D0983551

SSSP

Springer
Series in
Social
Psychology

SSSP

Action Control

From Cognition to Behavior

Edited by
Julius Kuhl and Jürgen Beckmann

Springer-Verlag
Berlin Heidelberg New York Tokyo

PD Dr. Julius Kuhl
Dr. Jürgen Beckmann

Max-Planck-Institut für psychologische Forschung
Leopoldstrasse 24, D-8000 München 40, F. R. Germany

With 19 Figures

ISBN 3-540-13445-X Springer-Verlag Berlin Heidelberg New York Tokyo
ISBN 0-387-13445-X Springer-Verlag New York Heidelberg Berlin Tokyo

© Springer-Verlag Berlin Heidelberg 1985
Printed in Germany

Typesetting, printing and binding: G. Appl, Wemding
2126/3140-543210

To Martha,
who is well acquainted with the limits
of action control

Preface

"It is not thought as such that can move anything, but thought which is for the sake of something and is practical." This discerning insight, which dates back more than 2000 years to Aristotle, seems to have been ignored by most psychologists. For more than 40 years theories of human action have assumed that cognition and action are merely two sides of the same coin. Approaches as different as S-O-R behaviorism, social learning theory, consistency theories, and expectancy-value theories of motivation and decision making have one thing in common: they all assume that "thought (or any other type of cognition) can move anything," that there is a direct path from cognition to behavior.

In recent years, we have become more and more aware of the complexities involved in the relationship between cognition and behavior. People do not always do what they intend to do. Aside from several nonpsychological factors capable of reducing cognition-behavior consistency, there seems to be a set of complex psychological mechanisms which intervene between action-related cognitions, such as beliefs, expectancies, values, and intentions, and the enactment of the behavior suggested by those cognitions.

In our recent research we have focused on *volitional* mechanismus which presumably enhance cognition-behavior consistency by supporting the *maintenance* of activated intentions and prevent them from being pushed aside by competing action tendencies.

Recently, many investigators from several subfields of psychology have discovered cognition-behavior inconsistencies. This led to several studies of the various factors contributing to observed discrepancies between cognition and behavior. Social psychologists have studied attitude-behavior inconsistencies. Clinical psychologists have become increasingly aware of the disruptive effects which self-regulatory deficits can have on an individual's ability to *behave* according to her/his preferences, feelings, and beliefs. Personality psychologists have studied various strategies employed by people when they find it difficult to maintain and enact an intention (e. g., in situations requiring delay of gratification or resistance to temptation). Finally, cognitive psychologists have constructed increasingly more complex computer models simulating problem-solving mechanisms, aiding in the enactment of intentions in difficult situations.

This book is an attempt to increase cross-fertilization across research areas concerned with the cognition-behavior relationship. Until now there have been very few attempts in this direction. Despite the common theme, the methods and theoretical assumptions differ considerably across these subfields. We believe that most of these differences are complementary rather than contradictory. The differences in theory and methodology arise from the fact that each approach focuses on a different facet of the problem. Although we have to confine ourselves, for the most part, to isolated facets of complex phenomena, it can be useful to "look across the border" occasionally and recognize the blind spots inherent in our own approach. Aside from its heuristic usefulness, cross-fertilization highlights the *interaction* between the various processes studied within isolated paradigms.

Preparing this book has taught us that there are some factors along the path from cognition to behavior which are difficult to control. Encouraging colleagues from quite different areas to contribute to a joint volume was not an easy task. Some authors had more problems in transforming their intention to participate in this project into the appropriate behavior than others did. All of us had to employ a substantial amount of action control to push aside the many competing action tendencies that beset our working day.

Several people helped us in this effort. Thanks are due Dr. Martin Irle for his encouragement and advice, and Hermann Hüttl and Britta Kotthaus for secretarial assistance. Part of the editorial work was made possible by the support of both editors by the German Science Foundation (DFG) and by a fellowship at the Center for Advanced Study in the Behavioral Sciences awarded to the senior editor. These individuals and institutions helped us to transform "thought as such" to "thought which is for the sake of something and is practical," to use Aristotle's definition.

Stanford and Munich   Julius Kuhl
March 1985   Jürgen Beckmann

Contents

Part III. Problem-Solving and Performance Control

9. Mechanisms of Control and Regulation in Problem Solving

10. Thinking and the Organization of Action

11. A Control-Systems Approach to the Self-Regulation of Action

12. From Cognition to Behavior: Perspectives for Future Research on Action Control

List of Contributors

Prof. Dr. Icek Ajzen
Department of Psychology, University of Massachusetts, Amherst, MA 01003,
U.S.A.

Dr. Jürgen Beckmann
Max-Planck-Institut für psychologische Forschung, Leopoldstrasse 24,
D-8000 München 40, F.R.Germany

Prof. Dr. Charles S.Carver
Department of Psychology, University of Miami, P.O.Box 248185,
Coral Gables, FL 33124, U.S.A.

Prof. Dr. Dietrich Dörner
Universität Bamberg, Lehrstuhl Psychologie II, Postfach 1549, D-8600 Bamberg,
F.R.Germany

Dipl.-Psych. Gunnar Friedrichsen
Fachbereich Pädagogik, Abteilung Psychologie,
Hochschule der Bundeswehr Hamburg, Postfach 700822,
D-2000 Hamburg 70, F.R.Germany

Dr. Peter M.Gollwitzer
Max-Planck-Institut für psychologische Forschung, Leopoldstrasse 24,
D-8000 München 40, F.R.Germany

Dipl.-Psych. Claudia Herrmann
Fachbereich Erziehungs- und Unterrichtswissenschaften, W7 der Freien
Universität Berlin, Habelschwerdter Allee 45, D-1000 Berlin 33, F.R.Germany

Prof. Dr. Martin Irle
Fakultät für Sozialwissenschaften der Universität Mannheim, A5,
D-6800 Mannheim 1, F.R.Germany

Yechiel Klar
Department of Psychology, Tel Aviv University, Ramat-Aviv, Tel Aviv, Israel

Prof. Dr. Rainer H. Kluwe
Fachbereich Pädagogik, Abteilung Psychologie, Hochschule der Bundeswehr Hamburg, Postfach 70 08 22, D-2000 Hamburg 70, F. R. Germany

Prof. Dr. Arie W. Kruglanski
Department of Psychology, Tel Aviv University, Ramat-Aviv, Tel Aviv, Israel

PD Dr. Julius Kuhl
Max-Planck-Institut für psychologische Forschung, Leopoldstrasse 24, D-8000 München 40, F. R. Germany

Prof. Dr. Michael F. Scheier
Department of Psychology, Carnegie-Mellon University, Pittsburgh, PA 15213, U. S. A.

Prof. Dr. Robert A. Wicklund
Fakultät für Psychologie und Sportwissenschaft der Universität Bielefeld, Abteilung Psychologie, Postfach 8640, D-4800 Bielefeld, F. R. Germany

Prof. Dr. Camille B. Wortman
Institute for Social Research, The University of Michigan, Ann Arbor, MI 48106, U. S. A.

Chapter 1
Introduction and Overview

Julius Kuhl and Jürgen Beckmann

Since its early beginnings, experimental psychology has focused on the cognitive mechanisms underlying the acquisition and representation of knowledge. Understanding human cognition seems the logical first step toward an explanation of human behavior. Once we know how people perceive their environments and how they judge their abilities to produce desired effects, we should be able to explain and predict their behavior. However, the link between cognition and behavior is less reliable than one might expect. People do not always perform in a manner consistent with their beliefs, values, attitudes, or intentions.

In his well-known study of attitude-behavior incongruence, La Pierre (1934) found that only 10% of the restaurants and hotels to which he sent a questionnaire indicated a willingness to serve "members of the Chinese race." However, when he had taken a Chinese couple to each of these establishments some time before sending out the questionnaire, only one out of 251 actually refused to serve them.

In fact, many investigators report disappointingly low attitude-behavior correlations across a great variety of situations (Wicker, 1969). Cognitive models of motivation that explain an individual's behavior on the basis of her or his expectancies and values do not have high predictive power. Although there are a large number of studies suggesting that expectancies and values have some effect on behavior (Ajzen & Fishbein, 1977; Atkinson, 1964; Heckhausen, 1977), the effects obtained are often weak and inconsistent across subjects and across situations (Kuhl, 1982).

How can one account for the disappointingly low cognition-behavior congruence? Are our theories insufficient, our measurement techniques inappropriate? In recent years considerable effort has been invested in overcoming several technical problems that might be the cause of an underestimation of the true cognition-behavior correspondence (Ajzen & Fishbein, 1980). In addition, several cognitive variables moderating cognition-behavior consistency have been investigated (Zanna, Higgins & Herman, 1982). Most psychologists seem to assume that, once all measurement problems are solved and all cognitive variables affecting behavior are discovered, they will be able to establish a close cognition-behavior relationship.

The present volume is prompted by the concern that this viewpoint may be too optimistic. There are good reasons to believe that it takes more than understanding cognition to account for behavior. Even the most reliable and comprehensive assessment of an individual's knowledge structures is insufficient to explain behavior. This point has been made very succinctly in Guthrie's (1935) criticism of Tolman's cognitive theory of behavior: "In his concern with what goes on in the rat's mind, Tolman has neglected to predict what the rat will do. So far as the theory is concerned the rat is left buried in thought; if it gets to the food-box at the end that is his concern, not the concern of the theory" (Guthrie, 1935, p. 172). To close the cognition-behavior gap we need to answer at least four questions: *First,* how do cognitive structures arouse motivational states? This question has been investigated by motivation psychologists for many years (e. g., Atkinson & Birch, 1978; Feather, 1982; Heckhausen, 1980; Weiner, 1980). The first two chapters address some of the issues related to this problem. *Second,* we need to understand the processes underlying temporal changes in those motivational states. This problem has been addressed by the *dynamics-of-action* theory (Atkinson & Birch, 1970; Kuhl & Atkinson, 1984).

The *third* problem, which is the major focus of this volume, relates to the mechanisms mediating the *formation* and *enactment* of intentions (see Part II below). In a given individual, many cognitive structures may be activated at the same time, and each of them may suggest several action alternatives. What mechanisms determine which alternative the actor actually intends to put into effect? What are the mechanisms that support the maintenance and enactment of an actor's current intention and protect it against competing action tendencies? Fifty years ago this problem was investigated thoroughly in the psychology of volition, or the "will". For several reasons (see Chapter 5), volitional processes have been neglected in the mainstream of psychological research on human action since then. Yet, all the above questions make it evident that there is a great need to study self-regulatory mechanisms that control the transition from cognition to action.

In this broad sense, the term *action control* not only comprises self-regulatory mechanisms for protecting a current intention. It also includes *performance control,* the fourth of the questions mentioned above. Performance control refers to mechanisms mediating the final execution of a sequence of behaviors. In fact, the well-known attempt by Miller, Galanter, and Pribram (1960) to close the cognition-behavior gap relates to performance control. According to Miller et al. (1960), the execution of an intended action is controlled by a system of hierarchically nested negative feedback loops (i. e., Test-Operate-Test-Exit Systems). Carver and Scheier (see Chapter 11, this volume) have elaborated this approach. The execution of plans of action often entails complex procedures to overcome obstacles. In this case, complex problem-solving activities have to be performed in order to find an action sequence that eventually leads to the desired goal state (see Chapters 9 and 10).

Overview of Chapters

This volume is divided into three parts. Part I contains three chapters that address the problem of how cognitive processes affect motivational states and action tendencies. The four chapters of Part II focus on self-regulatory processes that mediate the maintenance and enactment of intentions. Part III contains three chapters addressing issues of problem-solving and performance control. Since we do not believe that motivational, volitional and performance-control processes are independent of one another, we encouraged the authors to discuss their particular topics in the context of all three problems addressed in this volume.

For several decades, Expectancy × Value models have dominated motivational psychology. According to these models an actor intends to perform that action alternative which has the highest product of expectancy for achieving the aspired goal by the personal value (incentive) of that goal (Kuhl, 1982). However, the problem of transforming a motivational tendency or an intention into actual behavior is neglected by these models. Specifically, Expectancy × Value models seem to be based on an unrealistic assumption about human information-processing capacity. In most real-life situations, this capacity does not seem to be sufficient to allow for all of the calculations postulated by Expectancy × Value models. Furthermore, in many decision situations the time available is too short to consider all the information available. But, not taking into account the problems of capacity or time, there are many situations where a decision is not easily made because the given alternatives possess equal numbers of positive and negative aspects. To bridge the gap between the formation of an intention and its execution, we need to investigate volitional processes which (1) facilitate a decision among alternatives, (2) support the maintenance of an intentional state resulting from a decision, and (3) control the amount of action-related information processed in a given situation. Facilitation of decisions, maintenance of intentional states, and control of the extensiveness of information processing are factors that help overcome the difficulties of enacting intentions.

In Chapter 1, Ajzen suggests that the predictive power of traditional Expectancy × Value models decreases with the number of difficulties in enactment (see above) that enter the action sphere. Ajzen specifies three major sources of cognition-behavior inconsistency. The first source is a change in the initial intention before it is carried out. The second source is people's lack of confidence that the attainment of their behavioral goal is under their volitional control. Finally, whether a behavioral expectation formed on the basis of such an assessment of volitional control leads to actual goal attainment is contingent on the relation between people's confidence in their ability to exercise control over their own action and the extent to which they actually do control events in their lives.

Ajzen developed a theory of planned behavior which extends the traditional Expectancy × Value approach to incorporate these factors. In his original theory of reasoned action, a behavioral intention formed on the basis of attitudes toward a behavior and subjective norms was used as a predictor of a person's behavior. But, because of the factors mentioned above, every intended behavior is a goal, the attainment of which is usually subject to some degree of uncertainty. In many cases people might change their mind after forming an intention, thereby producing an inconsistency between the original intention and the behavior. Therefore, Ajzen concludes that it is not the behavioral intention which is the best predictor of behavior but the behavioral expectation formed after an intention has been formed. A behavioral expectation describes people's degree of confidence that they will be able to carry out a behavior once they intend to try it. Behavioral intentions and behavioral expectations differ whenever people anticipate that their intentions might change from the inception of an intention to the moment where an opportunity is given to perform the behavior or when they believe that attainment of their behavioral goal is not completely under their volitional control.

While Ajzen's approach, which aims to close the cognition-behavior gap, is still based on an expectancy-value model, Kruglanski and Klar (in the third chapter) propose a radical break with this tradition. According to these authors, an intention conceptualized as a particular kind of knowledge concerning personal goals, is usually deduced from certain (action) schemata. If certain preconditions specified in the schema are met, the respective intention or, if one is dealing with very simple cases, the action, will immediately succeed. The expectancy-value model is supposed to be just one instance of such an action schema. Since the possible number of schemas for action is, in principle, infinite, the question arises which schema will be selected. Kruglanski and Klar assume that this depends on an individual's socialization history and life experiences. But the acceptance of an action schema is also partly governed by certain motivational processes. Such motivational processes in the phase of intention formation were already discussed by James (1890) and Michotte and Prüm (1910) (see chapter 5 for more details).

Kruglanski and Klar suggest that a "need for structure" will soon put an end to the consideration of different action opportunities. If a person has a high need for structure he or she will relatively quickly "freeze" the strenghts of competing alternatives and accept an action schema. A similar notion of "freezing" action alternatives was proposed by Lewin (1951, 1952) who recognized that predecisional conflict must somehow be terminated. Kruglanski and Klar's conception of freezing is much more elaborate than Lewin's. Whereas Lewin saw freezing as merely a necessary condition for intention formation, Kruglanski and Klar specify the motivational determinants of freezing. A "need for structure" promotes freezing and a "fear of invalidity" inhibits freezing. Thus, according to Kruglanski and Klar, predecisional information processing is not as reality bound as rationalistic expectancy-value approaches suggest. Nevertheless, this at first sight irrational limitation of information processing involved in decision-making, may sometimes turn out to be more adequate for attaining a goal than would thorough elaboration of all available information, since the former results in quicker transformation of an intention into action than the latter. As Ajzen points out in Chapter 1, delay of the realization of an intention may result in a change of that intention.

In Chapter 4, Gollwitzer and Wicklund discuss the problem of how the activation of certain types of goals determines whether or not a cognitive state results in an action. In particular, they address, self-definitional goals (e. g., "I am a football fan") which – when activated – represent a very potent source for impeding or even blocking the attainment of various actual goals. The authors show that failing in one effort to gain a symbol for a self-definitional goal (e. g., a football fan not having enough time to watch the games) will lead to striving after and emphasizing other symbols (e. g., memorizing and discussing outcomes of games). Publicly stating that one has a particular intention (e. g., planning to watch a football game) may already provide such a symbol and thereby reduce the need to actually enact the intention. This creates an additional source of intention-behavior inconsistency.

Part II addresses the problem of which mechanisms an individual can employ in order to transform an intention into action. Furthermore, several factors which facilitate or impede the effectiveness of such self-regulatory processes are discussed. Chapter 5 gives a historical introduction to approaches dealing with this issue. Around the turn of the century a number of quite sophisticated models of self-regulatory processes had been developed in the tradition of the psychology of the will. This tradition has since been widely ignored in modern psychology. Nevertheless, the approaches taken by Ach, Michotte and Prüm, and James offer detailed descriptions of self-regulatory processes. Kuhl, who presents his theory of action control in Chapter 6, takes the work of Ach as a starting point. He emphasizes that the assumption that one can unequivocally derive an individual's behavior from his or her current intention is untenable. Not even a behavioral expectation as proposed by Ajzen (Chapter 1) will always suffice to accurately predict behavior. A crucial factor for cognition-behavior consistency, according to Kuhl, is the difficulty of enactment in relation to the efficiency of self-regulatory processes. Several factors are discussed which determine the difficulty of enactment. Kuhl then specifies a number of self-regulatory functions which facilitate volitional control in the face of such difficulty. Action versus state orientation seems to be a very potent determinant of the difficulty of enactment as well as of the various self-regulatory functions. A number of experimental results suggests that individual differences in and experimental manipulations of action and state orientations affect intention-behavior consistency.

Chapter 7 focuses on a specific type of shielding processes associated with Kuhl's construct of action orientation. According to Beckmann and Irle, the processes of dissonance reduction, described in Festinger's (1957) theory of cognitive dissonance, can be analyzed in terms of an action control model. These authors assume that dissonance reduction is employed as a self-regulatory process in the service of action control whenever the actual realization of an intention is jeopardized by cognitive inconsistency. They discuss a number of classical dissonance studies which corroborate their assumption. Furthermore, the authors report a number of recent experiments suggesting that dissonance reduction can be conceptualized as a self-regulatory process related to Kuhl's action-vs. state-orientation dimension. Dissonance reduction seems to be facilitated in the case of action orientation and impeded in the case of state orientation.

In Chapter 8, Herrmann and Wortman approach the problem of coping with stress from an action-control perspective. These authors discuss coping with life crises in the light of Kuhl's theory, especially focusing on the construct of action and state orientations. The experience of undesirable life events can block the enactment of a person's intentions. Whether or not this is actually the case seems to depend on whether the person engages in repetitive thought patterns about a lost object. Such repetitive thought patterns, according to Kuhl's theory, are characteristic of state orientation. Therefore, state orientation may impede the execution of intentions after the experience of critical life events. Up to this point state orientation has been discussed predominantly as being detrimental to cognition-behavior consistency, whereas action orientation was considered an adaptive mechanism. Herrmann and Wortman point out that this does not seem to hold in all cases. Immediate coping may indeed be obstructed by state orientation and facilitated by action orientation. But, long-term coping may be poorer with action orientation than with state orientation. Furthermore, premature coping efforts in the case of action orientation may result in failure which will lead to more distress than no coping efforts at all. Therefore, Herrmann and Wortman conclude that action orientation may impede the coping process if the resulting action does not alleviate the target problem. Conversely, in some cases the repeated examination of what has happened, as in the case of state orientation, will facilitate resolution of the crisis. However, in other cases where immediate activity is required, action orientation will be more advantageous for coping.

The chapters in part II primarily consider action control as a problem of maintaining an intention when one is faced with competing action alternatives. Part III focuses on the problems involved in the execution of an intention. One common problem is specifying a set of available action alternatives. This is rendered particularly difficult if goals and/or action alternatives are ill-defined.

Dörner (Chapter 10) discusses problem-solving in terms of a "trouble shooting" system developed to deal with the difficulties of enacting ill-defined intentions. In order to detect deficiencies in intentions and to implement the respective solutions, a controlling and regulating system is needed. Kluwe and Friedrichsen (Chapter 9) deal with cognitive activities that support the organization and the course of human thinking. They show that the uncertainty produced by a problem situation requires monitoring one's cognitive efforts. The resulting effects require decisions with regard to the course of one's own cognitive enterprise as well as its goals. Such "executive" and "metaplan" decisions can prevent people from executing erroneous or impracticable plans. As the authors point out, this might sometimes involve not only a change of the current action but also a shift to a different goal. Therefore, what is often considered to be cognition-behavior inconsistency may either be a mere shift from one type of action to another within the same plan, or a change of intention when it is determined that the goal is not attainable. Both cases are examples of maintaining action control. If, in the latter case, the intention were not abandoned, an individual's capacity for more promising actions would be blocked. This is detrimental to action control (see Chapter 6, Kuhl's concept of "degenerated intention").

Dörner (Chapter 10) focuses on the construction of the chain of operations leading to the goal. If appropriate schemas providing such a chain are not available, and consequently an automatic progression not feasible, heuristic processes will serve to produce a plan for action. Several factors which lead to deficient planning are discussed. One very potent determinant of planning activities seems to be feelings of competence or incompetence. People with high self-esteem apparently expose their own suppositions to critical examination whereas people with low self-esteem do not. Without critical checking, as emphasized by Dörner as well as by Kluwe and Friedrichsen, people run the risk of adopting erroneous plans which are detrimental to goal attainment and therefore result in discrepancies of people's cognitions (goals) and the results of their behavior.

In the last chapter in Part III, Carver and Scheier present their self-regulation approach toward performance control. These authors discuss control processes in terms of negative or discrepancy-reducing feedback loops. Current behavior is submitted to a continuous comparison process with salient behavioral standards. If a discrepancy is detected, attempts are made to bring one in line with the other. Kluwe, Friedrichsen, and Dörner, in their problem-solving approaches, acknowledge (just as Carver and Scheier do in their cybernetic-control analogy) multiple paths to the attainment of any given goal value. According to Carver and Scheier, when faced with an obstacle, people will first execute an expectancy assessment process concerning attainment of the actual goal. If the expectancy estimate is still favorable they will retry the activity with even more pronounced effort if their attention is focused on their own self. If expectancy assessment is unfavorable, people will disengage from the activity.

These chapters represent different aspects of the action control issue. Even though they address distinct phases in a course of action they overlap in several points because the various control processes are closely interrelated and the operation of one process may affect the organization and course of another process.

References

Ajzen, I., & Fishbein, M. (1977). Attitude-behavior relations: A theoretical analysis and review of empirical research. *Psychological Bulletin, 84*, 888–918.

Ajzen, I., & Fishbein, M. (1980). *Understanding attitudes and predicting social behavior.* Englewood Cliffs, N.J.: Prentice Hall.

Atkinson, J.W. (1964). *An introduction to motivation.* Princeton, N.J.: Van Nostrand.

Atkinson, J.W., & Birch, D. (1970). *The dynamics of action.* New York: Wiley.

Feather, N.T. (1982). Human values and the prediction of action: An expectancy-valence analysis. In N.T. Feather (Ed.) *Expectations and actions: Expectancy-value models in psychology.* Hillsdale, N.J.: Erlbaum.

Festinger, L. (1957). *A theory of cognitive dissonance.* Evanston, Ill.: Row & Peterson.

Guthrie, E.R. (1935). *The psychology of learning.* New York: Harper.

Heckhausen, H. (1977). Achievement motivation and its constructs: A cognitive model. *Motivation and Emotion, 1*, 283–329.

Heckhausen, H. (1980). *Motivation und Handeln.* Heidelberg: Springer.

James, W. (1980). *The principles of psychology.* (2 vols.) New York: Holt.

Kuhl, J. (1982). The expectancy-value approach within the theory of social motivation: Elaborations, extension, critique. In N.T. Feather (Ed.), *Expectations and actions: Expectancy-value models in psychology.* Hillsdale, N.J.: Erlbaum.

Kuhl, J., & Atkinson, J. W. (in press). Motivational determinants of decision time: An application of the dynamics of action. In J. Kuhl & J. W. Atkinson (Eds.), *Motivation, thought, and action*. New York: Praeger.

LaPierre, R. T. (1934). Attitudes vs. actions. *Social Forces, 13,* 230–237.

Lewin, K. (1951). *Field theory in social science*. New York: Harper & Row.

Lewin, K. (1952). Group decision and social change. In G. Swanson, T. Newcomb & E. Hartley (Eds.), *Readings in Social Psychology* (pp. 197–211). New York: Holt, Rinehart, & Winston.

Michotte, A., & Prüm, E. (1910). Etude expérimentale sur le choix volontaire et ses antécédents immédiats. *Travaux du Laboratoire de Psychologie expérimentale de l'Université de Louvain 1 (2)*.

Miller, G. A., Galanter, E., & Pribram, K. H. (1960). *Plans and the structure of behavior*. New York: Holt, Rinehart, & Winston.

Weiner, B. (1980). *Human motivation*. New York: Holt, Rinehart, & Winston.

Wicker, A. W. (1969). Attitudes versus actions: The relationship of verbal and overt behavioral responses to attitude objects. *Journal of Social Issues, 25,* 41–78.

Zanna, M. P., Higgins, E. T., & Herman, C. P. (Eds.) (1982). *Consistency in social behavior: The Ontario Symposium (Vol. 2)*. Hillsdale, N. J.: Erlbaum.

Part I

Cognitive and Motivational Determinants of Action

Chapter 2

From Intentions to Actions:
A Theory of Planned Behavior

Icek Ajzen

There appears to be general agreement among social psychologists that most human behavior is goal-directed (e. g., Heider, 1958; Lewin, 1951). Being neither capricious nor frivolous, human social behavior can best be described as following along lines of more or less well-formulated plans. Before attending a concert, for example, a person may extend an invitation to a date, purchase tickets, change into proper attire, call a cab, collect the date, and proceed to the concert hall. Most, if not all, of these activities will have been designed in advance; their execution occurs as the plan unfolds. To be sure, a certain sequence of actions can become so habitual or routine that it is performed almost automatically, as in the case of driving from home to work or playing the piano. Highly developed skills of this kind typically no longer require conscious formulation of a behavioral plan. Nevertheless, at least in general outline, we are normally well aware of the actions required to attain a certain goal. Consider such a relatively routine behavior as typing a letter. When setting this activity as a goal, we anticipate the need to locate a typewriter, insert a sheet of paper, adjust the margins, formulate words and sentences, strike the appropriate keys, and so forth. Some parts of the plan are more routine, and require less conscious thought than others, but without an explicit or implicit plan to guide the required sequence of acts, no letter would get typed.

Actions, then, are controlled by intentions, but not all intentions are carried out; some are abandoned altogether while others are revised to fit changing circumstances. The present chapter examines the relations between intentions and actions: the ways in which goals and plans guide behavior, and the factors that induce people to change their intentions, or prevent successful execution of the behavior. The first part of the chapter deals with prediction and explanation of behavior that is largely under a person's volitional control. A theory of reasoned action is decribed which traces the causal links from beliefs, through attitudes and intentions, to actual behavior. Relevant empirical research is reviewed, with particular emphasis on the intention-behavior link and the factors that may produce changes in behavioral intentions. The chapter's second part deals with a behavioral domain about which much less is known. There, an attempt is made to extend the theory of reasoned action to goal-directed behaviors over which an individual has only limited volitional control. First, internal and external factors that may influence volitional control are identified. Next, a behavior-goal unit is defined, and the theory of reasoned action is modified to enable it to predict and explain such goal-directed behavior. The modified theory, called "a theory of planned behavior," differs from the theory of reasoned action, in that it takes into account perceived as well as actual control over the behavior under consideration.

Predicting and Explaining Volitional Behavior: A Theory of Reasoned Action

A great many behaviors of everyday life may be considered under volitional control in the sense that people can easily perform these behaviors if they are inclined to do so. To illustrate, under normal circumstances most people can, if they so desire, watch the evening news on television, vote for the candidate of their choice in an election, buy toothpaste at a drug store, pray before going to bed, or donate blood to the Red Cross. The theory of reasoned action (Ajzen & Fishbein, 1980; Fishbein & Ajzen, 1975) is designed to predict volitional behaviors of this kind and to help us understand their psychological determinants.

As its name implies, the theory of reasoned action is based on the assumption that human beings usually behave in a sensible manner; that they take account of available information and implicitly or explicitly consider the implications of their actions. Consistent with its focus on volitional behaviors, the theory postulates that a person's *intention* to perform (or not to perform) a behavior is the immediate determinant of that action. Barring unforeseen events, people are expected to act in accordance with their intentions. Clearly, however, intentions can change over time; the longer the time interval, the greater the likelihood that unforeseen events will produce changes in intentions. It follows that accuracy of prediction will usually be an inverse function of the time interval between measurement of intention and observation of behavior.

Since we are interested in *understanding* human behavior, not merely in predicting it, we must next identify the determinants of intentions. According to the theory of reasoned action, a person's intention is a function of two basic determinants; one personal in nature and the other reflecting social influence. The personal factor is the individual's positive or negative evaluation of performing the behavior; this factor is termed *attitude toward the behavior.* Note that the theory of reasoned action is concerned with attitudes toward behaviors and not with the more traditional attitudes toward objects, people, or institutions. The second determinant of intention is the person's perception of the social pressures put on him to perform or not perform the behavior in question. Since it deals with perceived prescriptions, this factor is termed *subjective norm.* Generally speaking, people intend to perform a behavior when they evaluate it positively and when they believe that important others think they should perform it.

The theory assumes that the relative importance of these factors depends in part on the intention under investigation. For some intentions, attitudinal considerations may be more important than normative considerations, while for other intentions normative considerations may predominate. Frequently, both factors are important determinants of the intention. In addition, the relative weights of the attitudinal and normative factors may vary from one person to another. The discussion of the theory up to this point can be summarized symbolically as follows:

$$B \sim I \propto [w_1 A_B + w_2 SN] \tag{1}$$

In Equation 1, B is the behavior of interest, I is the person's intention to perform behavior B, A_B is the person's attitude toward performing behavior B, SN is the person's subjective norm concerning performance of behavior B, and w_1 and w_2 are empirically determined weighting parameters that reflect the relative importance of A_B and SN. The wavy line ($\sim$) in Equation 1 is inserted to suggest that intention is expected to predict behavior only if the intention has not changed prior to performance of the behavior[1]; and the intention itself is shown to be directly proportional to the weighted sum of attitude toward the behavior and subjective norm.

For many practical purposes, this level of explanation may be sufficient. However, for a more complete understanding of intentions it is necessary to explain why people hold certain attitudes and subjective norms. According to the theory of reasoned action, the attitude toward a behavior is determined by salient beliefs about that behavior. Each salient belief links the behavior with some valued outcome or other attribute. For example, a person may believe that "going on a low sodium diet" (behavior) "reduces blood pressure," "leads to a change in life style," "severely restricts the range of approved foods," and so forth (outcomes). The attitude toward the behavior is determined by the person's evaluation of the outcomes associated with the behavior and by the strength of these associations. Specifically, the evaluation of each salient outcome contributes to the attitude in proportion to the person's subjective probability that the behavior will produce the outcome in question. By multiplying belief strength and outcome evaluation, and summing the resulting products, we obtain an estimate of attitude toward the behavior based on the person's salient beliefs about that behavior.[2] This information-processing theory of attitude is presented symbolically in Equation 2, where A_B stands for attitude toward behavior B, b_i is the belief (subjective prob-

$$A_B \propto \sum_{i=1}^{n} b_i e_i \tag{2}$$

ability) that performing behavior B will lead to outcome i, e_i is the evaluation of outcome i, and the sum is over the n salient behavioral beliefs.

It can be seen that, generally speaking, a person who believes that performing

---

[1] In addition, the theory of reasoned action requires that intention and behavior be operationally defined so that they correspond in their target, action, context, and time elements (see Ajzen & Fishbein, 1977).

[2] Theories of a similar nature have been proposed by Edwards (1954), Rosenberg (1956), and others (see Feather, 1959).

a given behavior will lead to mostly positive outcomes will hold a favorable attitude toward performing the behavior while a person who believes that performing the behavior will lead to mostly negative outcomes will hold an unfavorable attitude. The beliefs that underlie a person's attitude toward the behavior are termed *behavioral beliefs*.

Subjective norms are also assumed to be a function of beliefs, but beliefs of a different kind, namely the person's beliefs that specific individuals or groups think he should or should not perform the behavior. These beliefs underlying the subjective norm are termed *normative beliefs*. Generally speaking, a person who believes that most referents with whom he ist motivated to comply think he should perform the behavior will perceive social pressure to do so. Conversely, a person who believes that most referents with whom he is motivated to comply think he should not perform the behavior will have a subjective norm that puts pressure on him to avoid performing the behavior. The relation between normative beliefs and subjective norm is expressed symbolically in Equation 3. Here,

$$SN \propto \sum_{j=1}^{n} b_j m_j \tag{3}$$

SN is the subjective norm, b_j is the normative belief concerning referent j, m_j is the person's motivation to comply with referent j, and n is the number of salient normative beliefs.

The above discussion of the theory of reasond action shows how volitional behavior can be explained in terms of a limited number of concepts. Through a series of intervening processes the theory traces the causes of behavior to the person's salient beliefs. Each successive step in this sequence from behavior to beliefs provides a more comprehensive account of the factors determining the behavior. At the initial level, behavior is assumed to be determined by intention. At the next level, these intentions are themselves explained in terms of attitudes toward the behavior and subjective norms. The third level explains these attitudes and subjective norms in terms of beliefs about the consequences of performing the behavior and about the normative expectations of relevant referents. In the final analysis, then, a person's behavior is explained by reference to his or her beliefs. Since people's beliefs represent the information (be it correct or incorrect) they have about their worlds, it follows that their behavior is ultimately determined by this information. Other, more distal factors, such as demographic characterictics or personality traits, are assumed to have no direct effects on behavior. According to the theory of reasond action, variables of this kind will be related to behavior if, and only if, they influence the beliefs that underlie the behavior's attitudinal or normative determinants.

Empirical Support

A considerable amont of evidence in support of the theory of reasoned action has accumulated in a variety of experimental and naturalistic settings. To provide a complete test of the relationships specified by the theory, it is necessary to elicit salient behavioral and normative beliefs in a pilot study, and use these beliefs, among other things, to construct a standard questionnaire. This questionnaire would contain measures of the following variables: (1) behavioral beliefs and outcome evaluations which are assumed to determine attitude toward the behavior (see Equation 2) and can be used to compute a belief-based estimate of this attitude; (2) normative beliefs and motivations to comply which underlie subjective norm (Equation 3) and can be used to compute an indirect measure of the normative component; (3) direct measures of attitude toward the behavior and subjective norm; and (4) intention to perform the behavior. Although various measurement procedures could be developed, in most applications of the theory, evaluative (e. g., "good – bad") and probablistic (e. g., "likely – unlikely") semantic differential scales have been employed. (See Ajzen & Fishbein, 1980, Appendices A and B, for detailed descriptions of measurement procedures used in tests of the theory.) The questionnaire thus constructed is administered to a sample of respondents whose actual behavior is subsequently recorded either by means of observation or, if direct observation is impractical, by means of self-reports.

Although complete applications of the theory require assessment of all variables from beliefs to overt behavior, many questions can be answered by investigating a more limited set of relationships. Thus, it is often sufficient to obtain direct measures of attitudes and subjective norms without assessing the underlying beliefs. In other cases, the intention-behavior relation is of little immediate concern; instead, the theory's ability to predict and explain intentions is at issue. In these instances, it is unnecessary to secure a measure of actual behavior. Viewed in combination, however, the available data span the whole range of relations from beliefs through attitudes, subjective norms, and intentions, to overt behavior.

Selected Research Findings

The extent to which the theory of reasoned action predicts behavioral intentions is usually evaluated by means of linear multiple regression analyses; the regression coefficients produced by these analyses serve as estimates of w_1 and w_2, the weights of the attitudinal and normative predictors (see Equation 1). In addition, whenever possible the intention-behavior correlation is reported, as are correlations between direct (A_B, SN) and indirect ($\sum b_i e_i$, $\sum b_j m_j$) measures of attitudes and subjective norms. Finally, the sample of respondents can be divided into those who did (or intented to) perform the behavior and those who did not. Differences in behavior (or intentions) can then be explained by examining patterns of differences in behavioral beliefs, outcome evaluations, normative beliefs, and motivations to comply. (See Ajzen & Fishbein, 1980, for examples.)

Table 2.1 presents a sample of research findings on the major relationships specified by the theory of reasoned action. It can be seen that the theory permits highly accurate prediction in a wide variety of behavioral domains. Generally speaking, people were found to act in accordance with their intentions. Strong intention-behavior correspondence was expected, given the largely volitional nature of the behaviors that served as criteria in these investigations. With one exception, all intention-behavior correlations exceeded 0.70. Interestingly, the exception ocurred in the prediction of having another child where the intention-behavior correlation, although significant, was only 0.55. Having another child is, of course, only partially under volitional control since fecundity, miscarriage, and other factors also mediate attainment of the goal. We shall return to this issue below.

The second column in Table 2.1 shows that, in each case, a linear combination of attitudes and subjective norms permitted highly accurate prediction of intentions. The relative importance of the two components is revealed by inspecting the next four columns. Except for reenlisting in the military, where the regression coefficient of subjective norm did not attain significance, both attitudes and subjective norms made significant contributions to the prediction of intentions. In seven of the nine studies listed in Table 2.1, the relative contribution of attitudes exceeded that of subjective norms; but in two cases, the pattern was reversed. Not surprisingly, women's decisions to have an abortion, and a couple's decision to have another child, were strongly affected by perceived expectations of important others.

Finally, the last two columns of Table 2.1 report relations between direct and belief-based measures of attitudes and subjective norms. It can be seen that the results tend to support Equations 2 and 3: Behavioral beliefs and outcome evaluations can be used to estimate attitudes toward a behavior (Column 7), while normative beliefs and motivations to comply provide estimates of subjective norms (Column 8).

Clearly then, the theory of reasoned action can afford highly accurate prediction of intentions and behaviors that are under volitional control. By examining closely the underlying belief structure, one can also gain a good understanding of the factors that ultimately determine a person's decision to perform or not to perform a given behavior. To illustrate, consider women's use of birth control pills. With respect to behavioral beliefs Ajzen and Fishbein (1980, pp. 141–142) summarized the important research findings as follows. "The major considerations that entered into the women's decisions to use or to not use the pill revolved around questions of physiological side effects, morality, and effectiveness. Although all women believed that using the pill leads to minor side effects (such as weight gain), they differed in their beliefs about severe consequences. The more certain a woman was that using the pill would not lead to such negative outcomes as blood clots and birth defects, the more likely she was to intend using the pill. Also associated with intentions to use the pill were beliefs that this was the best available method for preventing pregnancy." In addition, it was found that women intended to use the pill only if they had no strong moral objections. "On the normative side, the women's major concerns centered on the prescriptions of their husbands or boyfriends and doctors. They were highly motivated to comply with these referents, and women who believed that their husbands or boyfriends and doctors thought they should use the pill intended to

Table 2.1. Theory of reasoned action – sample of research findings

Criterion	Correlation	Multiple correlation	Correlations		Regression coefficients		Correlations	
	I–B	I–A_B; SN	I–A_B	I–SN	I–A_B	I–SN	A_B–$\sum b_i e_i$	SN–$\sum b_j m_j$
Cooperation in prisoner's dilemma game[a] (Ajzen, 1971)	0.82	0.82	0.75	0.69	0.53	0.40	–	–
Having another child[b] (Vinokur-Kaplan, 1978)	0.55	0.85	0.65	0.83	0.19	0.70	–	–
Choice of career orientation[c] (Ajzen & Fishbein, 1980)	–	0.86	0.83	0.64	0.67	0.29	0.81	0.83
Use of birth control pills (Ajzen & Fishbein, 1980)	0.85	0.89	0.81	0.68	0.64	0.41	0.79	0.60
Voting choice in 1976 presidential election (Ajzen & Fishbein, 1980)	0.80	0.83	0.81	0.71	0.61	0.27	0.79	0.73
Having an abortion[d] (Smetana & Adler, 1980)	0.96	0.76	0.50	0.69	0.21	0.46	0.58	–
Infant feeding[b] (Manstead, Proffitt, & Smart, 1983)	0.82	0.78	0.73	0.60	0.61	0.22	–	–
Smoking marijuana[a] (Ajzen, Timko, & White, 1982)	0.72	0.80	0.79	0.45	0.74	0.13	–	–
Reenlisting in the military (Shtilerman, 1982)	0.87	0.77	0.77	0.43	0.73	0.08*	0.64	0.41

[a] No beliefs were elicited or assessed in these studies.
[b] Only indirect (belief-based) measures of A_B and SN were obtained in these studies.
[c] Behavior was not assessed in this study.
[d] Only a belief-based measure of SN was obtained in this study.
* Not significant; all other coefficients $p < 0.05$.

do so. By the same token, women who believed that these two referents opposed their use of birth control pills formed intentions to not use them" (p. 142).

To summarize briefly, many behaviors of interest to social psychologists appear to be under volitional control and can be predicted with a high degree of accuracy from intentions to perform the behaviors in question. These intentions, in turn, appear to be based on personal attitudes toward the behaviors and perceived social norms. Attitudes are influenced by beliefs concerning a behavior's likely outcomes and evaluations of those outcomes, while subjective norms derive from normative beliefs regarding expectations of specific referent individuals or groups and motivations to comply with these referents. By examining differences in behavioral and normative beliefs (and in associated outcome evaluations and motivations to comply) we can go beyond prediction to provide a detailed explanation of volitional behavior.

The Intention-Behavior Relation

In light of the success achieved by the theory of reasoned action, it is important to keep its boundary conditions clearly in mind. For the most part, the theory's limitations have to do with the transition from verbal responses to actual behavior. The relations of beliefs, attitudes, and subjective norms to intentions are more clearly delineated than are the factors that determine whether or not the behavioral intention will be carried out. According to the theory of reasoned action, intention is the immediate antecedent of behavior. Strictly speaking, however, intentions can be expected to predict behavior only when two conditions are met. First, the measure of intention available to the investigator must reflect respondents' intentions as they exist just prior to performance of the behavior; and, second, the behavior must be under volitional control. As mentioned earlier, intentions may change over time, and any measure of intention obtained before the change took place cannot be expected accurately to predict behavior. This is largely a technical problem, however, since low predictive validity under such conditions merely reflects the reduced accuracy of the available measure of intention; it poses no challenge to the assumption that (current) intentions determine behavior. By way of contrast, the stipulation that behavior must be under volitional control imposes strict limitations on the theory's range of application; its ability to predict and explain human behavior will be greatly impaired whenever nonvolitional factors exert a strong influence on the behavior in question.

Changes in Intention

Many factors have been found to influence the stability of behavioral intentions. Examination of these factors sheds light on the ways and means by which it may be possible to prevent changes in intentions or modify predictions to take anticipated readjustments into account.

Effects in Time

Intentions change as time goes by. Some changes arguably occur simply as a function of time while others depend on the emergence of new information.

Salience of Beliefs. A goal's attractions and repulsions tend to be inversely proportional to psychological distance from the goal, and the avoidance gradient tends to be steeper than the approach gradient (Brown, 1948; Lewin, 1946, 1951; Miller, 1944). In a similar fashion, beliefs regarding a behavior's negative features, more so than its positive features, may become increasingly salient as the time of the behavior draws near. Thus, a person who is about to invest his hard-earned money in stocks may become increasingly concerned about the possibility that stock prices will decline in the future. If these shifts in evaluation reach the point at which the behavior's perceived disadvantages outweigh its perceived advantages, the individual is likely to reverse his intention and refrain from performing the behavior.

A different possibility was suggested by Semmer (personal communication): The conflicting behavioral tendencies produced by a goal's attractive and repulsive features may be resolved in favor of the more routinized responses. As the time for action approaches, people may fall back on familiar response patterns, that is, the probability of routine responses may increase, and the probability of novel responses may decline with the passage of time. Changes of this kind could help explain the difficulty of carring out a decision to refrain from such habitual behaviors as drinking or smoking.

New Information. Many changes in intentions, however, are the result of factors other than the mere passage of time. In fact, changes that appear at first glance to occur automatically may actually be mediated by internal processes (see Beckmann & Kuhl, 1984) or external factors. A multitude of unanticipated, and sometimes unforeseeable, events can disrupt the intention-behavior relation. A person's behavioral and normative beliefs are subject to change as events unfold and new information becomes available. Such changes may influence the person's attitude toward the behavior or his subjective norm and, as a result, produce a revised intention. To illustrate, consider a woman who intends to vote for the Democratic candidate in the forthcoming senate race. After her intention is assessed, she learns – by watching a television interview with the candidate a few days before the election – that he opposes abortion and equal rights for women. As a result, she "changes her mind": she forms new beliefs concerning the consequences of voting for the Democratic candidate, modifies her attitude toward this behavior, decides to vote for the Republican candidate instead, and actually does so in the election. Her voting choice corresponds to her most recent intention, but it could not have been predicted from the measure of intention obtained at an earlier point in time.

Several studies have demonstrated the disruptive effects of unforeseen events. For example, Songer-Nocks (1976a, 1976b) assessed intentions to choose the noncompetitive alternative at the beginning of a 20-trial, two-person experimental game. Half of the pairs of players were given feedback after each trial which informed them about the choices made by their partners and of the payoffs to each player. The other pairs were given no such information. Feedback concerning the partner's competitive or noncompetitive behavior may, of course, influence a player's own intentions regarding future moves. Consistent with this argument, Songer-Nocks reported that providing feedback significantly reduced the accuracy with which initial intentions predicted actual game behavior.

Somewhat more indirect evidence regarding the disruptive potential of unanticipated events is available from studies that have varied the amount of time between assessment of intention and observation of behavior. Since the probability of unforeseen events will tend to increase as time passes, we would expect to find stronger intention-behavior relations with short rather than long periods of delay. Sejwacz, Ajzen, and Fishbein (1980) obtained support for this prediction in a study of weight loss. A subsample of 24 college women indicated their intentions to perform eight weight-reducing behaviors (avoid snacking between meals, participate in sports on a regular basis, etc.) at the beginning of a two-month period, and again one month later. Correlations were computed between initial intentions and actual behavior over the two-month period, and between subsequent intentions and actual behavior during the final month. As expected, the intention-behavior correlations were stronger for the one-month period than for the two-month period. For example, the correlation between intention to avoid long periods of inactivity and actual performance of this behavior (as recorded by the women) was higher when the time period was one month ($r=0.72$) than when it was two months ($r=0.47$). Considering all eight behaviors, the average correlation increased from 0.51 for the two-month period to 0.67 for the one-month period.

Temporal delay between assessment of intention and observation of behavior thus tends to have a detrimental effect on behavioral prediction (see Hornik, 1970 and Fishbein & Coombs, 1974 for additional evidence in support of this conclusion). As time passes, there is an increase in the likelihood of unanticipated events and of concomitant changes in intentions. The result is a decline in the correlation between observed behavior and intentions assessed before the changes took place.

Confidence and Commitment

The discussion up to this point has dealt with rather drastic changes in intentions that lead to a reversal of behavioral plans. Intentions, however, vary in strength as well as direction, and changes can occur that would not be reflected in behavior. Consider, for example, a voter who assigns a probability of 0.85 to his intention to vote for the Republican candidate, and a probability of 0.15 to his intention to cast his vote for the Democratic candidate. Exposure to new information during the election compaign might reduce the perceived advantage of the Republican candidate, changing the strength of these intentions to 0.65 and 0.35, respectively; even so, the person would still be expected to vote for the Republican candidate.

As a general rule, when an intention is held with great confidence (i.e., when the intention is highly polarized), changes produced by new information will often be insufficient to reverse the planned course of action. In contrast, weak intentions to perform (or not to perform) a behavior carry less of a commitment; unanticipated events of relatively minor importance may influence such intentions enough to bring about a change of mind. It follows that intention-behavior correlations will usually be stronger when intentions are held with great, rather than little, confidence.

Sample and Warland (1973, Warland & Sample, 1973) as well as Fazio and Zanna (1978) have shown that attitudes held with high confidence are better predictors of behavior than are attitudes held with low confidence. Ajzen, White, and Timko (1982) examined more directly the effect of confidence on the intention-behavior relation. College students expressed their intentions to become members of a psychology subject pool (on a 7-point "likely-unlikely" scale), and indicated their confidence in their intentions on a 7-point scale that ranged from "extremely certain" to "not at all certain." At a later point in the experiment, they were given an opportunity to actually sign up for the subject pool. Using the median score on the certainty scale as a dividing point, respondents were partitioned into low and high confidence groups. As expected, the intention-behavior correlation was significantly stronger among respondents who had high confidence in their intentions ($r=0.73$) than among respondents with low confidence ($r=0.47$).[3]

It is interesting to note that the very act of stating an intention may induce hightened commitment to the behavior. In a series of experiments conducted by Sherman (1980), respondents who, in response to a question, predicted that they would act in a socially desirable manner were more likely to do so on a later occasion than were respondents who were not asked to predict their own behavior. Note, however, that the behaviors in question were of relatively little consequence: writing a counter-attitudinal essay, singing the national anthem over the telephone, and volunteering 3 hours to collect money for the American Cancer Society. Merely stating an intention may have much less of an effect on such personally significant behaviors as having an abortion or reenlisting in the military (see Table 2.1).

---

[3] In addition to greater stability, confident intentions (being more extreme) will tend to exhibit greater variance than intentions held with less confidence. Greater variability would also tend to increase the predictive validity of confident intentions. The standard deviation of intentions was, in fact, found to be significantly greater in the high ($SD=2.24$) than in the low ($SD=1.36$) confidence group. It is thus not clear whether improved prediction in the high confidence group was due to greater stability of intentions or whether it was simply a statistical artifact produced by a restriction of range in the low confidence condition.

Individual Differences

Some people change their intentions more readily than others. According to Snyder (1974, 1982), individuals differ in the extent to which their behavior is susceptible to situational cues as opposed to inner states or dispositions. He developed the "self-monitoring scale" to assess the tendency for a person's behavior to be guided by principle or inner disposition (low self-monitoring) on the one extreme, and by situational contingencies or pragmatism (high self-monitoring) on the other extreme. It stands to reason that the intentions of high self-monitoring individuals, who are sensitive to external cues, will readily be influenced by unanticipated events. The intentions of low self-monitors, however, should be less affected by external events, and should thus be relatively stable, since these individuals are sensitive primarily to internal states. We would thus expect stronger intention-behavior correlations among low as compared to high self-monitoring individuals.[4]

Data collected by Ajzen, Timko, and White (1982) support this prediction. The study was conducted at the time of the 1980 presidential election in the United States. Shortly before the election, college students completed questionnaires that, among other questions, included Snyder's self-monitoring scale and measures of intention to vote in the election and intention to smoke marijuana in the next 3 or 4 weeks. About 2 weeks after the election, participants were contacted by telephone and asked to report their behavior. As expected, individuals who scored below the median on the self-monitoring scale exhibited significantly stronger intention-behavior correlations than did individuals who scored above the median on the scale. The intention to vote predicted actual voting with a correlation of 0.59 for high self-monitors, and with a correlation of 0.82 for low self-monitors. The corresponding correlations with respect to the number of occasions on which the respondents reported to have smoked marijuana were 0.42 and 0.70.

Long-Range Prediction

It should be clear by now that many factors can influence the stability of intentions and hence the strength of the observed intention-behavior relation. One solution to this problem is to assess intentions immediately prior to observation of the behavior. Clearly, the shorter the delay, the less time and the fewer the opportunities for change. In practice, however, it may be neither feasible nor of much practical value to measure the intention in close temporal proximity of the behavior. Imagine, for example, that we are trying to predict behavior during such natural disasters as floods, tornados, earthquakes, or fires burning out of control.

---

[4] Snyder and Swann (1976) and Zanna, Olson, and Fazio (1980) have reported stronger *attitude*-behavior correlations for low self-monitors as compared to high self-monitors; but Zuckerman and Reis (1978) failed to obtain similar results.

It would be very difficult indeed to approach individuals fighting to save their lives or their posessions and ask them to state their intentions.

Even when possible, however, short-range predictions are often of little interest. Manufacturers of various consumer goods, from video games to automobiles, need to be able to anticipate buying behavior months or even years in advance; and banks, airlines, television companies, and other service organizations must predict the reactions of consumers long before offering a new type of service. Fortunately, long-range predictions of this kind are usually not concerned with the behavior of any given individual but rather with behavioral trends in relatively large segments of the population: how many people will buy a certain type of automobile in the course of a model year, the number of individuals who will volunteer for the various branches of the military by a certain target date, or the number of air travelers to be expected between two cities during a given time period. Aggregate intentions of this kind are apt to be much more stable than individual intentions. As we saw earlier, a multitude of unanticipated events can produce changes in the intentions of individuals: sudden illness or injury, a death in the family, a fortuitous win in the lottery, loss of one's job, an unexpected visit by a friend, and so forth. Of course, these are more or less random events that affect only some individuals at any given time. Their effects on intentions of different people are therefore likely to balance out, leaving the aggregate intention relatively unchanged. A young man who intends to enlist in the Navy may have a serious accident and make different plans for the future, but another man who had not considered joining the Navy may now intend to do so after applying unsuccessfully for a job in the civilian sector. On balance, the number of men intending to enlist in the Navy would remain unchanged.

Given that accuracy of behavioral prediction is influenced by the stability of intentions, the above discussion implies closer intention-behavior correspondence at the aggregate than at the individual level, especially in the case of long-range predictions. A good example is provided by research on family planning. In a study by Bumpass and Westoff (1969), women with two children were asked how many children they intended to have in their completed families. Six to ten years later (near the end of their reproductive periods) they were reinterviewed to obtain information about the number of children they actually had. Only 41% of the women had exactly the number of children they had planned; the remainder had more or fewer children than intended. On the average, however, the women's actual family size (3.3 children) was found to correspond precisely to the intended family size (also 3.3 children). Clearly, then, predictions at the aggregate level can be highly accurate even when the behavior of many individuals fails to correspond to their intentions.[5]

---

[5] The reason that Bumpass and Westoff (1969) found such close intention-behavior correspondence at the aggregate level is that the proportion of women who overproduced was almost exactly the same as the proportion of women who underproduced. Had there been a bias toward having more or fewer children than intended, the aggregate-level prediction would have been less accurate.

Plans, Goals, and Actions

The fact that intentions can change over time forces us to recognize their provisional nature. Strictly speaking, all an individual can say is that, *as of now,* he intends to perform a given behavior, and can assign a certain degree of confidence (subjective probability) to that intention. Assuming the behavior is under volitional control (and the person is prepared to exert maximum effort), failure to act in accordance with the intention would indicate that the person had a change of mind.

The Question of Volitional Control

Further complications enter the picture as we turn to behaviors that are not fully under volitional control. Failure to enact a behavior of this kind may occur either because of a change in intention or because performance of the behavior failed. A clear example is provided by the many smokers who intend to quit but either change their intentions or, when they do try, fail to achieve their goal. A number of investigators have in recent years turned their attention to the question of volitional control (e.g., Bandura, 1977, 1982; Kuhl, 1981, 1982; Semmer & Frese, 1979), and at least one attempt has been made to extend the theory of reasoned action to the prediction of partly nonvolitional behavior (Warshaw, Sheppard, & Hartwick, in press). The present discussion will draw on some of these analyses.

At first glance, the problem of volitional control may appear rather limited in scope. Its relevance is readily apparent whenever people try to overcome such powerful habits as smoking or drinking, or when they set their sights on such difficult to attain goals as marrying a millionaire or reducing weight. Closer scrutiny reveals, however, that even very mundane activities, which can usually be performed (or not performed) at will, are sometimes subject to the influence of factors beyond one's control. Consider, for example, a person who intends to spend the evening at home watching a movie on television. As she turns on the set, a puff of smoke indicates that its useful life is over, putting an end to her plans for the evening. Some behaviors are more likely to present problems of control than others, but we can never be absolutely certain that we will be in a position to carry our intentions. Viewed in this light it becomes clear that, strictly speaking, every intended behavior is a *goal* whose attainment is subject to some degree of uncertainty. We can thus speak of a behavior-goal unit; and the intention constitutes a plan of action in pursuit of the behavioral goal. We shall return to these issues. First, however, we must briefly consider some of the factors that influence volitional control over a behavioral goal.

Internal Factors

Many characteristics of an individual can influence successful performance of an intended behavior. Some of these internal factors are readily modified by training and experience while others are more resistant to change.

Individual Differences. At the most global level, we can conceive of differences among individuals in terms of their general ability to exercise control over their own actions. To assess such individual differences, one would have to collect information about the extent to which people manage to overcome difficulties of various kinds when attempting to perform different behaviors in a variety of settings. A person's volitional control over any given behavior may be related to this general control dimension.

To the best of my knowledge, no attempt has been made to define and assess actual behavioral control at a global level. The popular alternative is to consider people's *perceptions* of the extent to which they (as opposed to environmental factors) control events in their lives. Rotter (1966) has developed an internal-external locus of control scale designed to measure this generalized expectancy, and attempts have been made to relate scores on the scale to a variety of different behaviors (see, Lefcourt, 1982; Strickland, 1978).

Discussion of this research is beyond the scope of the present chapter but several related issues are worth considering. Generalized expectancies regarding locus of control are likely to influence behavior only to the extent that they have an impact on perceived control over the specific behavior in question. The general belief that external factors control most events in my life will have little effect on my decision to learn flying an airplane if I believe that I have control over success at this particular task.

Attribution of control (over the specific behavioral goal) to internal factors should, as a general rule, encourage attempts to perform the behavior. Whether such attempts actually succeed will, of course, depend on how realistic the person's attribution of control is. When people attribute control to internal factors, decide to perform the behavior, and proceed to implement their plans, they are likely to succeed only if they in fact have control over the behavior under consideration (see also Kuhl, 1982).

Beliefs in personal control, or lack of control, over behaviors and events should be related to perceived possession of various personal attributes and characteristics needed no perform the behaviors in question. We now turn to these more specific internal factors.

Information, Skills, and Abilities. A person who intends to perform a behavior may, upon trying to do so, discover that he lacks the needed information, skills, or abilities. Everyday life is replete with examples. We may intend to convert another person to our own political views, to help a child with his mathematics, or to repair a malfunctioning record player but fail in our attempts because we lack the required verbal and social skills, knowledge of mathematics, or mechanical aptitudes. To be sure, with experience we tend to acquire some appreciation of our abilities; yet new situations arise frequently, and failure to achieve our goals due to lack of requisite skills is the order of the day.

The idea that behavioral achievement is a function not only of intention (or motivation) but also of ability is of course not very original. Heider (1958) made it a cornerstone of his "naive analysis of action," and Jones and Davis (1965) incorporated it into their theory of correspondent inferences. According to Heider, a person is viewed as responsible for his action (i. e., the action is attributed to the person) if he is believed to have tried *and* to have had the ability to produce the observed effects. Evidence for this analysis in the context of achievement-related attributions has been provided by Weiner (1974) and his associates (e. g., Weiner, Frieze, Kukla, Reed, Rest, & Rosenbaum, 1971), as well as by other investigators (e. g., Anderson, 1974).

It seems self-evident that successful performance of an intended behavior is contingent on the presence of required information, skills, and abilities; and perhaps it is for this reason that few investigators have bothered to demonstrate empirically the interactive effect of intention and ability on actual behavioral performance.

Power of Will. Attainment of some behavioral goals requires what is commonly known as "will power" or "strength of character". Maintaining reduced weight, abstaining from alcohol or tobacco, and resisting temptations, such as going to a party instead of studying for an exam, are all familiar examples. People are often motivated to attain goals of this kind; their personal attitudes and intentions, however, may be less important than the degree to which they have control over their actions in the form of will power.

Ajzen, Averill, and Tirrell[6] collected some data of relevance in a preliminary study of temptation in several hypothetical situations. College students rated as particularly tempting the case of going to a party instead of studying. A brief paragraph outlined a scenario in which the respondents were studying on the night before an important exam and were invited to join a small party where they could get to know an attractive person they had wanted to meet for some time. Among other items, the questionnaire assessed (on a 7-point probability scale) the likelihood that the respondent would join the party (expected behavior), attitudes toward this behavior (on four evaluative semantic differential scales), and perceived control, that is, the perceived difficulty of declining the invitation (on a 7-point scale ranging from "very difficult" to "not at all difficult").

Perceived control was found to predict expected behavior significantly better $r=0.69$) than was the attitude toward the behavior ($r=0.25$).[7] Although this study used a hypothetical situation, its results suggest that people's ability to resist temptation (their willpower) may be an important determinant of certain types of behavior, over and above the influence of attitudes or intentions.

In his analysis of action control, Kuhl (1981, 1982) has introduced the some-

---

[6] Unpublished study, University of Massachusetts, 1982.
[7] The correlation of expected behavior with a measure of subjective norm (r = 0.24) was also significantly lower than its correlation with perceived control.

what related concept of state versus action orientation. A person's state or action orientation is viewed both as a relatively stable predisposition and as dependent on a variety of situational factors. Action-oriented individuals are assumed to focus their attention on action alternatives and to make use of their knowledge and abilities to control their performance. In contrast, state-oriented persons are likely to focus their attention on their thoughts and feelings (their present, past, or future state) rather than taking action consistent with their intentions. Kuhl (1982) has developed a scale to assess action orientation and, using this scale, has found higher intention-behavior correlations among action-oriented as opposed to state-oriented individuals. He has also investigated a variety of situational factors (e. g., a failure experience) that may result in state or action orientations, and thus influence execution of intended behavior (see also Mischel's, 1974, work on situational determinants of delay of gratification).

Emotions and Compulsions. Skills, abilities, and will power may present problems of control, but it is usually assumed that , at least in principle, these problems can be overcome. In contrast, some types of behavior are often viewed as controlled by forces that are largely beyond our control. People sometimes appear unable to cease thinking or dreaming about certain events, to stop stuttering, or to hold a tick in check. These compulsive behaviors are performed despite intentions and concerted efforts to the contrary.

Emotional behaviors seem to share some of the same characteristics. Individuals are often not held responsible for behaviors performed under stress or in the presence of strong emotions. We usually attribute little control to a person who is "overcome by emotion". Violent acts and poor performance are expected under such conditions, and there seems to be little a person can do about it. Some, however, have argued that emotional behaviors are not all that different from other types of behavior, and that their antecedents are very similar to those of nonemotional acts (e. g., Averill, 1980; Solomon, 1976). If this view is correct, then we should find relatively strong relations between intentions and emotional behavior, although problems of control may well prevail, especially in the case of intense emotional experiences.

In conclusion, as we move beyond intentions, various internal factors influence successful performance of an intended behavior. It may be fairly easy to gain control over some of these factors, as when we acquire the information or skills needed to perform a behavior. Other factors, such as intense emotions, stress, or compulsions, are more difficult to neutralize. Whatever the nature of the internal factor, however, it will tend to influence our control over the behavioral goal.

External Factors

Also impinging on a person's control over behavioral goals are external or situational factors. These factors will be discussed under two headings: (1) time and opportunity and (2) dependence on other people.

Time and Opportunity. It takes little imagination to appreciate the importance of circumstantial factors or opportunity. An intention to see a movie on a particular night, for example, cannot be carried through if tickets are sold out or the person is involved in a serious accident on the way to the theater. At first glance, lack of opportunity may appear equivalent to occurrence of unanticipated events that bring about changes in intentions, as discussed in an earlier section. While it is true that in the absence of appropriate circumstances people may come to change their intentions, there is an important difference between the two cases. When new information becomes available after a person has stated his intention, the new information may affect his salient beliefs about the behavior and thus lead to changes in attitudes, subjective norms, and intentions; at the end of this process the person is no longer interested in carrying out his original intention. By way of contrast, lack of opportunity disrupts an attempted behavior. Here, the person tries to carry out his intention but fails because circumstances prevent performance of the behavior. Although the immediate intention will be affected, the basic underlying determinants need not have changed.

Consider again the intention to see a particular movie on a given night. Reading a negative review or being told by a friend that the movie is not worth seeing may influence the person's beliefs such as to produce a more negative attitude toward the intended behavior and perhaps also a more negative subjective norm. As a result, the person may no longer intend to see the movie on the night in question or on any other night unless and until other events again cause him to change his mind.

Contrast this with the person who intends to see the movie, drives to the theater, but is told that there are no more tickets available. The environmental obstacle to performance of the behavior will force a change of plan; but it need not change the person's attitude toward seeing the movie or his subjective norm. Instead, it may merely cause the person to try again on a different night.

The question of time involves very similar considerations. A person who, on a given occasion, is unable to find the time required to plan and perform a behavior need not change his attitude, subjective norm, or intention; he can simply decide to perform the behavior an another occasion.

Dependence on Others. Whenever performance of a behavior depends on the actions of other people, there exists the potential for incomplete control over behavioral goals. A good example of behavioral interdependence is the case of cooperation. On can cooperate with another person only if that person is also willing to cooperate. Experimental studies of cooperation and competition in two-person games have provided ample evidence for this interdependence (see Rapoport & Chammah, 1965). For example, Ajzen and Fishbein (1970) reported correlations of 0.92 and 0.89 between cooperative strategy choices of the players in two prisoner's dilemma games. These high correlations suggest that a person's tendency to make cooperative choices depends on reciprocation by the other player.

A different example is provided by Fishbein's (1966) study of premarital sexual intercourse among undergraduate students. In this study it was found that intentions were significantly better predictors of behavior for females ($r=0.68$) than for males ($r=0.39$). Clearly, females in our society find it relatively easy to obtain the cooperation of males when they attempt to execute their intentions to engage in premarital sexual intercourse. By comparison, males often have greater difficulties in finding willing partners.

As in the case of time and opportunity, inability to carry out an intention because of dependence on others may have little effect on the underlying motivation. Often a person who encounters difficulties related to interpersonal dependence may be able to perform the desired behavior in cooperation with a different individual. Sometimes, however, this may not be possible as in the case of dependence on one's spouse. A wife's adamant refusal to have more children, for example, will usually cause the husband eventually to abandon his plan to enlarge the family, rather than shift his efforts to a different partner.

I have tried to show that time, opportunity, and dependence on others often lead only to temporary changes in intentions. When time is the constraining factor, the behavior may simply be delayed; when circumstances prevent performance of a behavior, the person may wait for a better opportunity; and when another person fails to cooperate, a more compliant partner may be sought. However, when repeated efforts to perform the behavior fail, more fundamental changes in intentions can be expected.

A Theory of Planned Behavior

The above discussion makes clear that many factors can obstruct the intention-behavior relation. Although volitional control is more likely to present a problem for some behaviors than for others, personal deficiencies and external obstacles can interfere with the performance of any behavior. Given the problem's ubiquity, a behavioral intention can best be interpreted as an intention to *try* performing a certain behavior. A father's plan to take his children fishing on the forthcoming weekend, for example, is best viewed as an intention to try to make time for this activity, to prepare the required equipment, secure a fishing license, and so forth. Successful performance of the intended behavior is contingent on the person's control over the various factors that may prevent it. Of course, the conscious realization that we can only try to perform a given behavior will arise primarily when questions of control over the behavior are salient. Thus, people say that they will try to quit smoking or lose weight, but that they intend to go to church on Sunday. Nevertheless, even the intention to attend Sunday worship services must be viewed as an intention to try since factors beyond the individual's control can prevent its successful execution.

These observations have important implications for the prediction of behavior from intentions. Strictly speaking, intentions can only be expected to predict a person's *attempt* to perform a behavior, not necessarily its actual performance. If our measure of intention fails to predict attempted behavior, it is possible that the intention changed after it was assessed (see the earlier discussion of the prediction of volitional behavior). However, if the intention does predict whether or not a person attempts to perform the behavior, but fails to predict attainment of the behavioral goal, it is likely that factors beyond the person's control prevented the person from carrying out his intention.[8]

Consider, for example, the study of weight reduction mentioned earlier (Sejwacz, Ajzen, & Fishbein, 1980). The college women in this study expressed their intentions to lose weight over a two-month period, and reported on their performance of various dietary behaviors and physical activities during that same period. The intention to lose weight was found to have a nonsignificant correlation of 0.16 with actual weight lost, but a significant correlation of 0.49 with an aggregate measure of behavior. It seems reasonable to argue that engaging in dietary behaviors and physical activities constitute attempts to reduce weight. Intentions were thus better predictors of attempts to reduce weight, than of actual changes in body weight. Clearly, losing weight depends not only on one's intention to do so, but also on other factors, such as will power and physiological variables that are only partly under volitional control. That the correlation between intentions and attempted behavior was no higher than 0.49 can perhaps be attributed to changes in intentions that may have occurred over the two-month period. Alternatively, it may have been due to a lack of will power, or to a state (as opposed to action) orientation (Kuhl, 1982).

Additional support for the present argument can be found in a study by Pomazal and Jaccard (1976). The intentions of college students to "donate blood at the upcoming blood drive" had a correlation of 0.46 with actual blood donations. However, it was noted that several participants in the study who came to donate blood were rejected for medical reasons or because of overcrowding. As expected, when showing up to donate blood (i.e., attempted behavior) was used as the behavioral criterion, the correlation with intentions increased to 0.59.

In short, behavioral intentions will often be better predictors of attempted than actual behavior. To insure accurate prediction in such instances, we would not only have to assess intentions but also obtain some estimate of the extent to which individuals are apt to exercise control over the behavior in question.

We are now ready to consider a possible expansion of the theory of reasoned action that will allow us to include consideration of nonvolitional factors as determinants of behavior. Equation 4 shows how the strength of a person's attempt to perform a behavior (B_t) interacts with the degree of his control (C) to deter-

$$B \propto B_t \cdot C \qquad\qquad\qquad (4)$$

mine the likelihood of actual performance of the behavior (B). The harder the person tries, and the greater his control over personal and external factors that may interfere, the greater the likelihood that he will attain his behavioral goal.[9] For some behaviors, a low level of effort (B_t) is sufficient, and successful perfor-

---

[8] Another possibility which cannot be discounted is that the person changed his intention while trying to perform the behavior.

[9] The notion of control is used here quite similar to Triandis's (1977) concept of "facilitative conditions" (F) in his model of interpersonal behavior. However, in Triandis's model, F interacts with intentions and habits to determine the likelihood of a behavior.

mance of the behavior depends largely on the level of control. Good examples are highly skilled activities such as typing or driving a race car. Increased effort on such tasks is less important than a high level of skill. For other behaviors, a minimal level of control is enough, and successful performance of the behavior varies with degree of effort. For instance, jogging 10 or 15 minutes every day requires only low levels of control; for most people, achievement of this behavioral goal depends largely on their willingness to try.

The question of control is often tied up with development of an adequate *plan* that will enable performance of the behavior. A plan usually consists of a set of intentions which, if carried out, are expected to result in the desired behavioral goal. It may also contain contingency plans; that is, alternative plans of action in case the intended sequence of behaviors is blocked. Often, plans are developed only in general outline: The initial behaviors may be clearly specified and later parts of the plan are to be developed, depending in part on the success of earlier actions (see Miller, Galanter, & Pribram, 1960). The behavioral attempt *(B_t)* constitutes the initiation of the plan; it is designed to overcome a perceived discrepancy between the present state and the desired goal. Successful execution depends on the adequacy of the plan itself, and on the various personal and external factors discussed earlier that may influence control over the behavior.

Consistent with the theory of reasoned action, the immediate determinant of a person's attempt to perform a behavior is his intention to try doing so *(I_t)*; and this intention is in turn a function of attitude toward trying *(A_t)* and subjective norm with regard to trying *(SN_t)*. These relations are expressed symbolically in Equation 5.

$$B_t \sim I_t \propto [w_1 A_t + w_2 SN_t] \tag{5}$$

Here, as in Equation 1, the wavy line between B_t and I_t indicates that expressed intentions to try performing a behavior can change before the behavioral attempt is observed, and w_1 and w_2 are empirically determined weights for the two predictors of I_t. Thus, the more favorable a person's attitude toward trying to perform a behavior, and the more he believes that important others think he should try, the stronger his intention to try.

We must go beyond the theory of reasoned action, however, when we consider the determinants of A_t, the attitude toward trying to perform a behavior. Clearly, the attitude toward trying and succeeding (i.e., the attitude toward the behavior) will usually differ from the attitude toward trying and failing. Whenever the possibility of failure is contemplated, therefore, the attitude toward trying will be determined not only by the attitude toward (successful) performance of the behavior *(A_s)* but also by the attitude toward a failed attempt *(A_f)*. This idea is expressed in Equation 6, where the attitudes toward behavioral success and

$$A_t \propto [p_s A_s + p_f A_f]; \ p_s + p_f = 1.0 \tag{6}$$

failure are weighted by the respective subjective probabilities of these events (p_s and p_f). The subjective probabilities of success and failure should be related to beliefs about the presence or absence of personal and external factors discussed earlier that may facilitate or inhibit behavioral performance. This formulation is structurally similar to Lewin's treatment of level of aspiration (Lewin, Dembo, Festinger, & Sears, 1944). According to Lewin's analysis, a goal's valence is equal to the product of the valence of achieving success times the subjective probability of success, minus the product of the valence of failure times the subjective probability of failure. If we substitute "behavioral attempt" for "goal" and "attitude" for "valence," then Equation 6 is equivalent to Lewin's formulation.[10]

Inspection of Equation 6 shows that the attitude toward trying to perform a behavior is equal to the attitude toward the behavior (A_s) when success is certain $(p_s = 1$ and $p_f = 0)$; that is, when the possibility of failure does not enter a person's considerations.

As was true of attitude toward a behavior, attitudes toward successful and unsuccessful behavioral attempts can be viewed as determined by underlying beliefs. Thus, A_s should be a function of salient beliefs concerning the likely consequences of successfully performing the behavior; these beliefs will be much the same as the beliefs about performing a volitional behavior where success is implicitly assumed. In contrast, A_f should be determined by salient beliefs concerning the likely outcomes of a failed behavioral attempt; these beliefs would normally not enter a person's considerations in the case of a behavior viewed as volitional.

It is possible to extend a similar analysis to the determinants of subjective norms. Thus, a person's belief that most important referents would approve or disapprove of his attempting to perform a given behavior might be viewed as a function of two prior subjective norms: one applying to a successful attempt, the other to an unsuccessful attempt. However, this distinction appears less relevant in the case of subjective norms than in the case of attitudes toward trying. When a person believes that important referents think he should try to perform a behavior, this subjective norm will in most cases be independent of success or failure; it has to do more with the social desirability of trying than with the likelihood of success. To be sure, the person may believe that, after the fact, his important others will react very differently to success and failure, but these are *behavioral* beliefs regarding the consequences of a successful or unsuccessful attempt, not normative beliefs. They should thus influence the attitude toward trying, but not the subjective norm with respect to a behavioral attempt.

In the case of subjective norm, a simpler model is proposed. The subjective norm for trying to perform a behavior (SN_t) is viewed as a function of the subjective norm concerning (successful) performance of the behavior (SN), multiplied by the subjective probability of success attributed to the referents (p_r), as shown in Equation 7.

$$SN_t \propto p_r SN \tag{7}$$

---

[10] Warshaw, Sheppard, and Hartwick (in press) have also proposed a similar formulation. In their analysis, attitude toward pursuing a goal is determined by attitude toward pursuit with success and attitude toward pursuit with failure, each weighted by the expectancy (probability) of success or failure, respectively. In addition, they proposed that attitudes toward actions involved in pursuit of the goal also be considered. The present analysis assumes that these latter attitudes are reflected in attitudes toward successful and unsuccessful behavioral attempts.

In other words, important referents are viewed as recommending an attempt when they approve of the behavior and believe the attempt is likely to succeed. For example, a woman may believe that her husband, children, and best friends think she should get a job. If she also believes that, according to these important referents she has a good chance of finding a job, she will come to form the belief that her important others think she should try to get a job.

As to the beliefs that underlie subjective norms toward trying to perform a behavior, two possible approaches can be suggested. One is to consider the perceived normative expectations of specific referents with respect to a behavioral attempt. A second approach would be to follow Equations 7 and 3. That is, one would assess subjective norm regarding the behavior (SN), and multiply this variable by the referents' perceived probability of success (p_r), as suggested in Equation 7. To understand the determinants of SN, one would elicit salient normative beliefs regarding specific referents, multiply each by the corresponding motivation to comply, and sum the products. This analysis was described earlier and is summarized in Equation 3.

The discussion of the theory of planned behavior up to this point is presented schematically in Fig. 2.1. The one new feature in Fig. 2.1 is the introduction of BE, behavioral expectation. In their extension of the theory of reasoned action, Warshaw, Sheppard, and Hartwick (in press) stressed the distinction between what a person *intends* to do (I) and what he expects he actually *will* do (BE). Behavioral expectation thus refers to a person's estimate of the likelihood that he actually will perform a certain behavior. Generally speaking, people will expect to perform a behavior if they intend to try it (I_t) and if they believe (have a high subjective probability) that they can control it (b_c), as shown in Equation 8.

$$BE \propto b_c I_t \tag{8}$$

Behavioral intentions and expectations will tend to differ whenever respondents anticipate that their intentions might change or when they believe that attainment of their behavioral goal is not completely under their volitional control. According to Warshaw, Sheppard, and Hartwick (in press), behavioral expectations are therefore likely to predict actual behavior more accurately than are behavioral intentions.

Inspection of Fig. 2.1 shows that, according to the present analysis, BE would be expected to correlate with *attempted* behavior (B_t) since these two variables

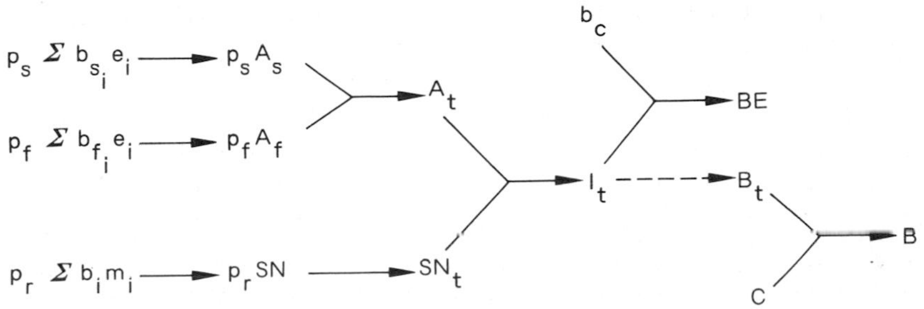

Fig. 2.1. Schematic presentation of the theory of planned behavior

are influenced by some of the same factors. The correlation between behavioral expectation and *actual* behavior *(B)*, however, will depend on correspondence between the person's belief in his control over the behavior *(b_c)* and the degree of his actual control *(C)*. To the extent that the person's assessment of his skills, will-power, and other requisite personal factors, and of the presence of favorable or unfavorable external factors is realistic, his behavioral expectation will predict his actual behavior. However, when his assessment fails to accurately reflect real-ity, prediction of behavior from behavioral expectation will suffer. In any case, it must be kept in mind that unlike behavioral intentions, behavioral expectations may have no *causal* effect on actual behavior.

From a practical point of view, it will be very important to identify factors that correlate with realistic perception of behavioral control. In addition to past expe-rience, these factors may include confidence in one's subjective judgment of control, availability of a detailed plan of action, and general self-knowledge. To the extent that perceived control is likely to be realistic, it can serve as an esti-mate of actual control and, together with intention to try, it can be used to predict the probability of a successful behavioral attempt.

Subjective perceptions of control may, of course, influence attempts to per-form behavior regardless of their accuracy. A person who has a pessimistic view of his control over the behavior may never try and may thus fail to find out that he was wrong. As a result, perceived control will usually correlate with behavior-al performance. Again, however, this correlation will tend to be strong only when perceived control corresponds reasonably well to actual control.

Bandura's (1977, 1982) work on "self-efficacy" provides support for the rela-tion between perceived control and behavioral performance, usually in the form of overcoming certain phobias or strong habits. For example, in one study (Ban-dura, Adams, & Beyer, 1977), adult snake phobics following a period of treat-ment rated the strength of their expectations that they could perform various behaviors in relation to snakes. Correlations between these estimates of self-effi-cacy (behavioral control) and subsequent behavior toward a snake were 0.83 and 0.84 for two groups of participants who had received performance-based as op-posed to vicarious treatment, respectively. Experiences during treatment (and perhaps prior to treatment) seemed to have created quite accurate perceptions of self-efficacy. Since all participants were there to try to overcome their phobias, attainment of the behavioral goal was dependent primarily on their (perceived) control over the behavior involved.

Another issue worth considering at this point is the influence of past behavior on present performance. Some investigators have suggested that past perfor-mance of a behavior exerts an influence on present behavior that is independent of behavioral intentions, attitudes, or subjective norms. For example, Bentler and Speckart (1979) reported a study in which college students' use of alcohol, marijuana, and hard drugs at one point in time made a significant contribution independent of intentions to the prediction of the performance of these behav-iors two weeks later. According to the present analysis, past performance of a be-havior may be correlated with control over the behavior *(C)*. In the case of addic-tive behaviors, such as drinking alcohol or using drugs, frequent past behavior is likely to be associated with lack of control. A person who is addicted may intend not to take hard drugs or drink excessively, but lacks the control to achieve his behavioral goal. As a result of its correlation with control, past behavior would in fact be expected to influence present behavior over and above the effect of intention (see Equation 4). According to the theory of planned behavior, past performance should have no independent effect on present behavior only when a person has complete control, that is, when dealing with a volitional behavior. Drinking and use of drugs clearly do not qualify.

Examination of Equations 4 through 8 and Fig. 2.1 shows that the theory of reasoned action discussed at the beginning of this chapter is a special case of the theory of planned behavior. The special case occurs when the subjective probability of success or perceived control (p_s or b_c) and actual degree of control over the behavioral goal (C) are at their maxima. This can be seen clearly if both are scaled from 0 to 1. When p_s and C equal 1, the likelihood of a behavioral attempt is equivalent to the likelihood of performing the behavior (see Equation 4), attitude toward trying reduces to attitude toward the behavior (Equation 6), and subjective norm with respect to trying is the same as subjective norm regarding performance of the behavior (Equation 7). In addition, when b_c equals 1, behavioral expectation becomes equivalent to behavioral intention (Equation 8). Of course, when p_s or b_c are equal to 1 (failure is not even considered as a possibility) and when actual control (C) is perfect, the behavior is under complete volitional control and the theory of reasoned action applies. However, when the possibility of failure is salient and actual control is limited, then it becomes necessary to go beyond the theory of reasoned action. It is here that the theory of planned behavior will prove most useful.

Summary and Conclusions

Successful performance of social behavior was shown to depend on the degree of control a person has over internal and external factors that may interfere with the execution of an intended action. The extent to which attainment of a behavioral goal depends on skills, abilities, will power, or opportunity varies, however, from behavior to behavior. When factors of this kind exert a negligible influence on successful performance of a behavior and the possibility of failure is not a salient consideration, the behavior may be said to be under volitional control; the immediate and only determinant of such a behavior is the intention to perform it. A measure of intention is thus expected to permit accurate prediction of volitional behavior, unless the intention changes after it is assessed but before the behavior is observed. Evidence for strong intention-behavior relations has been obtained in numerous applications of Fishbein and Ajzen's (1975; Ajzen & Fishbein, 1980) theory of reasoned action. In addition, this research has shown that it is possible to explain differences in intentions and behavior in terms of attitudes toward the behavior and subjective norms, and then to trace the determinants of these variables to the underlying behavioral and normative beliefs.

The theory of reasoned action applies to behaviors that are under volitional control. Its predictive accuracy diminishes when the behavior is influenced by factor over which at least some people have only limited control. A theory of planned behavior was proposed which expands the theory of reasoned action and permits it to deal with behaviors of this kind. According to the proposed theory, social behavior follows more or less well developed plans. The success of an attempt to execute the behavioral plan depends not only on the effort invested (the strength of the attempt) but also on the person's control over other factors, such as requisite information, skills, and abilities, including possession of a workable plan, willpower, presence of mind, time, opportunity, and so forth.

Whether or not an attempt is made to perform a given behavior and the strength of that attempt are determined in an immediate sense by the intention to try performing the behavior. This intention is in turn a function of two factors: the attitude toward trying and the subjective norm with regard to trying. The attitude toward trying is based on two separate attitudes, one toward a successful behavioral attempt and one toward an unsuccessful attempt, each weighted by the subjective probability of the event in question. Finally, these two attitudes are determined by salient beliefs regarding the consequences of a successful or unsuccessful behavioral attempt and by evaluations of these consequences.

In a somewhat similar manner, the subjective norm with respect to a behavioral attempt is viewed as based on the subjective norm regarding (successful) performance of the behavior, weighted by judgments of the likelihood of success as attributed to important social referents.

According to the theory of planned behavior, therefore, the considerations that, in the final analyses, enter into the determination of a behavioral attempt include beliefs about the likely consequences of success and failure, the perceived probabilities of success and failure, normative beliefs regarding important referents, and motivations to comply with these referents. Generally speaking, a person will attempt to perform a behavior if he believes that the advantages of success (weighted by the likelihood of success) outweigh the disadvantages of failure (weighted by the likelihood of failure), and if he believes that referents with whom he is motivated to comply think he should try to perform the behavior. He will be successful in his attempt if he has sufficient control over internal and external factors which, in addition to effort, also influence attainment of the behavioral goal.

The theory of reasoned action was shown to represent a special case of the proposed theory of planned behavior. The two theories are identical when the subjective probability of success and the degree of control over internal and external factors reach their maximal values. When this is the case, we are dealing with purely volitional behavior to which the theory of reasoned action can be directly applied. When subjective probabilities of success and actual control are less than perfect, however, we enter the domain of the theory of planned behavior.

References

Ajzen, I. (1971). Attitudinal vs. normative messages: An investigation of the differential effects of persuasive communication on behavior. *Sociometry, 34,* 263–280.

Ajzen, I. & Fishbein, M. (1970). The prediction of behavior from attudinal and normative variables. *Journal of Experimental Social Psychology, G,* 466–487.

Ajzen, I., & Fishbein, M. (1977). Attitude-behavior relations: A theoretical analysis and review of empirical research. *Psychological Bulletin, 84,* 888–918.

Ajzen, I., & Fishbein, M. (1980). *Understanding attitudes and predicting social behavior.* Englewood-Cliffs, N.J.: Prentice-Hall.

Ajzen, I., Timko, C., & White, J.B. (1982) Self-monitoring and the attitude-behavior relation. *Journal of Personality and Social Psychology, 42,* 426–435.

Ajzen, I., White, J.B., & Timko, C. (1982). Direct experience, confidence, and the attitude-behavior relation. Unpublished manuscript, University of Massachusetts.

Anderson, N.H. (1974). Cognitive algebra: Integration theory applied to social attribution. In L. Berkowitz (Ed.), *Advances in experimental social psychology,* Vol.7. New York: Academic Press.

Averill, J.R. (1980). A constructivist view of emotion. In R. Plutchik & H. Kellerman (Eds.), *Emotion: Theory, research, and experience. Volume 1: Theories of emotion.* New York: Academic Press.

Bandura, A. (1977). Self-efficacy: Toward a unifying theory of behavioral change. *Psychological Review. 84,* 191–215.

Bandura, A. (1982). Self-efficacy mechanism in human agency. *American Psychologist, 37,* 122–147.

Bandura, A., Adams, N.E., & Beyer, J. (1977). Cognitive processes mediating behavioral change. *Journal of Personality and Social Psychology, 35,* 125–239.

Beckmann, J., & Kuhl, J. (1984). Deforming information to gain action-control: Functional aspects of human information processing in decision making. *Journal of Research in Personality, 18,* 224–237.

Bentler, P.M., & Speckart, G. (1979). Models of attitude-behavior relations. *Psychological Review, 86,* 452–464.

Brown, J.S. (1948). Gradients of approach and avoidance responses and their relation to motivation. *Journal of Comparative and Physiological Psychology, 41,* 450–465.

Bumpass, L., & Westoff, C. (1969). *The later years of child bearing.* Princeton, NJ: Princeton University Press.

Edwards, W. (1954). The theory of decision making. *Psychological Bulletin, 51,* 380–417.

Fazio, R.J., & Zanna, M. (1978). Attitudinal qualities relating to the strength of the attitude-behavior relationship. *Journal of Experimental Social Psychology, 14,* 398–408.

Feather, N.T. (1959). Subjective probability and decision under uncertainty. *Psychological Review, 66,* 150–164.

Fishbein, M. (1966). Sexual behavior and propositional control. Paper read at the Psychonomic Society meetings.

Fishbein, M., & Ajzen, I. (1975). *Belief, attitude, intention, and behavior: An introduction to theory and research.* Reading, Mass.: Addison-Wesley.

Fishbein, M., & Coombs, F.S. (1974). Basis for decision: An attitudinal analysis of voting behavior. *Journal of Applied Social Psychology, 4,* 95–124.

Heider, F. (1958). *The psychology of interpersonal relations.* New York: Wiley.

Hornik, J.A. (1970). Two approaches to individual differences in an expanded Prisoner's Dilemma game. Unpublished master's thesis, University of Illinois.

Jones, E.E., & Davis, K.E. (1965). From acts to dispositions: The attribution process in person perception. In L. Berkowitz (Ed.), *Advances in experimental social psychology,* Vol.2. New York: Academic Press.

Kuhl, J. (1981). Motivational and functional helplessness: The moderating effect of state versus action orientation. *Journal of Personality and Social Psychology, 40,* 155–170.

Kuhl, J. (1982). Action- vs. state-orientation as a mediator between motivation and action. In W. Hacker, W. Volpert, & M. von Cranach (Eds.), *Cognitive and motivational aspects of action.* Amsterdam, NL: North-Holland Publishing Co.

Lefcourt, H.M. (1982). *Locus of control: Current trends in theory and research* (2nd Ed.). Hillsdale, NJ: Erlbaum.

Lewin, K. (1946). Action research and minority problems. *Journal of Social Issues, 2*, 34–46.

Lewin, K. (1951). *Field theory in social science.* New York: Harper.

Lewin, K., Dembo, T., Festinger, L., & Sears, P.S. (1944). Level of aspiration. In J.McV. Hunt (Ed.), *Personality and the behavior disorders.* New York: Ronald Press.

Manstead, A.S.R., Proffitt, C., & Smart, J.L. (1983). Predicting and understanding mothers' infant feeding intentions and behavior: Testing the theory of reasoned action. *Journal of Personality and Social Psychology, 44,* 657–671.

Miller, G.A., Galanter, E., & Pribram, K.H. (1960). *Plans and the structure of behavior.* New York: Holt.

Miller, N.E. (1944). Experimental studies of conflict. In J.McV. Hunt (Ed.), *Personality and the behavior disorders.* New York: Ronald Press.

Mischel, W. (1974. Processes in delay of gratification. In L.Berkowitz (Ed.), *Advances in experimental social psychology* (Vol.7). New York: Academic Press.

Pomazal, R.J., & Jaccard, J.J. (1976). An informational approach to altruistic behavior. *Journal of Personality and Social Psychology, 33,* 317–326.

Rapoport, A., & Chammah, A.M. (1965). *Prisoner's dilemma: A study on conflict and cooperation.* Ann Arbor: University of Michigan Press.

Rosenberg, M.J. (1956). Cognitive structure and attitudinal affect. *Journal of Abnormal and Social Psychology, 53,* 367–372.

Rotter J.B. (1966). Generalized expectancies for internal versus external control of reinforcement. *Psychological Monographs, 80* (1, Whole No.609).

Sample, J., & Warland, R. (1973). Attitude and prediction of behavior. *(Social Forces, 51,* 292–303.

Semmer, N., & Frese, M. (1979). Handlungstheoretische Implikationen für kognitive Therapie. In N.Hoffman (Ed.), *Grundlagen kognitiver Therapie: Theoretische Modelle und praktische Anwendungen.* Bern: Huber.

Sejwacz, D., Ajzen, I., & Fishbein, M. (1980). Predicting and understanding weight loss: Intentions, behaviors, and outcomes, In I.Ajzen & M.Fishbein, *Understanding attitudes and predicting social behavior.* Englewood Cliffs, N.J.: Prentice-Hall.

Sherman, S.J. (1980). On the self-erasing nature of errors of prediction. *Journal of Personality and Social Psychology, 39,* 211–221.

Shtilerman, M. (1982). The influence of decision making process on the relationship between attitudes, intentions and behavior. Unpublished master's thesis. Tel-Aviv University.

Smetana, J.G., & Adler, N.E. (1980). Fishbein's value x expectancy model: An examination of some assumptions. *Personality and Social Psychology Bulletin, 6,* 89–96.

Snyder, M. (1974). Self-monitoring of expressive behavior. *Journal of Personality and Social Psychology, 30,* 526–537.

Snyder, M. (1982). When believing means doing: Creating links between attitudes and behavior. In M.P.Zanna, C.P.Herman, & E.T.Higgins (Eds.), *Consistency in social behavior: The Ontario Symposium,* (Vol.2). Hillsdale, N.J.: Erlbaum.

Snyder, M., & Swann, W.B., Jr. (1976). When actions reflect attitudes: The politics of impression management. *Journal of Personality and Social Psychology, 34,* 1034–1042.

Solomon, R.C. (1976). *The passions.* New York: Anchor Press.

Songer-Nocks, E. (1976). Situational factors affecting the weighting of predictor components in the Fishbein model. *Journal of Experimental Social Psychology, 12,* 59–69. (a)

Songer-Nocks, E. (1976). Reply to Fishbein and Ajzen. *Journal of Experimental Social Psychology, 12,* 585–590. (b)

Strickland, B.R. (1978). Internal-external expectancies and health-related behavior. *Journal of Consulting and Clinical Psychology, 46,* 1192–1211.

Triandis, H.C. (1977). *Interpersonal behavior.* Monterey, CA: Brooks/Cole.

Vinokur-Kaplan, D. (1978). To have – or not to have – another child: Family planning attitudes, intentions, and behavior. *Journal of Applied Social Psychology, 8,* 29–46.

Warland, R.H., & Sample, J. (1973). Response certainty as a moderator variable in attitude measurement. *Rural Sociology, 38,* 174–186.

Warshaw, P. R., Sheppard, B. H., & Hartwick, J. (in press). A general theory of intention and behavioral self-prediction. In Bagozzi (Ed.), *Advances in Marketing Communication*. Greenwich, CT: JAI Press

Weiner, B. (1974). *Archievement motivation and attribution theory*. Morristown, N.J.: General Learning Press.

Weiner, B., Frieze, I., Kukla, A., Reed, L., Rest, S., & Rosenbaum, R. M. (1971). *Perceiving the causes of success and failure*. New York: General Learning Press.

Zanna, M. P., Olson, J. M., & Fazio, R. H. (1980). Attitude-behavior consistency: An individual difference perspective. *Journal of Personality and Social Psychology, 38,* 432–440.

Zuckerman, M., & Reis, H. T. (1978). A comparison of three models for predicting altruistic behavior. *Journal of Personality and Social Psychology, 36,* 498–510.

Chapter 3

Knowing What to Do: On the Epistemology of Actions

Arie W. Kruglanski and Yechiel Klar

Do people know what they are doing? This admittedly simplistic phrasing conceals some questions of fundamental significance to a psychological theory of actions: Are human actions thoughtful and rational or are they often mindless and automatic? Are they consciously determined or do they frequently stem from unconscious forces inaccessible to human cognizance? In the present chapter we deal with some of the foregoing problems using as our frame of reference a theory of lay epistemology developed recently by Kruglanski and his colleagues (see, Kruglanski, 1980; Kruglanski & Ajzen, 1983; Kruglanski, Baldwin, & Towson, in press; Kruglanski & Klar, 1982, Bar-Tal, Yarkin, & Bar-Tal, 1982). Our central thesis will be that voluntary actions by definition are based on intentions which essentially are types of knowledge about what it is that a person wishes to do. Thus, a general theory about the acquisition of all knowledge should be of help in elucidating how a particular knowledge is acquired, in this instance, the knowledge of one's intentions.

Given that the process of action control is comprised of (1) intention formation and (2) action formation (out of the intention), the present paper deals, mainly, with the first part. However, we will also apply our analysis to the intention-action relations. We begin with a brief outline of our epistemic theory stressing especially those assumptions and arguments that are of particular relevance to the problem of actions. We then apply the theory to the following major topics in action theory: (1) the expectancy-value approach to human motivation, (2) the attitude-behavior relation, (3) mindless versus thoughtful actions, and (4) the unconscious determinants of actions.

A Theory of Lay Epistemology

Our theory of lay epistemology addresses the process whereby people acquire all their knowledge. For instance, one might *know* that "it is daytime now", that "the Earth is round" or that one "intends to go to the theatre." From our standpoint all the above propositions represent different types of knowledge and are reached via the same basic process. We find it useful to consider knowledge in terms of its two aspects; its contents and the confidence one has in the contents in question. For instance, one may have little confidence that "it is raining now," and a great deal of confidence that "it will be dark here by 10 p. m.," and so forth.

The distinction between contents and confidence suggests the two basic functions that the epistemic process has to fulfill. First, the contents of knowledge need to be generated somehow. Thus, we need to have a phase of *cognition generation*. Secondly a given degree of confidence must be bestowed on the generated contents. Thus, we need to have the phase of *cognition validation*. These separate phases of the epistemic process are further discussed in what follows.

Cognition Generation. Where do the contents of our knowledge come from? In some respects the origins of knowledge are quite enigmatic. Cognitions get somehow generated in our minds. They seem to crystalize out of the "stream of consciousness" (James, 1890), depending as it may on attentional shifts, the salience of ambient stimuli (Taylor & Fiske, 1978; Rumelhart & Ortony, 1977), and the momentary mental availability of various ideas (Tversky & Kahneman, 1974). Besides being influenced by transient factors like saliency and availability, people's capacity to generate cognitions on a specific topic may depend on their intelligence and creativity.

Furthermore, peoples' *motivational states* may determine their tendency to generate ideas on a given topic. Three major epistemic motivations have been identified. The need for cognitive structure, the fear of invalidity and the need for specific conclusions. The *need for cognitive structure* is a desire to have some knowledge on a given topic; to have any firm knowledge as opposed to ambiguity, doubt, or confusion. A need for structure could stem from various sources: intolerance for ambiguity (Frenkel-Brunswick, 1949; Smock, 1955), lack of time, the obligation to reach a decision, or the need for immediate action. Arousal of the need for structure in a given domain may make one inclined to quickly generate a pertinent hypothesis, refrain from probing it too deeply, and confronting it with further, potentially inconsistent evidence, alternative hypotheses, or competing action tendencies. The *fear of invalidity* stems from an anticipated cost of a mistaken judgment. Its arousal may make one inclined to generate numerous alternative hypotheses on a given topic and prudently attend to all the relevant evidence one can find. Finally, a *conclusional need* is the need to uphold a conclusion with a particular content. Any wish or motivation relating to that particular content might underly a conclusional need (e. g., to possess a certain aptitude, be in good physical condition, be loved and respected, have a bank account that is not badly in the red, and so forth). A conclusional need might be aroused by a commitment to a particular idea or to a certain course of action. The preference for desirable conclusions may dispose a person to generate alternatives to an undesirable hypothesis or, to refrain from generating alternatives to a desirable hypothesis. Sometimes, it may lead one to a state of ambiguity if the only available unambiguous information seemed painful or unpleasant (see Snyder & Wicklund, 1981); at other times it might lead one to bolster a preferred alternative (Beckmann & Kuhl, 1984).

Cognition Validation: The Problem of Proof. Our theory of lay epistemology assumes that the process of proving or validating our knowledge is invariably deductive. The knower begins with a conditional (if-then) premise interrelating two concepts. When the occasion arises to instantiate or affirm the antecedent term of the conditional, the consequent term can be deductively "proven." For instance, a person might initially believe that "if drops of water are falling from the sky . . . then it is raining." This person might then conclude, upon observing falling drops on some subsequent occasion that it is raining.

In proving a given idea we thus transmit our subjective confidence from our premises to our conclusions. Of course, persons may vary greatly in the types of premises to which they subscribe. Accordingly, they will go about proving their ideas by collecting vastly different categories of evidence. A cultist, for example, might believe that a statement is worth believing only if it is endorsed by the guru; an inductivist might believe that if very many ravens are black then all ravens are, and an experimental scientist might believe that a hypothesis is true only if a specific empirical outcome is observed in a controlled experiment. Note that while the contents of the above premises are rather different, they all share the conditional (if-then) form and are validated in an identically deductive way (e. g., by confirming their antecedent term): The experimentalist may accept the hypothesis if he or she experimentally observes the requisite datum, the inductivist will accept that all ravens are black upon encountering an abundance of black ravens and the cultist may accept a statement if it is endorsed by the guru. In sum, any concept can be conditionally linked with any other concept in a person's phenomenology. Thus, peoples' premises may differ and so may the evidence that they consider relevant or legitimate. However, the deductive process of proving one's ideas is nonetheless identical in all instances.

"Unfounded" Knowledge. Not every single bit of confident knowledge needs to be linked in a person's awareness with other knowledge from which it may have been deduced. Once a given proposition has been accepted as true a person may forget its evidential basis. For instance, we may be certain that the area under a circle is the largest for a given perimeter or that Earth rotates around the sun – without quite remembering the deductive support for our beliefs. In this sense persons may not be conscious of the reasons for their behavior, which (if pushed a bit) could be taken to mean that some of our behaviours may have "unconscious" determinants.

Furthermore, some of our beliefs might never have been deducted from evidence. A person might sometimes "just know" that something is the case, have a strong intuition, or a premonition that it is true without experiencing a need to prove such knowledge in any way. In lay epistemic terms, this could occur because of strong motivational forces suppressing the generation of alternative hypotheses and preventing the recognition of evidence inconsistent with one's current beliefs. Alternatively, a person may lack the capacity to come up with alternatives to a given hypothesis, that is, be unable to imagine to him or herself that things could be different than presently conceived. An individual need not be reflexively aware of these extralogical underpinnings of his or her knowledge.

In this sense, too, we may speak of unconsciously, or nonconsciously, determined actions.

A Threat to Confidence: A Logical Inconsistency. Just as consistency with (i. e., deducibility from) a firmly accepted premise subjectively validates (or proves) a given conclusion, an inconsistency with a firmly believed proposition subjectively disproves a hypothesis. An inconsistency between two equally believable propositions brings about the experience of doubt; such doubt can be removed only if the inconsistency is resolved. In turn, resolution of an inconsistency is accomplished by denying one of the contradictory cognitions, that is deciding which of the two is false (cf. Kruglanski & Klar, 1982). Thus, given two inconsistent cognitions, the knower is likely to give up or deny the one held with the lesser degree of confidence. In case of an inconsistency between a "hypothesis" and a "fact," we usually abandon the hypothesis because the fact label denotes a more confidently held cognition than a hypothesis.

In the case of action-relevant cognitions, doubt (caused by the incompatibility of two or more equally compelling action alternatives) will hardly result in action; at least, action based on doubt will be very fragile or hesitant (see, Beckmann & Irle, Chapter 7 this volume). In addition, the screening of an intended action might be endangered (see, Kuhl, Chapter 6, this volume). This is referred to in what follows as epistemic "freezing" and "unfreezing."

The Tentativeness of Knowledge: Freezing and Unfreezing Influences. Any currently believed proposition may be subsequently undermined and rejected. This may happen if the knower encounterred a seemingly undeniable bit of knowledge inconsistent with the belief in question. For instance, one might have believed all one's life that the Earth is round or that milk is good for babies, yet readily abandon such beliefs after their validity was denied by respectable scientific authorities or by trusted media of communication. In principle, it is possible to engender plausible rival alternatives to any judgment, conclusion, or decision tendency. This means that for all those numerous beliefs to which we are currently committed there may exist plausible alternatives which, thus far, we have not managed to generate. In the language of lay epistemology, we have "frozen" on our current beliefs because of factors having to do with our capacity and/or motivation to generate alternative hypotheses on a given topic.

Some of the postulated motivational influences on the freezing and unfreezing of beliefs were studied in a recent series of experiments by Kruglanski and Freund (1983). In these experiments predictions derived from the epistemic theory were applied to such seemingly disparate phenomena as primacy effects in impression formation (cf., Luchins, 1957), ethnic stereotyping (cf., Hamilton, 1979), and anchorage of numerical estimates in initial values (cf., Tversky & Kahneman, 1974). All of these phenomena can be interpreted as instances of epistemic freezing. Primacy effects reflect the crystallization of impressions after scanning only the early information about a person and remaining insensitive to later information. Similarly, stereotyping may reflect the basing of judgments on preexisting categories and an insensitivity to fresh evidence. Finally, anchoring may reflect the persisting impact of initial estimates and the relative inefficacy of subsequent evidence in occasioning a revision of those estimates.

Based on this line of reasoning, Kruglanski and Freund (1983) conducted three experiments in which need for structure, manipulated by degrees of time pressure was orthogonally crossed with need for validity manipulated by expected loss of face in the case of a judgmental inaccuracy. In the experiment dealing with ethnic stereotyping, for example, students at a teacher's seminary were asked to evaluate a composition supposedly written by a local eighth grader. Stereotyping was assessed as the degree to which evaluations seemed to be based on the child's ascribed ethnic background, rather than on examination of the essay. Subjects were given either a full hour (low need for structure) or only ten minutes (high need for structure) to complete their grading. In the low need for validity condition, subjects were assured that the purpose of the research was not to assess in any way the correctness of their judgments, but rather to identify possible individual differences in evaluative style. In the high need for validity condition, subjects were told that they would be asked to defend their grading before a group of peers. The results of this experiment lent strong support to the lay epistemic analysis with freezing being significantly stronger under high versus low need for structure, and significantly weaker under high versus low need for validity. These results were replicated in two other experiments in which primacy effects and numerical anchoring were the judgmental phenomena under study.

The Searchlight Metaphor. Our preceding portrayal of the epistemic process fits well with Karl Popper's (1973) "search-light" metaphor of human consciousness. Our mental capacity is limited so that at any given time we can consciously keep track of (or focus on) a highly restricted portion of our conceptual repertoire. The "search-light" of cognizance can move about, rapidly illuminating different subsets of notions at different moments. Thus, we may lose track of the deductive bases of our conclusions and become unaware of reasons for our firm beliefs. Furthermore, we may accept propositions as valid without realizing that they contradict other firm beliefs of which we are not momentarily cognizant. It follows that the doubt-inducing effect of cognitive contradictions may hold only for contradictions of which a person is consciously aware. Our total repertoire of beliefs can be fraught with inconsistencies which we may not have the opportunity to recognize.

An Epistemic Analysis of Actions

Any action begins with an *intention* which is then transformed into an external event like raising an arm, walking, or turning on the light. Accordingly, two issues require explanation: (1) How is an intention formed? and (2) How is it transformed into an action? The present chapter is mostly concerned with the former question. But let us begin by briefly considering the latter. An intention may be conceived of as some desired state which is compared with the outcomes of various activities that may be performed at a given time. Thus, we postulate a feedback loop (cf., Miller, Galanter, & Pribram, 1960) between outcomes produced by a person and his or her intentions. When the outcomes approximate the intention to a sufficiently close degree, a cognitive link is established between the activities and the intention. Simply, a person learns that if intention i is to be realized, activity a needs to be executed. This activity-intention link may then be committed to memory so that next time the person intends the action he or she knows how to produce it, knows what component activities are needed to realize the intention, and how to produce those component activities.

Admittedly, the foregoing analysis is question begging in a sense. It transmits the problem backwards since one still needs to explain how the component activities are acquired. But the regression need not be infinite if one is willing to assume that by the time of birth, the infant is equipped with a variety of activities which he or she knows how to execute. A baby may "know" how to cry, how to turn its head or move its trunk or limbs. Those inborn activities may then be used as components in the acquisition of more advanced behavioral skills via a cybernetic process described previously (see for example: Hoocker, 1943).

Be that as it may, our primary concern in this chapter is with a somewhat earlier phase of action production: the formation of an intention. As we stressed at the outset, we view an intention as a kind of knowledge. When we say that person p intends to execute action a we mean that p knows or is aware that a is what he or she wishes at the moment. Just as one may know that it is daytime now or that one is sitting at one's desk and writing, one may also know what it is that one is *wishing* to do at any given moment. This view of intention as a mental or cognitive event (i. e. as a specific content of knowledge) is not unique to the present analysis. In much the same vein, Hume (1738, p. 399) for example, wrote: "By the will to act I mean nothing but the internal impression we feel and are conscious of when we knowingly give rise to any new motion of our body or new perception of our mind . . ."

Conceiving of an intention as a specific knowledge content suggests that the formation and revision of intentions may be governed by the same rules that govern the acquisition of all knowledge. Our theory of lay epistemology attempts to characterize just such a process. Therefore, it should be relevant to the topic of intentions as well. A few examples should suffice to show that this way of thinking about intentions is quite plausible.

Action Schemata. Our theory suggests that, as with other types of knowledge, intentions can be deduced from certain schemata to which an individual may subscribe. We refer to those kinds of schemata as action schemata. More formally speaking, an action schema is a premise, that is, a conditional statement whose consequent term is an intention. As we emphasized earlier there seem to be no apriori constraints on the contents of conditional statements, so that any antecedent term can be conditionally joined with any consequent term. This has the important implication that a behavioral intention can be deduced from virtually any antecedent, depending on the specific if-then linkages that might have formed in a given person's phenomenology. For instance, a person may employ an action schema whereby "if the leader issues a command, then I intend to execute it," or "if the action serves a very worthwhile goal, and it is within my ability to perform it, then I intend to do so" (Atkinson & Birch, 1970), or "if I performed action X in the past then I intend X now as well" (cf., Fazio & Zanna, 1980). The phenomenon of modeling (Bandura, 1969) can be explicated in terms of specific action schemata from which behavioral intentions are deduced. For instance, a small child may proceed from the premise whereby if an older sibling or a parent is performing a given action then the child intends to do the very same thing as well. Such an analysis should have implications for ways of affecting modelling. For instance, the tendency to model should be weakened if a child was made to abandon the premises from which the modelling intention was deduced, and so forth. In sum, action schemata may function as decision rules guiding the person on how to act on various occasions.

Factors of Saliency and Availability. Our analysis suggests that the *availability* and *saliency* of various ideas should affect the crystallization of intentions. This proposition seems well substantiated by everyday experience: the sight of an attractive object (inducing its momentary saliency and availability) may evoke an intention to acquire it, a phenomenon well recognized in the selling and advertising professions. Thus, a central aim in advertising is getting the potential buyers to easily recall a product's name. If the name "Mr. Clean," for instance, readily comes to mind, a person may well ask for the product when in need of a detergent.

The Freezing and Unfreezing of Intentions. According to our approach, intentional stability or instability may be understood by reference to the phenomena of epistemic freezing and unfreezing. In other words, an intention would most likely be stable under conditions known to effect freezing; For instance, under a conclusional need with which the intention was congruent or, under a heightened need for structure. Thus, when under time pressure to act quickly (e. g., at an auction sale where several persons are competing for a given item) an individual may experience a strong need for structure concerning his or her intentions. Under these circumstances, intentions (e. g., whether to bid for the object on sale) may crystallize quickly and without many fluctuations.

An intention would be likely to shift under an unfreezing force like a strong conclusional need with which the intention is incongruous. For example, an intention to indulge in a midnight snack might quickly be abandoned if the individual was reminded of the potentially disastrous consequences of the intended action for his or her figure. An intentional shift also may be occasioned by a heightened need for validity or a fear of invalidity. Thus, a youth may change the decision to enroll in a given course of studies upon being suddenly confronted with the question of whether that is what she or he *really* wants to do, as well as being reminded of the potentially costly consequences of making an erroneous judgment in the matter.

To summarize, we have argued that intentions can be conceived of as specific contents of knowledge so that a general epistemic theory about the acquisition of all knowledge should be relevant to the process of intention formation. From this perspective we have suggested that intentions can be deductively derived from various action schemata to which a person may subscribe, and which that person may invoke at a given moment. This, in turn, implies that formation of specific intentions may depend on the saliency and/or the mental availability for the individual of particular schemata. Furthermore, the stability and instability of intentions may be explicated by reference to our notion of epistemic freezing. This notion has considerable significance for the problem of action control: If intentions cannot be stabilized, the individual will not be able to control motivation, that is, stick with the intention until it is carried out. An intention may stabilize quickly if the individual is under a high need for structure, and/or if the intention is congruent with a salient conclusional need. Similarly, an intention could shift or be unfrozen under a heightened fear of invalidity and/or if its content is incongruous with a conclusional need. Epistemic needs might also determine what kind of action schemata will be employed in order to: 1) prevent a desirable intention from being skipped (conclusional needs), 2) avoid uncertainty and ambiguity which may block immediate action (need for structure), or 3) refrain from expensive mistakes (need for validity). This concludes our epistemic analysis of actions. In what follows we apply the analysis to several major topics in contemporary action theory.

Rational and Less Rational Action Tendencies

A unique feature of the epistemic interpretation of actions is that it offers an integrative approach which promises to bridge the dichotomy between rational and irrational actions. Proponents of this dichotomy typically assume that, by and large, actions are purposive, that is, are goal oriented, reasoned, or calculated – in short, they are rational in the sense of being effectively oriented toward the attainment of the actor's objectives. However, alongside rational actions, mention is often made of various less rational behavioral tendencies. In those cases reference is made to habitual, overlearned, automatic, uncontrollable, unconscious, or mindless actions.

According to our proposal, both types of action, rational and irrational, represent, the operation of action schemata, albeit of different content. In the following section, we interpret in this particular way the popular expectancy-X-value model of human action.

The Expectancy-Value Model

The expectancy-value model (cf., Atkinson & Birch, 1970, 1974; Feather, 1982) is by far the most influential cognitive approach to the explanation of actions. As Kuhl (1982, p.126) put it: "there seems to be implicit consensus among many theorists concerned with social motivation that expectancy-value theory is for motivation theory what evolution theory is for biology; a firm universally accepted foundation for all theories of specific phenomena to build upon ... Consequently, expectancy-value theories are often then basis rather than the target of empirical investigations . . ."

The basic contention of the expectancy-X-value model is that the tendency to act is determined by two parameters; the *value* of the goal mediated by the behavior and the *subjective* probability or *expectancy* that the behavior will actually mediate the goal. The expectancy and value factors are assumed to combine multiplicatively in determining behavior; if either the expectancy or the value is at zero magnitude, no behavior is expected to occur. The expectancy-value model was conceived of as a universalistic model for explaining goal-directed, rational actions but to be largely inapplicable to "automatic" behaviors which are determined more by forces of habit or inertia (cf., Atkinson & Birch, 1974) than by rational choice.

But recently, important objections were raised concerning the universalistic claims of the expectancy times value formulation. Kuhl (1982) in particular pointed out that even if subjects do take into account expectancy and value parameters they need not combine them multiplicatively. Rather, they could do so in terms of several "logical" models. A subjective probability may not be a phenomenologically representative construct. Thus, people may "think", that is, reason deductively from verbal propositions rather than "calculate" before deciding on a given course of action (cf., Kuhl, 1982, p. 152). For example, a person could employ a schema whereby if the goal is attainable (sufficient expectancy) and desired (sufficient value), the action will be performed. Yet a different model could be the disjunctive model whereby *if* either the expectancy *or* the value reached a threshold of sufficiency, the action would be undertaken. Finally, a person could employ a "pure-expectancy," or a "pure-value" model, that is, consider just one of the two parameters and engage in a behavior if it exceeded some critical level. Indeed, in a recent study Kuhl (1977) obtained support for the possibility that different people may behave according to different models. Failure-oriented subjects, for example, behaved significantly more according to a pure expectancy model than subjects low in fear of failure. (cf., Kuhl, 1982, p. 151).

A further restriction on the universality of the expectancy-X-value model is suggested by Kuhl's variable of action control, which is a tendency determined by interaction of personal and situational factors depicting a meta-cognitive orientation towards actions. Two basic orientations are distinguished: a state orientation and an action orientation. A person is action oriented when his or her action is simultaneously focused on (1) some aspect of the present state, (2) some aspect of a future state, (3) a discrepancy between the present and the future state, and (4) at least one action alternative that might remove the discrepancy. By contrast, state orientation is described as a motivational tendency that instigates cognitive activity for its own sake, that is, without perceiving it as instrumental to future action.

Kuhl and Beckmann (1983) hypothesized and found that depending upon their type of action orientation, subjects tended to employ a different model when deciding on a course of action. „All state-oriented subjects based their choices on the complex conjunctive (expectancy-X-value) rule, whereas most action-oriented subjects based their choices on the simpler expectancy rule"

(Kuhl, 1982, p. 154). According to Kuhl "(these) results provide additional support to the assumption that information about expectancy and value is not sufficient for predicting action. The construct of state versus action orientation seems to convey additional explanatory value ..."

To summarize, criticisms in preceding sections cast serious doubts on the universality of the expectancy-X-value model. Firstly, the model leaves out the entire domain of habitual (overlearned) behaviors that surely constitute a large proportion of everyday activities. Secondly, Kuhl's research demonstrates that different persons may employ different expectancy-X-value models and their choice of a particular model may depend on their state versus action orientation.

It should be noted that while critical of the multiplicative model, Kuhl does accept the basic idea that the parameters of expectancy and value (even if in need of differentiation and supplementation) are of a special relevance to human action. Our epistemic analysis suggests a different possibility as outlined in the section to follow.

Expectancy-Value Models as Specific Action Schemata. It is possible to view the various expectancy-value models (or: desision rules) discussed by Kuhl as representing specific "action schemata." Each such "expectancy-value" schema when adjudged as situationally applicable may be used by the individual for deducing a behavioral intention. Thus, we accept Kuhl's proposal that various expectancy-value models may be conceived of as logical algorithms of a conditional form. "If I am able to reach a goal, via performing X and I want to reach this goal then I will enact X" is an example of a particular action schema phrased in expectancy and value terms as are the remaining models identified by Kuhl.

But if all those models essentially represent the contents of specific action schemata and if the list of possible such schemata is endless then the expectancy-value formulation should abdicate its centrality as an explanatory paradigm for human action. The various expectancy-value models can now be conceived of as a mere sample out of a potentially endless population of action schemata. In a sense then, our present analysis carries one step further the critique initiated by Kuhl (1982a). According to this critique, the multiplicative expectancy-X-value formula may not be as universal as was originally supposed and there may exist other ways in which different persons use the expectancy and/or value concepts in order to derive their intentions. We are suggesting, further, that often behavioral intentions may be deduced from schemata based on concepts other than expectancy or value (see Kuhl, 1983, p. 106f., for a similar view). For instance, in a recent experimental study conducted by Kuhl (1982b) it turned out that expectancy and value information explained *leisure* time activities of sixth graders but not *routine* or *obligatory* activities. Examples of the latter type of activities are: doing homework, brushing one's teeth, helping mother do the shopping ..." (Kuhl, 1982, p. 155). According to the present conception, the latter activities could simply be deduced from action schemata formulated in terms other than expectancy or value (cf. Kuhl, in press).

As we indicated earlier, the conception of action schemata allows an integration of behaviors previously conceived of as habitual, automatic, or externally controlled with those conceived of as rational, calculated, or, goal oriented. Simply, the different types of behavior may follow from different action schemata that people may employ in various circumstances. For example, the habitual or highly overlearned behavior of stopping at a red light, or turning left at the crossroads could be explained in terms of a specific action schema that a person might invoke. For instance: "If I see a red light I bring the car to a halt", "If I see the intersection of Main and 5th I turn left", and so forth. According to the present analysis, the process whereby a given action schema would be invoked, adjudged as applicable or modified would be the same regardless of the type or content of a specific action schema (e.g., regardless of whether it be of a "rational," or a "habitual" content). We are referring to the general epistemic process described earlier and assumed applicable to all possible schemata or knowledge contents, transcending schemata specifically concerned with "actions."

It should also be clear that we consider as somewhat futile any attempt to enumerate or catalogue the various possible types of action schemata that people may employ. Such a list of possible schemata is open-ended in principle. Depending on their unique socialization histories, or life experiences, different persons could derive their behavioral intentions from vastly different schemata. Instead of attempting to construct an exhaustive list of action schemata we propose an approach combining the *nomothetic* and *ideographic* levels of analysis. The nomothetic level concerns the general epistemic process by which action schemata are invoked, evaluated, and applied. The ideographic part concerns the contents of specific action schemata that should be particularly identified for each separate person/situation combination. Jointly the two approaches may allow us to predict and/or control the occurrence of particular actions in specific circumstances. ·

The Attitude-Behavior Relation

Earlier, we outlined arguments against an attempt to predict subjects' behavior from the contents of specific schemata (e.g., the expectancy × value schema) without first checking whether this particular schema is actually used by the subjects in the particular situation. From this perspective, it is of interest to examine briefly the social-psychological program of predicting behavior from attitudes. The concept of attitude has been among the most central ones in modern social psychology. An attitude is often defined as an evaluative (pro-con) response toward an object. The interest of social psychologists in the study of attitudes can be used to predict (approach-avoidance type) behavior toward objects (cf. Kiesler, Collins & Miller, 1969). Thus, it came as a disappointing surprise to learn that the empirical evidence for the attitude behavior relation is weak and unimpressive (cf. Wicker, 1969). A significant advance toward explaining such evidence and clarifying the attitude-behavior issue was accomplished by the works of Fishbein and Ajzen (1975), and Aizen and Fishbein (1977, 1980). These authors argued that attitudes toward an object are general, whereas behaviors toward an object are specific and there seems to exist no good reason for assuming that a general attitude would necessarily lead to a specific behavior. Suppose you had a very positive attitude toward movies. From this it hardly follows that on the night of January 23rd you will attend a particular movie by Peckinpah which is playing at the Odeon.

Ajzen and Fishbein further argued that the appropriate construct from which behavior is predictable is the behavioral intention. Indeed, research evidence cited by Ajzen and Fishbein (1977) supports their analysis by yielding impressively high intention-behavior correlations. Ajzen's and Fishbein's (1980) theory has concerned the determinants of behavioral intentions. These were assumed to be: (1) attitude toward the behavior and (2) subjective norm toward the behavior. An attitude toward a behavior "... is determined by salient beliefs about that behavior. Each salient belief links the behavior with some valued outcome ... The attitude toward the behavior is determined by the person's evaluation of the outcome associated with the behavior and by the strength of these associations ... By multiplying belief strength and outcome evaluation and summing the resulting products, we obtain an estimate of attitude toward the behavior ..." (see Ajzen, Chapter 2, this volume).

Subjective norms concerning a behavior "... are also assumed to be a function of beliefs ... that specific individuals or groups think he should or should not perform the behavior ... (thus) a person who believes that most referents with whom he is motivated to comply think he should perform the behavior will perceive social pressure to do so ..." (p. 5).

An Epistemic Reanalysis of the Attitude-Behavior Problem. According to the present conception, an attitude is a specific type of knowledge; a knowledge that a given object is "good" or "bad", "worthy" or "unworthy," and so forth. We thus concur in Ajzen and Fishbein's observation that no necessary relation should be expected between an attitude and a behavior. An attitude represents a content of knowledge or a specific proposition ("University is a worthwhile institution", "movies are fun"), and as we stressed earlier, no epistemic content is *necessarily* a component of an action schema. For instance, a person might accept the proposition that "the work of Picasso is very good" without linking it to the intention of "visiting a Picasso exhibit at the Museum of Modern Art." Furthermore, according to the present analysis, an absence of a link between an attitude and a behavior need not stem from a generality-specificity difference between the two. We reiterate that any content of knowledge or proposition can be linked with an intention and this includes propositions phrased in highly abstract terms. For example, a person may believe that "if a given behavior represents social responsibility I shall try to execute it." Clearly, the social responsibility concept is rather general or abstract yet the schema in which it is contained can be used as a premise from which numerous specific actions are deduced. Thus, a person may donate money to a charity drive or take part in a street demonstration upon identifying the behaviors in question as signifying social responsibility.

It is noteworthy that Ajzen and Fishbein, while critical of the view that intentions to behave are determined by attitudes toward objects, do not give up the quest for intentions' universal determinants which in their framework are: (1) an attitude toward the behavior and (2) a subjective norm concerning the behavior. In commenting on this proposal it is good to note, first, that an attitude and a subjective norm toward a behavior need not be conceived of as independent of one another. Specifically, the attitude toward a behavior has to do with the behavior's likely consequences and the subjective norm clearly alludes to one such important consequence: being approved of by significant others. It follows that in Ajzen's and Fishbein's system behavioral intention is assumed to derive from just one determinant, an attitude toward a behavior, in turn determined partially by a subjective norm, but also by other expected consequences of the behavior.

Furthermore, recall that for Ajzen and Fishbein an attitude toward a behavior is determined by the *value* of the behavior's consequences and the *expectancy* that the behavior will mediate those particular consequences. In other words, Ajzen's and Fishbein's model of the determinants of intentions turns out to be a particular version of the expectancy × value model; a point acknowledged by Ajzen (Chapter 2, this volume). But if so, critiques of the expectancy × value models outlined earlier are applicable to Ajzen's and Fishbein's theory as well. Thus, the strong empirical support for a relation between attitudes toward behaviors and behavioral intentions (cf. Ajzen & Fishbein, 1977) could be restricted to circumstances in which subjects happen to invoke an expectancy × value schema; no comparable support for this relation should be expected in cases where subjects may employ alternative action schemata phrased in other than expectancy or value terms.

To sum up, according to the present conception, while "attitude toward an object" need not be necessarily linked to a behavioral intention, neither need "attitude toward a behavior": An "attitude toward a behavior" is an "expectancy × value" schema to which there may exist various schematic alternatives, as we have seen.

Variables Moderating the Intention-Behavior Relation

Do persons invariably act in accordance with their intentions? In a sense, the question is meaningless. After all, actions are intentional by definition, so whenever there is action there must be intention as well. But it is still possible to ask whether one may hope to predict future actions from previously measured intentions. This already is a reasonable question since intentions may shift and an intention measured at Time 1 may turn out to be a poor predictor of an action performed at Time 2 (cf. Ajzen, Chapter 2, this volume). In this context let us examine two factors proposed as moderators of the intention-behavior relation: Kuhl's action-control dimension and Snyder's self-monitoring variable.

Action Orientation. According to Kuhl (1982), state-oriented subjects may more often act in disregard of their intentions than action-oriented subjects. For instance, when engaged in a given task, state-oriented subjects may be readily given to mediation concerning the causes of their performance rather than carrying on with the task as would action-oriented subjects. According to the present interpretation, state-oriented subjects may have a systematic preference for the activity of *contemplation,* whereas action-oriented subjects may prefer activities with a definite *completion* component. Thus, we need not posit that state-oriented subjects often act in disregard of their intentions. Rather, state-oriented subjects may be more prone to form an intention to meditate than action-oriented subjects.

The intention to meditate could be deduced in the usual way from an appropriate action schema. For instance, in the course of socialization a person could be brought up to believe that "if something unexpected occurs it is necessary to thoroughly investigate it;" such a schema could frequently give rise to a meditative intention. One implication of the above analysis is that state-oriented subjects intending to meditate would be *less* likely to disregard their intention than action-oriented subjects intending to do so. This prediction qualifies Kuhl's (1982) original generalization concerning the higher tendency of action-versus state-oriented subjects to enact their intentions (but see Kuhl's (1984) distinction between self-generated and other intentions).

Furthermore, if action-oriented subjects are strongly motivated toward task completion, they should strive to arrive quickly at *cognitive structures* which may enable them to act and complete their assignments. Such subjects should therefore exhibit a fast freezing process, a conclusion supported by Kuhl's and Beckmann's (1983) finding that action-oriented subjects tend to initiate action on the basis of minimal information. Finally, action-oriented subjects should manifest a preference for specific conclusions, in particular those which allow them to act fast and with confidence (for instance, conclusions which imply that they have adequate ability to perform the task). Admittedly, the above predictions are at present speculative and require further empirical probing to establish their plausibility.

Snyder's "Self-Monitoring" Variable. According to Snyder (1974, 1982) the dimension of self-monitoring is relevant to the attitude-behavior relation. Low self-monitoring subjects are assumed to guide their behavioral choices on the basis of relevant inner states. High self-monitors, on the other hand, are assumed to base their choices mainly on situational cues. From the above analysis Snyder (in press) derived the prediction that low monitors should regard their attitudes as more relevant guides to action than high monitors. Ajzen, Timko, and White (1982, see Ajzen, Chapter 2, this volume) reasoned further that both low and high monitors may regard attitudes (toward behavior) as relevant to their actions, but that high self-monitors should be more likely than low-self-monitors to shift their intentions on the basis of situational cues. This prediction was supported by Ajzen's et al. (1982) findings.

A somewhat different analysis of the self-monitoring variable is suggested by the present epistemic theory. According to this interpretation, both high and low self-monitors may be acting in accordance with their action schemata; however, the contents of dominant action schemata may differ for the two types of subjects. The low monitors may frequently invoke a schema whereby "only if *I* feel like doing something – I do it," whereas the high monitors may invoke a schema whereby "only if something is situationally appropriate – I do it." It is noteworthy that both inner "feelings" and situational norms may be subject to fluctuations, and at unknown rates. In cases where one's own feelings about a given behavior may fluctuate at quicker rates than the situational norms, high self-monitors may be more stable in their intentions than low self-monitors. However, when "norms" may shift more quickly than feelings, intentions of low self-monitors may exhibit the greater stability, just as found by Ajzen et al. (1982).

Mindless vs. Mindful Actions

The utilitarian expectancy-value model considered earlier is generally viewed as representing a rationalistic, economic, or calculating approach to action. By contrast, habitual or inertia-based actions are typically considered as less thoughtful or rational. A somewhat similar distinction between more and less thoughtful actions appears in Langer's (1978) recent work on mindlessness. To Langer, "mindless" actions occur when people fail to process all the information contained in the situation, and proceed, instead, on the basis of minimal information allowing them to enact a script (Schank & Abelson, 1977). For instance, in one particularly striking demonstration of the mindlessness phenomenon (Langer, Blank & Chanowitz, 1978) people about to use a copying machine in the library of the CUNY Graduate Center were approached by a person who asked them to use the machine first. In one condition *(Request-Only)* the experimenter said: Excuse me, I have 5 (20) pages may I use the zerox machine? In the *Placebic Information* condition the experimenter said: Excuse me, I have 5 (20) pages. May I use the xerox machine because I have to make copies? Finally, in the *Real Information* condition he said: Excuse me, I have 5 (20) pages. May I the xerox machine because I am in a rush?

When the requested favor was small rather than large, that is, when the number of pages to be photocopied was 5 rather than 20, the percentage of complying subjects was as high in the *Placebic Information* condition as in the *Real Information* condition which was significantly higher than the percentage of compliance in the *Request-Only* condition. However, when the requested favor was large (the 20 page condition), the percentage of complying subjects was as low in the *Placebic Information* condition as in the *Request-Only* condition, which was significantly lower than the percentage in the *Real Information* condition.

Langer et al. (1978) interpreted these findings in terms of the notion of scripting or mindlessness: when the favor requested is small subjects respond to a script whereby a request for a favor is complied with when it is accompanied by a phrase serving as a reason or a justification for the request. Under these circumstances the subject does not process all the available information which, if only attended, would clearly suggest that "having to make copies" does not actually justify using a machine out of turn. By contrast, when the favor requested is large ignoring the information is potentially effortful; this makes it worthwhile to think about the information given, or engage in a mindful action.

But our lay-epistemic framework affords a somewhat different way of viewing Langer's et al. results. According to this interpretation, subjects' behavior in the Langer et al., study was not utterly mindless or mindful. In all the conditions they were processing at least some of the information given, but not all the available information, before they had decided (Or, in our terms, before they had frozen their epistemic process). The mindless subjects attended to the information that the favor needed was small and decided to comply. The mindful subjects attended additionally to the information that the favor requestor is in a rush. Had the favor been of a greater value, subjects might have asked for more information in order to comply (e.g., what made the copying so urgent? Is the matter at all justifiable?). Hence, it is the *amount* of processing and its thoroughness, not the processing itself that distinguishes between the two groups.

It is of interest that our epistemic theory specifies the conditions under which persons would seek more or less information in adjudging a schema's applicability. The amount of information sought should depend on the individual's epistemic motivations, the needs for structure, for validity, and for conclusional contents.

For example, under *high need for structure,* subjects might be motivated to reach a decision or to act without being concerned too much with the veridicality of the decision or action. Consequently, they will process less information before freezing the epistemic process. Conversely, under *high fear of invalidity,* subjects will be concerned with the costs of an erroneous decision – (it migth be a consequential error to comply with an unjustified request or to reject a justified one, so it is worthwhile to consider the facts again. Subjects will thus process more information in this condition.

Under high conclusional need the amount of processing will depend on the congruence between the need at issue and the requested action. For example, if the price of complying is high, that is, if compliance bears unpleasant consequences, as in the Langer study, subjects will ask for more information before accepting the "justified request" schema. More so than when the price is low, as with the small favor. If the subject's desired conclusion is "I am a kind and considerate person" he will attend to less information before accepting the applicability of the schema. Less so than if the cherished conclusion is "I am a cautious person; nobody is going to pull one over on me . . .".

More generally, according to our epistemic analysis an action is neither mindless nor mindful, thoughtless nor fully rational. Rather than implying such dichotomous distinctions we suggest, instead, that depending on the conditions, an action schema may be accepted after either a long or a relatively brief informational search. What Langer refers to as mindful action usually is undertaken after a more extensive search than a mindless action. But even in the case of the most mindful action, not all of the relevant information can be attended to. Since there is a potentially endless amount of information relevant to any schema, the acceptance of a schema is necessarily based on limited information only.

Furthermore, calling an action mindful or rational implies that the actor's behavioral decision was "right" in some sense, or that the schema from which the action was derived is valid or truly applicable. But actually, no schema is correct or applicable in an absolute sense (cf. Kruglanski & Ajzen, 1983). A schema considered applicable at Time 1 could be revoked later at Time 2, after the individual had considered a further sample of relevant evidence. For example, the judgment that a reason given for a decision was acceptable or justifiable could be revoked after one had considered the possibility that the speaker was lying.

Unconsciously Determined Actions

An important issue that a theory of actions must address is the problem of unconsciously determined behavior. One way of viewing the issue is in terms of the possibility that people may occasionally be unaware of their behaviors' determinants (Nisbett & Wilson, 1977). Such a possibility is quite compatible with our epistemic analysis. For instance, our "searchlight" metaphor of human consciousness is compatible with the idea that persons may very quickly forget the conceptual basis (action schemata) from which their intentions may have been deduced. A child might initially deduce the intention to do something from observing a parent doing it. However, after a while this deductive basis of the action may be forgotten and the child may remain with a largely unexplained (or a wrongly explained) urge to engage in a given behavior.

Alternatively, an action schema may be accepted *because* of its (conclusional) desirability, but a person may be unaware of such a determination. Under such circumstances, an individual could simply experience confidence in the schema and an unwillingness to further test its validity. But a person need not be reflexively aware that his or her confidence has anything to do with the schema's (conclusional) desirability.

To illustrate how the foregoing sources of "unconscious determination" may work in a particular case consider a verbal conditioning experiment reported by Bowers (1981) in which the subject, a young lady, displayed in the course of time a substantial increase in a preference for landscape paintings, the reinforced category of response, over portraits – representing the unreinforced category. After the conditioning trials were over, the experimenter inquired about the subject's perception of the experiment. It turned out that she was aware of having generally preferred landscapes and she also noticed the experimenter's systematic approval for landscapes. At the same time she strongly denied that her preferences had anything to do with the experimenter's approval. As she expressed it: "I picked the landscapes because I liked them better than the portraits. Besides, you only said "good: after I made my selection, so what you said could not possibly have influenced my choice of pictures." Thus, the subject in this case did have access to an important determinant of her behavior, the experimenter's reinforcing responses for landscapes – yet she failed to relate the approval to her own landscape preferences despite the demonstrable relation between the two.

How could this have happened? Rather simply indeed: After having made the first reinforced choice the subject might have reasoned that it must have been well justified if the experimenter also shared it. But this deductive link may well have been forgotten; instead, the subject seemed to remember well the temporal priority of her expressed preference for landscapes over that of the experimenter. In and of itself such priority is consistent with the idea that the subjects' preferences were "internally" determined rather than being shaped or influenced externally by the experimenter. Of course, the subject *could* have considered the possibility that the trials were not independent and that the experimenter's reaction on a given trial could well have affected her reactions on the next trial. But such possibility *need not* have occurred to the subject or, if it did, it might have been quickly discarded because of its undesirable implication that the subject lacks firm aesthetic standards, a conclusion at odds with a self-concept of a worldly, cultured person.

The *mystique* of behavior's unconscious determinants may well have to do with the primitive "contents" of unconscious forces in the Freudian paradigm (notably, sex and aggression). But if we interpret unconsciousness as simply a lack of awareness (Erdelyi, 1974) concerning the determinants of some of our actions the phenomenon is readily compatible with a view of the epistemic process in which the deductive bases of our conclusions can be quickly forgotten and in which we may not be reflexively aware of the motivational underpinnings of our tendencies to accept or reject specific conclusions.

Conclusion

In the preceding pages we attempted to demonstrate that a theory of lay epistemology is relevant to the understanding of human actions. Our basic contention has been that intentions are specific contents of knowledge, the knowledge of what it is that one wishes to do. If so, intentions should be governed by the same rules that govern the acquisition of any knowledge. They should be deducible from other kinds of knowledge or action schemata that the person may subscribe to. They can be frozen and unfrozen depending on the mental availability of alternative intentions and the magnitude of the various epistemic needs mentioned earlier, the need for cognitive structure, the fear of invalidity, and the need for specific conclusions.

Our epistemic analysis of intention formation was applied to various major topics in action theory. It was noted that the expectancy-value model of human motivation may merely represent (the content of) one action schema among many and that habitual, overlearned, or mindless actions may be essentially subject to the same epistemic rules as thoughtful, calculated, or rational actions. Finally, the problem of unconsciously determined actions was considered from the present epistemic perspective. According to this viewpoint, persons may often forget the deductive bases of their intentions or be reflexively unaware of the motivational underpinnings of their conclusions.

The analysis outlined in the foregoing pages is already consistent with a plethora of findings in the domain of research on actions, such as data controverting the universality of the expectancy-value model (Wicker, 1969). However, most of the implications and predictions of the present epistemic analysis have not yet been specifically tested. We are referring, for example, to predictions concerning conditions (1) under which one action schema (say, expectancy × value) rather than another schema would serve as a basis for behavioral decisions, (2) under which actions would be mindless rather than mindful, and (3) under which persons would be conscious or unconscious of their actions' determinants. The latter questions represent possible directions of future research that could be profitably guided by an epistemic theory of actions.

References

Ajzen, I., & Fishbein, M. (1977). Attitude-behavior relations: A theoretical analysis and review of empirical research. *Psychological Bulletin, 84,* 888–918.

Ajzen, I., & Fishbein, M. (1980). *Understanding attitudes and predicting social behavior.* Englewood-Ciffs, N. J.: Prentice-Hall.

Ajzen, I., Timko, C., & White, J. B. (1982). Self-monitoring and the attitude-behavior relation. *Journal of Personality and Social Psychology, 1982, 42.*

Atkinson, J. W., & Birch, D. (1970). *The dynamics of action,* New York: Wiley.

Atkinson, J. W., & Birch, D. (1974). The dynamics of achievement oriented activity. In J. W. Atkinson, & J. O. Raynor (Eds.): *Motivation and achievement.* (pp. 271–326). New York: Wiley.

Bandura, A. (1969). *Principles of behavior modification.* New York: Holt, Rinehart & Winston.

Bar-Tal, D., Yarkin, K., & Bar-Tal Y. (1982). *Planing and performing interpersonal interaction: A cognitive motivational approach.* Unpublished manuscript, Vanderbilt University.

Beckmann, J., & Kuhl, J. (1984). Altering information to gain action control: Functional aspects of human information processing in decision making. *Journal of Research in Personality, 18,* 224–237

Bowers, K. S. (1981). Knowing more than we can say leads to saying more than we can know: On being implicitly informed. In D. Magnusson (Ed.), *Toward a psychology of situations.* Hillside, N. J.: Erlbaum.

Erdelyi, M. H. (1974). A new look at the New Look: Perceptual defense and vigilance. *Psychological Review, 81,* 1–25.

Fazio, R. H., & Zanna, M. P. (1980). Direct experience and attitude behavior consistency. In L. Berkowitz (Ed.), *Advances in experimental social psychology,* Vol. 14, New York: Academic Press.

Feather, N. (Ed.) (1982). *Expectations and actions: Expectancy-value models in psychology.* Hillsdale, N. J.: Erlbaum.

Fishbein, M., & Ajzen, I. (1975). *Belief, attitude, intention and behavior: An introduction to theory and research.* Reading, Ma., Addison-Wesley.

Frenkel-Brunswick, E. (1949). Intolerance of ambiguity as emotional and perceptual personality variable. *Journal of Personality, 18,* 103–143.

Hamilton, D. L. (1979). A cognitive attributional analysis of stereotyping. In L. Berkowitz (Ed.) *Advances in experimental social psychology,* Vol. 12, New York: Academic Press.

Hoocker, D. (1943). Reflex activities in the human fetus. In R. G. Backer. J. S. Koonin of H. F. Wright (Eds.) *Child behavior and development.* 17–28, New York: McGraw-Hill.

Hume, D. (1938). *A Treatise of Human Nature.* Selby-Bigge. (first ed. 1738).

James, W. (1890). *The principles of psychology.* New York: Holt.

Kiesler, C. A., Collins, B. E., & Miller, N. (1969). *Attitude change,* New York, Wiley.

Kruglanski, A. W. (1980). Lay epistemo-logic-process and contents. *Psychological Review, 87,* 70–87.

Kruglanski, A. W., & Ajzen, I. (1983). Bias and error in human judgment. *European Journal of Social Psychology,* 13, 1–44.

Kruglanski, A. W., Baldwin, M. W., & Towson, S. (in press). The lay epistemic process in social cognition. In D. Frey, & M. Irle (Eds.) *Theories in social psychology,* Vol. 2, Bern: Huber.

Kruglanski, A. W., & Freund, I. (1983). The freezing and unfreezing of lay inferences: Effects on impressional primacy, ethnic stereotyping and numerical anchoring. *Journal of Experimental Social Psychology, 19,* 448–468.

Kruglanski, A. W., Klar, J. (1982). A view from a bridge: Synthesizing the consistency and the attribution paradigms from a lay epistemic perspective. Unpublished manuscript, Tel-Aviv, University.

Kuhl, J. (1977). *Mess- und prozeßtheoretische Analysen einiger Person- und Situations-Parameter der Leistungsmotivation* Bonn: Bouvier.

Kuhl, J. (1982a). The expectancy-value approach within the theory of social motivation. Elaborations, extensions, critique. In N. T. Feather (Ed.), Expectations and actions: *Expectancy-value models in psychology.* Hillsdale, N. J., Erlbaum.

Kuhl, J. (1982b). Action vs. state orientation as a mediator between motivation and action. In: W. Hacker, W. Volper, & M. von Cranach (Eds.), *Cognitive and motivational aspects of action.* Amsterdam: North Holland Publishing Co.

Kuhl, J. (In press). Motivation and information processing: A new look at decision – making, dynamic conflict, and action control. In: R. M. Sorrentino & E. T. Higgins (Eds.), *The handbook of social behavior.* New York: Guilford Press.

Kuhl, J., & Beckmann, J. (1983). Handlungskontrolle und Umfang der Informationsverarbeitung: Wahl einer vereinfachten (nicht optimalen) Entscheidungsregel zugunsten rascher Handlungsbereitschaft. *Zeitschrift für Sozialpsychologie, 14,* 241–250.

Langer, E. (1978). Rethinking the role of thought in social interaction. In J. Harvey, W. Ickes, & R. Kidd (Eds.), *New directions in attribution research.* Hillsdale, N. J.: Erlbaum.

Langer, E., Blank, A., & Chanowitz, B. (1978). The mindlessness of ostensibly thoughtful action: The role of placebic information in interpersonal interaction. *Journal of Personality and Social Psychology, 36,* 635–642.

Luchins, A. S. (1957). Experimental attempts to minimize the impact of first impressions. In C. A. Hovland (Ed.), *The order of presentation in persuasion.* New Havem, Conn.: Yale University Press.

Miller, G. A., Galanter, E., & Pribram, K. H. (1960). *Plans and the structure of behavior.* New York: Holt, Rinehart and Winston.

Nisbett, R. E., & Wilson, T. D. (1977). Telling more than we can know; Verbal reports on mental processes. *Psychological Review, 84,* 231–259.

Popper, K. R. (1973). *Objective Knowledge: An evolutionary approach.* Oxford: Clarendon.

Rumelhart, D. E., & Ortony, A. (1977). The representation of knowledge in memory In R. C. Anderson, R. J. Spiro, & W. E. Montague (Eds.), *Schooling and the acquisition of knowledge.* Hillsdale, N. J.: Erlbaum.

Schank, R., & Abelson, R. P. (1977). *Scripts, plans, goals, and understanding.* Hillsdale, N. J.: Erlbaum.

Smock, D. C. (1955). The ininfluence of psychological stress on the "intolerance of ambiguity." *Journal of Abnormal and Social Psychology, 50,:* 177; 182.

Snyder, M. (1974). Self-monitoring of expressive behavior. *Journal of Personality and Social Psychology, 30,* 526–537.

Snyder, M. (1982). When believing means doing? Creating links between attitudes and behavior. In M. P. Zanna, E. T. Higgins, & C. P. Herman (Eds.), *Consistency in social behavior.* The Ontario Symposium, (vol. 2) Hillsdale, N. J.: Erlbaum.

Snyder, M. L., & Wicklund, R. A. (1981). Attribute ambiguity. In J. H. Harvey, W. Ickes, and R. F. Kidd (Eds.), *New Direction in Attribution Research, (Vol. 3),* Hillsdale, N. J.: Erlbaum.

Taylor, S. E., & Fiske, S. T. (1978). Salience, attention and attribution: Top of the head phenomena. In L. Berkowitz (Ed.), *Advances in experimental social psychology,* Vol. 11. New York: Academic Press.

Tversky, A., & Kahneman, D. (1974). Judgment under uncertainty: Heuristics and biases. *Science, 185,* 1124–1131.

Wicker, A. W. (1969). Attitudes versus actions: The relationship of overt and behavioral responses to attitude objects. *Journal of Social Issues, 25,* 41–78.

Chapter 4

The Pursuit of Self-Defining Goals*

Peter M. Gollwitzer and Robert A. Wicklund

The cognition-behavior relation central here is the human pursuit of self-definitions. The striving after such self-definitions as child-rearer, parent, musician, or humanitarian is treated as a goal-oriented enterprise, such that the cognized goal (e. g., to be a humanitarian) brings forth numerous behaviors directed toward the individual's trying to realize that self-defining goal. In the course of spelling out some of the dynamics of self-completion processes, we will show how the pursuit of self-completion has the side-effect of interfering with a variety of other types of cognition-behavior relations. For example, the relation between attitude and behavior, between intention and behavior, and even that between situational cues for behavior and actual behavior can all suffer demise or even elimination owing to the person's pursuit of a self-defining goal.

A Central Distinction: Self-Defining vs. Non-Self-Defining Goals

Before we raise the issue of cognition-behavior relations as affected by the pursuit of self-defining goals, it is important to lay a theoretical groundwork. In the following pages we will illustrate what we have in mind with *self-defining goal,* and we will also specify the workings of the self-completion process. To start with an example: Two students are undertaking their first major exercise in the laboratory of an experimental psychology class. The class project is to train a pigeon to manifest a particular superstitious behavior – namely that of turning around three times and walking backward toward the food source prior to being fed. Each student has read B. F. Skinner, has already witnessed several shaping-up experiments, and is ready to go ahead with this first exercise.

The objective, commonly agreed-upon goal of the exercise is clear and dis-

* The writing of this chapter was facilitated by a stipend to the second author from the Alexander von Humboldt-Stiftung, Bonn-Bad Godesberg, and by NSF Grant BNS 7913828 to Melvin L. Snyder and Robert A. Wicklund.

tinct: train the pigeon to execute the superstitious behavior with great reliability. Certainly this is a goal for both students of the present example, but let us make a further assumption about the psychology of one of the students. Suppose that the orientation of one of them is only peripherally on the objective goal (the teacher's goal), and more centrally on the goal of being a *psychologist*. From our theoretical background (Wicklund & Gollwitzer, 1981, 1982) this self-defining goal is pursued by means of numerous possible, culturally agreed-upon symbols. Thus, the student's pursuit of the self-definition *psychologist* may be marked by attempts to put his name on publications, association with recognized psychologists, a collection of psychology books and journals, a temporary job as instructor, and any other indication of this seemingly professional, culturally-agreed upon self-definition. To the extent that symbols of one's psychologist status are lacking, the dynamics postulated by the self-completion notion imply that the person will then pursue alternative symbols of the self-definition.

As it happens, neither student is very successful in reaching the objective goal set by the professor. The first student's pigeon learns to peck the floor, the second student's pigeon learns to sit down and cackle. Given that the first student's orientation in the entire situation has to do only with the *non*-self-defining goal, we might expect that student simply to feel frustrated, and perhaps to try other methods to train the bird to perform the superstitious behavior. On the other hand, the reaction of the aspiring psychologist – the one oriented primarily toward a self-defining goal – may well be much different. Assuming that an accomplished feat of animal training is nothing more than one possible symbol of being a psychologist, the student can readily resort to alternative symbolic routes to completing the self-definition. In fact, the theory implies that faltering in one effort to gain a symbol (bird-training) will lead to striving after and emphasizing other symbols, such as a high grade point average in psychology courses, profound knowledge of great figures in psychology, and even attempts to teach others how to shape up superstitions in birds. While it may seem unlikely that the incompetent would set out to teach others within the area of his incompetence, the reader is referred to the section entitled "Lack of education and attempted influence" for an illustration of such effects.

A Theory of Symbolic Self-Completion

Symbolic self-completion theory (Wicklund & Gollwitzer, 1981, 1982) provides a body of ideas on how people pursue the self-defining goals to which they aspire. The core assumption of the theory is that indicators of a self-definition are mutually substitutable for one another. The central observation to be made about the human, within the context of a notion of self-completion, is that central flaws in the person's training or performance regarding the self-defining goal are covered over by what the theory calls *self-symbolizing efforts*.

Historical Background

The idea of substitution is central to the thinking of Lewin (1926) and several of his students. Their analysis of goal-oriented behavior and interrupted activities is the conceptual background of self-completion theory. According to Lewin, a tension system is set up once people aspire to goals. This tension can be reduced as soon as the individual undertakes efforts to approach the goal. When the individual is interrupted in these efforts, the tension system is expected to propel the individual to a substitute activity that aims at tension reduction.

In support of Lewin's theorizing, his students, in particular Mahler (1933) and Lissner (1933), found that subjects' interest in resuming an interrupted, original activity was sharply curtailed when subjects were given the opportunity to complete a substitute activity. In other words, people can substitute for a shortcoming regarding one goal by successfully approaching another goal. What are the ramifications of this phenomenon? First, Ovsiankina (1928) points to the fact that a tension buildup can only be expected when the individual is personally involved in the activity that potentially leads to goal attainment. Accordingly, one would expect that only the interruption of involving activities makes people search for substitutes. Second, Lissner (1933) and Mahler (1933) claimed that activities function as substitutes only when they are related to the original activity that is interrupted. Henle (1944) specifies that this relation has to do with superordinate personal goals, such as being creative, musical, intelligent, and so forth. This means that an activity can substitute for another only when both activities lead to the same personal goal. Finally, Mahler (1933) pointed out that substitute activities are more effective in reducing tension related to interrupted goal striving when the substitute activity becomes a social fact, that is, when it is noticed by others.

With the Lewinian school as a background, the theory of symbolic self-completion can be summarized, using the concepts of *commitment* to self-defining goals, *symbols* of completeness, and *social reality.*

Commitment to Self-Defining Goals

With the term *self-definition* we refer to a conception of one's self as having a readiness to enact certain classes of behavior. If the self-definition is being a "jogger," for instance, then the activities deal with actual running, wearing appropriate clothes, associating with runners, and so forth. It is not necessary that these activities are actually carried out; rather, the individual claims to have the potential to carry them out. Accordingly, a self-definition is to be construed as an ideal, or goal.

Commitment to a self-definition means that the individual aspires to this "ideal" condition, wherein all of the qualities appropriate to the self-definition are embodied. The processes we shall discuss below, which have primarily to do with the mutual substitutability of symbols appropriate to particular self-definitions, should be observable only among individuals who are clearly committed.

In Lewin's language, a goal-specific tension exists only as long as the person is involved psychologically in the goal-pursuit. Once the personal commitment to the goal has been abandoned, all of the interruption effects (e.g., Ovsiankina, 1928; Zeigarnik, 1927) and substitution effects (e.g., Lissner, 1933; Mahler, 1933) are no longer observable.

Symbols of Completeness

Symbols are the building blocks of self-definitions, and the construction and preservation of a self-definition depends heavily on the person's use and possession of relevant symbols of completeness. A symbol can be a word, behavior, or a physical entity that potentially signals to others one's self-definitional attainment. One should not speak of a single, unequivocal symbol of the attainment of a self-definition. Rather, each self-definition may be viewed as composed of a set of symbols. People learn about these alternative indicators of self-definitions through interactions with others (cf. Cooley, 1902; Mead, 1934), and in turn, once the individual displays the symbol, others react as though the person embodies that self-definition.

The symbols of any given self-definition can take a variety of forms. At a simple but important level there are self-descriptions (e.g., a person who comforts another who suffers some emotional problem may refer to himself as "psychologist"), that is, highly literal and direct indications that one possesses the self-definition in question. Of course, the human is not solely dependent on these kinds of open self-characterizations. There are numerous comparatively subtle indicators of a self-definition, many of which are describable as status symbols. Having a diploma from a graduate school is a broadly recognized symbol of a person's self-definition, and it will propel the person toward a sense of completeness. Similarly, titles, official occupational positions, and membership in select interest groups are all socially evolved mechanisms for providing the individual with indicators, or markers of possessing an aspired-to self-definition.

As is illustrated in our research (below), an important dimension associated with symbols of completeness is the relative durability of the symbol. For instance, one's education, professional position, or relevant inherited traits would be highly durable symbols – not readily amenable to alteration by the incomplete person. On the other hand, acting as though one possesses the self-definition (trying to teach relevant skills to others; associating with experts), describing the self in ways that would further one's sense of completeness, and various other efforts, would fall toward the malleable end of the durable-to-malleable continuum. It is easy to see the importance of this dimension when one sets out to make clear predictions from the theory: Our starting point has characteristically been one of placing subjects into a position in which they are lacking a specified durable symbol. Because they cannot readily alter their standing on that symbol, it is then necessary for them to turn to other sources of symbolic support – namely, specifiable symbols that are highly malleable. The workings of this method will be seen in most of the research to be reported below.

Social Reality

From our theoretical position, the symbol in itself does not suffice to generate a sense of completeness. The symbols associated with any given self-definition serve a communicative function, no matter whether the symbol consists of a self-description, acting out the role of someone possessing the self-definition, acquiring materials or objects appropriate to the self-definition, or anything else. Theoretically, they signal to the community or society that one does indeed possess the self-definition; this indication of having attained a positive self-definitional status firms up one's sense of completeness.

There are at least two pieces of research that bear directly on this point. Mahler (1933) found that a substitute task served to reduce the tension associated with an original task only under special social circumstances. Quite independent of whether subjects knew that they had the right answer on the substitute task, it was found that the task reduced tension only to the degree that the experimenter took notice of the subject's completing the task. In short, the attainment of an intellectual goal (in the case of Mahler's experiment) was mediated directly by social circumstances. More direct to our theoretical point is a study by Gollwitzer (1981), which demonstrates that self-symbolizing efforts that are noticed by others are especially effective in enhancing self-definitional completeness.

Self-Symbolizing: The Cognition-Behavior Relation

When do people begin to reflect about their self-definitional standing? The answer we suggest here is hinted at in the writings of Mead (1934) and Shibutani (1961), who state that disruptions of social behavior steer the person's attention in the direction of the community. The interrupted person is said to take the perspective of the "generalized other" – which means being attuned to the values of society. We have pointed out above that the array of symbols associated with a given self-definition is defined within the community. Accordingly, individuals who encounter a hindrance in working toward a self-definition should then evaluate the self in line with the way society would view their extent of completeness. Thus, falling short with regard to one symbol will be experienced as missing an important facet of the symbolic array that defines a "complete" self-definition.

The hindrance of self-definitional progress can come about in a number of different ways: (1) One kind of disruption is the failure to complete an ongoing self-symbolizing act (as for instance, the psychology student of the above example who failed to demonstrate his animal-training capability). (2) Another possibility is direct evaluation from others, that is, other people pointing to the lack of symbolic support for one's self-definition. (3) A further source of disruption stems from social comparison with people who are more advanced regarding the self-definition in question. Such disruptions are expected to instigate self-reflection, making self-defining individuals evaluate their standing vis-à-vis the symbolic array that the community associates with a complete self-definition.

But there is more to these disruptive events than the instigation of self-reflection. As Lewin (1926) and his students (Lissner, 1933; Mahler, 1933; Ovsiankina, 1928) showed, the interruption of goal-oriented activity also results in a tension state. In the case of self-defining goals, this tension state can fuel self-symboliz-ing behavior. In other words, if a person's progress towards self-completion has been brought to a halt, the person will then be acutely aware of the falling short of completeness, and the tension state will propel the individual in the direction of substituting an alternative symbol for the symbolic lack associated with the disruptive experience. Another way of viewing this substitution idea is that those who face failure with respect to one symbol of completeness – generally a highly durable symbol – are not forced to abandon the quest for completing a self-defi-nition. Instead, they can (and are motivated to) move on immediately to one of the other numerous self-symbolizing routes.

The cognition-behavior relation of self-symbolizing activities now becomes evident: The cognition that guides self-symbolizing activities is not solely the ex-perience of failure or falling short. Rather, the individual tries to keep self-sym-bolizing activities in line with the claim of a complete self-definition. The experi-ence of falling short only serves to instigate a self-reflection that directs the individual's attention towards the whole array of symbols implied by a "com-plete" self-definition, *and* to provide for a tension state that motivates the indi-vidual to take one of the available alternative routes to completeness.

Two Research Examples

We are suggesting that people committed to a self-definition will bring their self-symbolizing behavior into line with the claim of completeness. This idea can be tested in a very straightforward manner. First, the individual is made to think about (or to experience) a symbolic weakness with respect to an aspired-to self-definition. In our research paradigms this weakness generally has to do with a highly durable symbol, that is, a symbol which is not readily attainable or alter-able. Then the person is offered an easily attainable alternative symbolic route to completeness; and finally, it is observed whether the individual strives for com-pleteness via the use of this symbolic route. It is predicted that the individual tries to substitute for the lack, thus projecting a picture of completeness.

Lack of Education and Attempted Influence. In the first study conducted under the rubric of symbolic self-completion (Wicklund & Gollwitzer, 1981, Study 1), male and female undergraduates were first asked to report a skill or an area of special knowledge which they considered to be of special interest. Subjects listed such skills as playing tennis, swimming, speaking Spanish, and playing piano. Then they were asked to indicate the amount of formal education (in months) they had received in the activity areas they had mentioned. Finally, subjects were asked to write an instructional essay that potentially could motivate other people to get involved in the subjects' favorite activity area. When subjects were finished with this essay they were asked to fill out a form that allowed them to specify how many different groups of people should receive their essay. Subjects chose from 12 potential target groups, such as high school students, undergraduates, foreign students, and so forth. The only additional information subjects had about these potential audiences of their instructional efforts was that these groups of people were part of the psychology department's subject pool.

Subjects who were acutely aware of their lack of education – having just indicated a relatively insufficient number of years of formal education on a questionnaire – can be called *incomplete* on the symbolic dimension *education*. The opportunity to teach or persuade others within the subject's own self-definitional area offers an opportunity to regain a degree of completeness, in that teaching or persuading (1) places the subject into the role of expert musician, athlete, or whatever self-defining dimension is central, and (2) also serves to generate a potentially broader social reality for one's self-definitional status. In short, having persuaded others within one's own self-defining area constitutes a symbol of completeness in that area. In support of this line of thought, we observed a negative correlation between amount of formal education and the number of different groups to which subjects wanted to send their essays ($r = -0.34$, $p < 0.01$).

Lack of Education and Positivity of Self-Descriptions. In an experiment by Gollwitzer, Wicklund, and Hilton (1982, Study 1) the same format was followed as in the study just presented. However, instead of influencing others, subjects were asked to come up with self-descriptions which could be varied in terms of their positivity. More specifically, subjects had to make a public statement that described the quality of their performance on a fictitious test measuring capability in their self-definitional area. As stated above, self-descriptions are an easily accessible and – if noticed by others – a highly effective form of self-symbolizing. Accordingly, subjects who fall short with respect to the amount of formal education can use positive self-descriptions as a means of approaching completeness. As a consequence, their self-descriptions should *not* reflect the educational shortcoming. On the contrary, they should be self-aggrandizing, thereby implying a complete self-definition. In line with these ideas we found a negative correlation ($r = -0.27$, $p < 0.02$) between subjects' educational background (months of formal education) and positivity of self-descriptions. Thus, the study reflects the paradoxical effect that people who begin with a weak educational background deliver self-descriptions that lay claim to a particularly positive status regarding the aspired-to self-definition.

The Role of Commitment to a Self-Definition

There is an important prerequisite for the self-symbolizing process to begin. The phenomenon that people compensate for a shortcoming (instead of behaving in a way which is consistent with it) cannot be expected in every case. Rather, it is necessary that the individual be committed to the self-definition, in the sense of showing a continued striving for a complete self-definition. For the person who has given up striving for progress as psychologist, musician, humanitarian, or any other self-definition, disruption of progress cannot occur and thus a sense of falling short, that is, the negative evaluative state of incompleteness, cannot develop. For such a person, having to admit to a shortcoming is no longer experienced as disruptive; consequently, no tension state builds up that could propel the individual to self-symbolizing efforts. Ovsiankina's (1928) finding, that those who are interrupted while trying to solve an uninvolving task tend not to resume the task, supports this line of thought.

In each of the two studies reported above there were also participants who, although willing to indicate an activity area of special interest (e. g., playing piano, speaking Spanish), indicated that they had not pursued any of the relevant activities recently. In other words, they had stopped striving for progress regarding the self-definition to which they once aspired. When looking only at these subjects ($N = 42$), whom we designated as *noncommitted,* we found in the first study that educational background (months of formal education) and the number of target groups these subjects wanted to influence tended to be positively related ($r = 0.11$, *ns*). In the second study, amount of education and positivity of self-descriptions for noncommitted subjects ($N = 12$) showed a similar pattern ($r = 0.32$, *ns*).

These data suggest that noncommitted individuals tend to bring subsequent behavior into line with the strength of their educational backgrounds: Noncommitted subjects with only a few years of formal education tend to be less interested in influencing numerous target groups, and, in addition, they tend to make self-descriptions that are comparatively modest. Thus, noncommitted individuals do not seem to influence others or self-aggrandize in proportion to their incompleteness. Instead, they tend to engage in these activities in a manner that is consistent with what they know about themselves regarding the strength of their educational background; that is, a weak educational background leads to reduced influence attempts and modest self-descriptions, whereas a strong educational background leads to enhanced influence attempts and to positive self-descriptions.

In other words, the behavior of noncommitted subjects tends to be related to their educational background in a 1:1 manner. For committed subjects, on the other hand, the educational background obtained serves the function of supporting one's claim of a complete self-definition. When the committed individual falls short, self-symbolizing activities are triggered. Accordingly, for committed subjects, a weak educational background leads to enhanced influence attempts and to positive self-descriptions – a result that is just opposite to what is found with noncommitted subjects.

The Relation Between Self-Report and Behavior

One cognition-behavior relation with which social psychologists and personologists have traditionally been concerned is that between a self-report that describes an inner quality (attitude or trait) and behavior that is presumably predicted by that inner quality. It would be ideal for the psychology of attitude measurement, as well as for personality psychology, if the connection between a person's self-report and that same person's behavior were a direct one. But it is easy to document the absence of such correlations. Wicker (1969), reviewing research on correlations between overt behavior and behavior-relevant attitudes, lists numerous examples.

In a study by Freeman and Aatov (1960), for instance, college students' attitudes toward cheating were measured on four projective tests, as well as on a more direct self-report measure. The correlations between each of these measures and actual cheating behavior varied between 0.10 and −0.19. Wicker (1969) also reports an example from his own work. There, students' attitudes toward research were measured. Subjects had to rate concepts such as *scientific research, psychological research* and *participating as a subject in psychological research* on semantic differential, evaluative scales. The behavioral measures were obtained one to four weeks after the attitude assessment. There were four levels of behavior, corresponding to the steps in the recruiting process, with "stated unwillingness to participate" and "stated willingness, appointment scheduled, and appearance at the experiment" at the opposing ends of the behavioral rating scale. The correlation coefficients relating attitudes and participation behavior were as follows: scientific research, −0.04; psychological research, 0.06; and participation as a subject in psychological research, 0.17.

Factors Interfering with Self-Report Validity

Why do psychologists find these low correlations between self-report and behavior? The problem at hand seems rather simple: A person is asked a number of straightforward questions (e.g., "Do you detest cheating?" or "How highly do you regard scientific research?") and then a sample of relevant behavior is obtained. The respondent is then found to engage in cheating, or not. In regard to the second question asked, the respondent is found to make a contribution to scientific research or not. Given the clarity of the problem, one should expect psychologists to be able to produce correlations that are of a considerably greater magnitude than 0.30. So what are the factors that might be responsible for the usually obtained low correlations that hardly ever exceed 0.30?

Four very common disturbing factors, discussed in some detail by Wicker, are reviewed below. It is important to note that these factors have often been seen as central in the psychologist's attempt to maximize self-report validity; however, none of them is particularly theoretically deep. That is, none of them addresses the psychological variables that move a person toward consistency between behavior and self-report. In large part, these factors may be seen as methodological prerequisites for obtaining self-report – behavior congruency.

1. Multiple Influences on Behavior. Another term for this factor is perhaps the "proportion of variance accounted for" problem. The classic report of LaPierre (1934) may serve as an example. In this study, hotel managers' attitudes towards Chinese were assessed via a self-report measure, and it was observed whether these hotel managers rented a room to a travelling Chinese couple. Clearly, the prejudice toward Chinese may account for very little of a hotel manager's behavior toward Chinese, given that they are paying customers. In other words, the major factor is the economic incentive; the minor factor is one's unfavorable attitude toward the customer. There is, of course, no a priori reason why any given attitude or trait should be expected to account for all of the variance in behavior. This problem is ultimately a practical one.

2. Measurement Issues. This problem receives a great deal of attention in research involving the validation of personality measures. Commonly the discussion centers around the reliability of the self-report measurement instrument (Jackson, 1982). On the other hand, Epstein (1979) has raised the question of reliable measurement of the criterion-behavior. He argues that single observations of behavior are always unreliable in the sense that they lack stability. An action performed on a given occasion may not be repeated under different circumstances or at a different point in time. Epstein reports increased attitude-behavior correlations when the measurement of the criterion-behavior is based on repeated observations.

Surely, an unreliable measure will not help the researcher's efforts to find a high correlation between self-report and behavior. At the same time, many highly reliable measures still do not show the expected correlation between self-report and behavior (cf. Gibbons, 1978; Hormuth, 1982). Thus, unreliable measurement is by no means the sole problem. However, we do not wish to belabor the point that unreliability and measurement-based obstacles to high validity should be dealt with.

3. Specificity of the Attitude Object. Wicker (1969) pointed out that many instances of inconsistency may be due to the fact that the attitude object is usually defined very generally while the behavioral response is considered at a highly specific level. For instance, suppose a sample of people is given the item "Do you love sports?" Subsequently, their spending of money for sports equipment is monitored, and correlated with the "like sports" item. If the correlation found is close to 0.00, the specificity hypotheses would see the explanation in the fact that the object of attitude was simply "sports;" there was no differentiation between watching sports on TV, teaching sports, engaging in sports, and the specific type of athletic activity also remained unspecified.

An attitude toward an object does not imply one single type of behavior but a whole array of behaviors toward that object (Ajzen, 1982; Ajzen & Fishbein, 1980; Fishbein & Ajzen, 1975). Therefore, relations between general attitudes and behavior could only be observed when the whole array of behaviors toward the attitude object is taken into account. However, if we are interested in predicting one specific behavior, such as whether a certain individual will or will not buy a certain piece of sports equipment, one should measure the attitude towards the behavior in question, that is, the individual's attitude towards buying this piece of equipment instead of the individual's liking of sports. When attitudes towards specific behaviors regarding the attitude object are measured, attitude-behavior correlations are obtained that are in the 0.60 to 0.80 range (Ajzen,

1982); a marked difference to the low correlations that are observed when general attitudes (e. g., liking) toward the attitude object are measured.

4. Can the Behavior be Manifested? Wicker's point here is that there may be physical limits on whether the behavior that is of interest to the psychologist can even be performed. A respondent's positive or negative attitude toward Gypsies can hardly be manifested when there are no Gypsies in the vicinity; a self-described personality trait "extravert" cannot be realized on a behavioral level unless there is adequate opportunity for the person to mix socially. This problem, quite clearly, is entirely a methodological concern. With this, and the other three considerations as a background, we will now turn to what can be regarded as psychological factors basic to congruency between self-report and behavior.

Psychological Roots of Congruency Between Self-Report and Behavior

Abelson (1982) comes to the conclusion that the psychological state that increases self-report and behavior consistency is that of clarity about the behaviorally relevant inner, personal quality. To support this conclusion he refers to Snyder's (1979) work on self-monitoring. There it could be shown that people who are more in touch with their attitudes (so-called "low" self-monitors) show better attitude-behavior consistency than people who are more sensitive to situational cues (so-called "high" self-monitors). Abelson sees the same principle at work in the research reported by Scheier, Buss, and Buss (1978): People who were high on a scale that measures self-consciousness, that is, a disposition to direct attention toward one's self, were found to show stronger attitude-behavior relations than people who were low on self-consciousness. Finally, he refers to Fazio and Zanna (1978) to support his argument. These researchers conducted a number of studies which show that direct behavioral experience with the attitude object can increase the attitude-behavior relation. Fazio and Zanna (1978) attribute this effect of direct experience to the following process: "During a direct experience, because of more information and better focus, the individual forms a relatively clear, strong, and confident attitude . . . which better predicts behavior."

However, researchers in the tradition of dissonance theory might question Abelson's contention that salience or clarity of one's attitudes provides for better attitude-behavior consistency. Pallak, Sogin, and Cook (1974) found that salience or clarity of one's attitudes does not hinder people from changing their attitudes after they have engaged in counter-attitudinal behavior. When they varied salience by manipulating subjects' extremity of the initial attitude, it turned out that subjects with more extreme initial attitudes showed more attitude change in the direction of counter-attitudinal behavior than did subjects with moderate initial attitudes. From the perspective that clarity makes for strong initial attitude-behavior relations, this result is hard to interpret; from the perspective of dissonance theory this result is a consequence of heightened dissonance between the counter-attitudinal behavior and the salient, extreme attitude. Green (1974) found results parallel to Pallak, Sogin, and Cook, although in this study the internal, personal quality was not an attitude, but that of a physiological need (i. e., thirst).

Abelson's suggestion that the psychological state that favors attitude-behavior consistency is one of clarity about the internal, personal quality obviously falls short. Rather, research in the tradition of dissonance theory suggests that psychological variables that are of a more motivational nature also affect such consistency. We will keep this in mind when we discuss the self-report and behavior consistency issue from the perspective of self-completion theory.

Self-Report and Behavior: Consistency with What?

It seems reasonable to assume that any given behavior is associated with at least a few relevant self-related cognitions. Thus, a thorough analysis of the self-report and behavior-consistency issue has to address the question of "consistency with what?", that is, to which aspect of the person's self is the observed behavior related?

Dissonance researchers have tackled this question in the context of the individual's changing cognitions to reduce dissonance. There the question was: Which type of cognition – out of a few possible choices – will people bring into line with the behavior to which they have committed themselves (Götz-Marchand, Götz, & Irle, 1974; Beckmann & Irle, Chapter 8, this volume). A parallel issue arose in objective self-awareness theory (Duval & Wicklund, 1972; Wicklund, 1975). There the issue of "consistency with what?" was addressed in the realm of self-report validity research (Wicklund, 1980, 1982; Wicklund & Frey, 1980) and is thus particularly relevant for the present discussion.

Self-awareness theory states that self-focused individuals come to realize their personal inconsistencies, or inadequacies, on whatever dimension of the self is salient in the condition of self-focused attention. The result of this self-evaluative condition is a motivational consequence: If it is difficult to remove oneself from the self-aware condition, then the person can be expected to show an increase in consistency. In other words, individuals will then behave so as to conform to their moral principles, attitudes, beliefs, or whatever aspect of the self that is focused on.[1]

---

[1] While it may be tempting to suggest that the condition of self-focus vs. that of outward focus parallels Kuhl's postulated dimension of action orientation vs. state orientation (cf. Carver & Scheier, Chapt. 11 this volume), it does not seem theoretically profitable to propose a general correspondence between self-focus and action (or between self-focus and passivity). The reason for this lies in certain components of the dynamics – sequential processes – that are part of the workings of self-awareness. The kinds of conditions that have generally been employed to induce self-focused attention would force the person, at least at the outset, to cease ongoing activities and to stop in order to evaluate one's progress or condition vis-á-vis personal standards. Whether self-focused attention then leads the person toward action in the sense of abiding by those standards depends on whether the actions are possible, and on whether circumstances readily permit such appropriate actions. When not, then avoidance of the self-focusing circumstances is postulated. Such avoidance might variously be labeled as action orientation or as state orientation, in that what is being (actively) avoided is the state of having to focus on one's incompetent or unprepared self.

In a study by Gibbons (1978), for instance, subjects first filled out the Mosher sex guilt inventory, then read an erotic passage from a paperback book, and finally rated the passage on the criteria of: arousing, enjoyable, and well-written. About half of the subjects sat before a mirror, a classic self-focus inducing device, while they made this tri-partite rating of the pornographic passage. The resulting correlation coefficients between the scale score and the three dependent measures were 0.45, 0.74, and 0.58 respectively. The results in the control condition that was run without placing subjects in front of a mirror were 0.10, 0.20, and −0.23. In short, the results suggest that subjects behave much more in line with their attitudes or morals when behavior takes place under self-awareness conditions.

Prior to the Gibbons research, Carver (1975) published a pair of studies showing that physical punishment (electric shock) given to another person is especially predictable from a punishment-relevant self-report measure when the punishment is carried out under self-focusing conditions. Shortly thereafter, Scheier (1976) reported an experiment which used the same format as the Carver studies. However, Scheier also angered some of his subjects and found among this group no correlation between an earlier stated belief in the value of punishment and subsequently administered punishment. Instead, the effect of self-awareness was enhancement in the level of punishment, accompanied by an increment in anger.

What happened? Did subjects under self-awareness conditions in the Scheier study fail to show consistency? Wicklund (1980, 1982; Wicklund & Frey, 1980) suggests that whenever the individual carries more than one self-aspect that is relevant to the behavior in question, then − when the individual is made self-aware − these standards, goals, or end points may come into conflict with each other. In such a case the individual's striving to be consistent may take the form of bringing the behavior into line with that "end point" that comes to the fore given self-focused attention. Wicklund draws on William James (1910) to find an answer to the question of which "end point" or which aspect of the self will come to the fore when a person is made self-aware. James had suggested that immediate psychological states such as emotions and desires have a more seizing quality than self-components having to do with a person's social norms, and that volitional decisions are more central to a person's self than intellectual processes. Accordingly, the Scheier results do not simply demonstrate the individual's inconsistency with prior self-report; rather they demonstrate consistency with another aspect of the person's self − that of the emotional experience of anger.

Thus, objective self-awareness theory teaches us that what might, on the surface, look like an example of inconsistency may still be a case of consistency. Even if behavior contradicts self-reports, people might still act on a consistency principle, that is, they simply bring their behavior into line with a different aspect of the self than the one reflected in their self-reports. This idea is crucial for the understanding of self-reports in the case of the self-defining human.

The Self-Symbolizing Function of Self-Reports

The relation between self-related cognitions and behavior takes on a distinct quality within the context of self-completion processes. To be sure, the psychologist's construal of *cognition* as well as of *behavior* must be modified and reconsidered when dealing with a person who is committed to a self-definition. Within the arena of self-report validity, the psychologist's perspective (i. e., the assumed relation between self-related cognitions and behavior) is usually like this: It is hoped that the respondent shows some behavior that stems directly from the self-related cognition.

Another way of saying this is that the psychologist hopes that certain, easily specifiable behaviors stem directly from the attitudes, traits, or abilities that are sampled via self-report.

When we refer to the self-defining individual as being governed by a set of very special dynamics we mean the following: The individual's self-related cognitions, insofar as they are pertinent to a self-definition to which the person aspires, may fall in the service of the self-defining goal to which the individual aspires.

Self-related cognitions are then not simply descriptive summary statements of the person's past (positive or negative) behaviors and experiences. Instead, these self-related cognitions are to be seen as oriented towards building a sense of completeness regarding the aspired-to self-definition.

When the individual expresses these self-related cognitions via self-reports, such self-reports are then not necessarily coordinated to the individual's subsequent behaviors in a simple 1:1 relation. This is because these self-reports have a transitory quality and do not have to reflect a stable underlying condition. Rather, they are related to the waxing and waning of the individual's self-defining needs. To the extent that the person is incomplete (lacking self-definitional symbols), self-descriptions should strongly reflect the person's efforts to gain completeness. To the extent that the person does possess a degree of completeness, self-descriptions will then be less affected by self-definitional needs. These phenomena are illustrated rather directly in an experiment by Gollwitzer (1981).

Varying Social Reality for Positive Self-Descriptions. Female subjects committed to the self-defining goal of raising a family were asked to respond to eight questions that either required self-descriptive answers (e.g., "How well do you get along with children?") or did not require self-descriptive answers (e.g., "What is the average number of children in an American family?"). The self-related questions could easily be answered with positive self-descriptions, that is, subjects could say any conceivable positive thing about themselves, whereas the knowledge-related items were difficult to twist into self-descriptive material. Subjects wrote their answers to these questions on small paper slips, put these slips into prepared envelopes, and finally sealed the envelopes. The experimenter carried these envelopes to a presumed partner subject who allegedly was told to comment on at least four of the subject's answers. In fact, depending on the experimental condition, the experimenter herself commented either on the self-descriptive answers *(Social Reality* condition*)* or on the non-self-descriptive answers *(No Social Reality* condition*)*.[2]

The theoretical meaning of what has happened thus far is the following: Subjects who have gathered up a social reality for their self-descriptions should have moved in the direction of self-definitional completeness, according to the view of the theory. They have offered up a symbol – the relatively simplistic positive self-description – for public consumption, and were given feedback that the symbol was recognized by another person. Accordingly, a gain in the sense of completeness should have resulted. Those in the *No Social Reality* condition, in contrast, should not have experienced any such gain since their positive self-descriptions were not recognized by another person.

In all conditions, subjects were then given a second chance to describe themselves. Adler (1912), and in a more experimental form Kelley (1951), have suggested that people with upward-oriented aspirations can pursue these aspirations psychologically by presenting themselves as similar to people who already have prestige and recognition. Congruent with that suggestion, the symbolic route offered to subjects here was the opportunity to describe their personality as similar or dissimilar to relevant, successful others (i. e., to successful mothers).

This was done in the following manner: In all conditions, subjects were introduced to a presumed personality psychologist. He presented a semantic differential type of personality questionnaire that already carried check marks connected by straight lines. The experimenter explained that this profile represented the "ideal" personality for a mother, that is, successful mothers have a personality profile similar to this ideal. In fact, the experimenter had drawn this line onto the questionnaire so that it described a person with five positive and five negative traits. Finally, subjects were asked to indicate their own personality traits on this questionnaire (Fig. 4.1).

Subjects who had already won a measure of completeness *(Actual Social Reality* condition*)* should have had less reason to self-symbolize on the profile measure; and this is what happened (as shown in Fig. 4.2). On a measure of dissimilarity to the ideal profile (the sum of the squared differences between the ideal check mark and the subject's check mark for each single item on the personality questionnaire), *Actual Social Reality* subjects were significantly more dissimilar from the ideal than *No Social Reality* subjects.

The most blatant aspect of the results is that the tendency to perceive oneself as more, or less similar to an expert is not necessarily a stable tendency. The degree to which subjects claimed similarity to the expert was very highly dependent on the preexisting sense of completeness, in that the subjects *without* an immedi-

---

[2] There was also a condition in which subjects had good reason to except acknowledgement for their positive self-descriptions, but where the actual social reality had not materialized by the time of the dependent measure. In this *Expected Social Reality* condition, the experimenter led the subject to believe that the partner subject was late but would, at some later point in time comment on the self-descriptions of the subject. Since the results for this condition and for the *Actual Social Reality* condition were identical, we will not discuss the results of the *Expected Social Reality* condition further. The reader is referred to Gollwitzer (1981) and Wicklund and Gollwitzer (1982, Chapter 5) for a more complete description of the experiment.

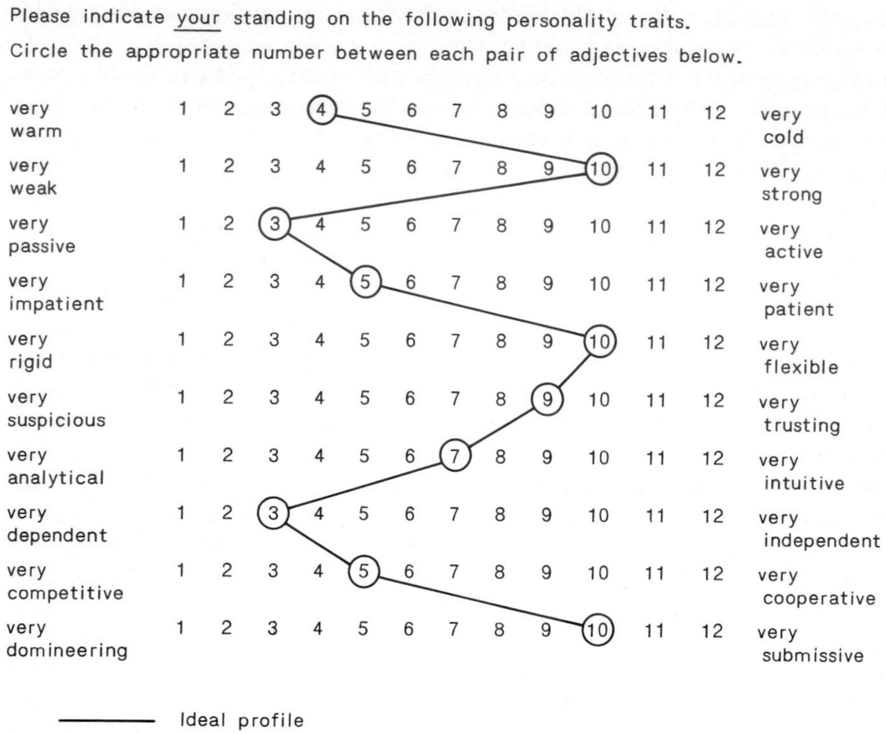

Please indicate <u>your</u> standing on the following personality traits.
Circle the appropriate number between each pair of adjectives below.

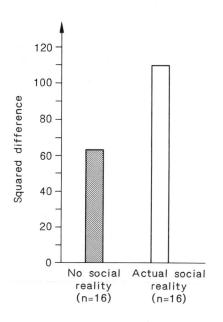

very warm	1	2	3	④	5	6	7	8	9	10	11	12	very cold
very weak	1	2	3	4	5	6	7	8	9	⑩	11	12	very strong
very passive	1	2	③	4	5	6	7	8	9	10	11	12	very active
very impatient	1	2	3	4	⑤	6	7	8	9	10	11	12	very patient
very rigid	1	2	3	4	5	6	7	8	9	⑩	11	12	very flexible
very suspicious	1	2	3	4	5	6	7	8	⑨	10	11	12	very trusting
very analytical	1	2	3	4	5	6	⑦	8	9	10	11	12	very intuitive
very dependent	1	2	③	4	5	6	7	8	9	10	11	12	very independent
very competitive	1	2	3	4	⑤	6	7	8	9	10	11	12	very cooperative
very domineering	1	2	3	4	5	6	7	8	9	⑩	11	12	very submissive

——————— Ideal profile

Fig. 4.1. Personality profile questionnaire with ideal profile drawn in

Fig. 4.2. Mean dissimilarity of own personality profile to the ideal profile (Gollwitzer, 1981)

ately preexisting social reality (on the basis of the earlier self-description) showed a definite propensity to characterize themselves as similar to experts. In short, these self-characterizations that came forth in the dependent variable were in no way the stable end product of a long behavioral history, or of the exact knowledge of one's training or experience, or any such thing. Rather, perceived similarity to the expert was largely at the whim of the momentary condition of incompleteness.

Second, a further question has to do with whether the subjects' personality characterizations are valid by the criterion of any related behavior or index of ability that is potentially observable by the investigator. In this particular instance, since we know that subjects were randomized across conditions, the personality characterization probably has a validity of zero by the criterion of any index we would care to name. But we can go one step further with the validity question. *Actual Social Reality* subjects were given acknowledgement for their child-rearing-relevant self-descriptions. Thus, at the very least, they had the concordance of the community that they were indeed potentially capable mothers. Logically, then, they would have proceeded to view themselves as somewhat more expert than subjects in the *No Social Reality* condition, and this means that on the dependent measure they would have placed themselves closer to the accomplished mothers whose profiles were displayed. The fact that the *opposite* results were obtained says that the results describe a "reverse validity," such that subjects whose social reality indicates to them a certain competence are the least likely to openly characterize themselves as experts. The reason for this, as proposed earlier, has to do with the interference of self-defining needs. And, of course, there is no inconsistency here when we view the situation from the standpoint of the person who is working toward completeness in a self-definition. The more incomplete the individual is (e. g., without social reality), the more the person is expected to strive after further symbols and social reality.

Varying Social Reality for Ambitions. The present experiment (Gollwitzer & Mendez, 1983) uses much the same format as the above study, but deals with varying degrees of social reality for self-definitionally relevant ambitions. Normally one would expect that those who indicate high aspirations, and who are then recognized for those aspirations, would go on to characterize themselves as relatively capable in the relevant activity area. That is, a superficial consistency would be expected between perceived similarity to an expert and the extent of social recognition the person has had for his statements of aspiration. But, as in the previous study, it can be shown that the gaining of socal reality – this time for a stated ambition – can lower one's tendency to describe oneself as similar to experts or professionals.

The present study differed from the previous one in just three aspects: First, the participants were female undergraduates committed to the self-defining goal of "female professional." Second, the self-related questions subjects had to answer during the first part of the experiment asked them to write down intended attainments rather than actual attainments. These questions read like this: "What type of position do you plan to occupy in your professional career?"

or "How much money do you plan to make on your first job?" Finally, the check marks subjects found on the personality questionnaire presented to them by the second experimenter were this time described as representing the "ideal" personality for a female professional. The format of the check marks and the resulting profile were the same as in the earlier study.

The results – self-ascribed dissimilarity to the expert – were strikingly parallel to the prior study. Again, the dissimilarity measure revealed that *Actual Social Reality* subjects reported their personalities to be more dissimilar to the ideal personality than *No Social Reality* subjects. The present study suggests that whenever individuals committed to a specific self-definition tell others their ambitions, an increased sense of completeness will result (Fig. 4.3). The consequence is the suppression of further striving toward the self-defining goal of professional woman.

This finding makes very good sense from the standpoint of the person striving toward a self-defining goal. Coming to the dependent variable phase of the present study with an absence of social reality, the person should be particularly motivated to gain further symbols and social reality – hence the result that *No Social Reality* subjects claimed the greatest similarity to the profile of the accomplished professional. Interestingly, the results would be nonsensical from the viewpoint of a superficial consistency (or self-report validity) view of the situation. One would think that stating an aspiration, and then receiving social acknowledgement for that aspiration, would goad the person on to feelings of expertise and potential accomplishment. But of course the results took the opposite form: These supposedly goaded-on subjects were the modest ones when it

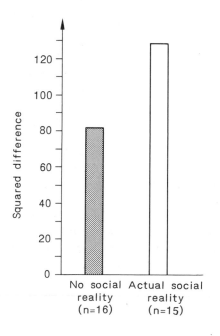

Fig. 4.3. Mean dissimilarity of own personality profile to the ideal profile (Gollwitzer and Mendez, 1983)

came to characterizing their similarity to the expert. In short, the observer who fails to see the self-definitional issues involved in the present paradigm will have to note another example of inconsistency.

The present pair of experiments, as well as the two studies reported earlier where subjects' incompleteness was caused by a lack of educational background, should caution psychologists from expecting consistent self-report and behavior relations given that self-reports are in the service of self-defining needs. There are two reasons for this: First, self-descriptions that serve self-definitional needs tend to reflect self-definitional attainments in a reverse manner. That is, people with a weak symbolic background try to claim strengths, whereas people with a strong symbolic background can afford to show modesty. Consequently, any subsequent performance may contradict these self-descriptions, insofar as the performance of individuals with minor self-definitional attainments will necessarily tend to show faults, whereas the performance of the more advanced individuals should reflect their higher level of self-definitional attainment.

Second, self-descriptions that serve self-definitional needs change the urgency of these needs. That is, a positive self-description that attains social reality reduces the self-definitional need which originally called forth the positive self-description. As a consequence, subsequent behavior will not reflect the claims made via the positive self-description, since no motivation remains to fuel activities that are in line with these claims.

The reader might think at this point that the picture of the self-defining human we are drawing is one of ultimate inconsistency. However, this particular picture holds only if one takes the traditional view of consistency, that is, if one expects self-report and behavior to relate in a 1:1 manner. As soon as one asks the question of "consistency with what?" – as suggested in the context of self-awareness theory (Wicklund, 1982) – a picture of consistency results: Self-defining individuals bring self-reports and/or behaviors into line with the self-definitional needs they presently experience.

The studies presented also suggest that – in the case of the self-defining human – clarity about one's self-definitional attainments does not guarantee self-reports that favor behavioral consistency. In both experiments, subjects had ample time to think about their self-definitional attainments (Experiment 1) or self-definitional aspirations (Experiment 2). Still, subjects' self-definitional needs were affecting their self-reports. That is, subjects who were made more complete via a social reality manipulation presented themselves less positively than their comparatively less complete counterparts. It appears, then, that Abelson's (1982) contention that clarity about one's internal condition makes for more accurate, and thus behaviorally more consistent self-reports, ignores the potential effects of motivational tendencies on these self-reports. In the present pair of studies, self-definitional needs could easily change self-reports in the self-complimentary direction even though subjects were given the possibility of reaching clarity about the facet of the self on which they were asked to report.

Self-Symbolizing: The Interference with Goals That Are Not Self-Defining

So far we have analyzed the working of self-completion processes within the realm of the self-report validity issue. We have asked the question of whether self-reports in the service of self-definitional needs are followed by behavior in a consistent manner. Thus, the cognition-behavior relation we have addressed is one where the cognition is self-related and qualifies as a symbol when expressed via self-report; in addition, the implied behavior also qualified as a symbol of completeness. Because both the self-related cognition and the implied behavior are embedded in a self-definitional theme, we have postulated a set of dynamics that are better described as consistency with self-definitional needs rather than as self-report and behavior inconsistency.

We now want to address a more complex cognition-behavior relation, where the behavioral part does not only qualify as an indicator of self-definitional attainment, but is also related to a non-self-defining goal. The following example illustrates what we have in mind: Suppose that a male is asked if he is attracted to a certain female, and whether he would like to get to know her. Once he has indicated his liking for the female, he is given a chance to act on his feeling of attraction to her. One would expect the male individual to match his courtship behavior to the degree of attraction he experiences to the female. But what if the female asks for a very particular type of courtship behavior that is also relevant to the self-definitional needs of the male? It is under these conditions that the male's attraction to the female might conflict with the person's self-definitional concerns.

The present experiment (Gollwitzer & Wicklund, 1985) was designed along this line of thought. Subjects were put in a situation where they could either describe themselves so as to conform to the requests of a member of the opposite sex to whom they felt attracted, or else come up with self-descriptions that were serving self-definitional needs. Male subjects committed to various artistic, academic, and athletic self-definitions were first given false personality feedback to manipulate their self-definitional completeness: subjects were informed that they had, or did not have, the personality qualities frequently found with successful people in their fields of expertise. In a second, presumably unrelated part of the experiment, subjects were made to expect a conversation with an attractive female undergraduate ("Debbie"). After subjects had read a description of Debbie's appearance, attraction measures were taken. Then, subjects were given a chance to ingratiate themselves with Debbie, in that they found that Debbie would like to know more about her male partners before meeting them. The information she required was related to subjects' standing in their self-definitional activity areas. Before subjects wrote down this information, they were informed about certain preferences regarding the form this self-descriptive information should take: Debbie either liked males who kept these self-descriptions very modest, or preferred males who described themselves very positively.

In terms of the issue of the relation between cognition and behavior we were looking at the following: First, there is the non-self-defining theme of making friends with an attractive woman. The resultant courtship behavior can be assumed to be controlled by two "interpersonal" cognitions: the self-related cognition of feeling attracted to this woman, and the cognitions the subject entertains regarding the courtship preferences this attractive woman holds. Second, there is a self-definitional theme created for those subjects who had received negative personality feedback. These subjects should have been oriented towards bringing their courtship behavior into line with their claim of completeness regarding the aspired-to self-definition. The important question, then is this: Will the "interpersonal" cognitions lose their grip on a subject's behavior (i. e., self-descriptions in the present study) when self-definitional needs are aroused?

In general, one would expect subjects to follow Debbie's preference to the degree they were attracted to her. However, self-descriptions in one's self-definitional area are also symbols of completeness. Subjects who had received negative personality feedback should, therefore, have implemented these self-descriptions as a means of recovering completeness. For the group of subjects that had received positive personality feedback, self-definitional needs were at least partly abated, thus the main factor influencing subjects' self-descriptions should have been their liking for Debbie.

When computing a correlation between subjects's liking for Debbie and their willingness to follow her self-descriptive preferences, we found high consistency among subjects who had received positive personality feedback ($r=0.49$, $p<0.01$); for subjects who had received negative personality feedback, no such consistency was found ($r=-0.12$, ns). The self-definitional concerns aroused by the negative personality feedback evidently hindered subjects from being attuned to their liking for Debbie and/or from acting on their liking for Debbie. Furthermore, when it was evident to subjects that Debbie preferred modest males, the relatively incomplete subjects were incapable of bringing themselves to be modest (in sharp contrast to the relatively complete group). Even when it was obvious that only modesty could be successful in the interaction, incomplete subjects proceeded to describe themselves as at the 66th percentile in their respective self-definitional areas, while the more complete subjects lowered their self-descriptions all the way to the 43rd percentile.

These results indicate that when self-definitional needs were aroused, interpersonal cognitions lost control over the subjects' behavior towards the target person. Subjects' self-related cognitions of feeling attracted to her, as well as the cognized behavioral requests of the target person, were hardly reflected in subjects' behaviors. Thus, self-definitional goals can upset the cognition-behavior relation of non-self-defining goals whenever the behavioral side of this relation is relevant to self-definitional claims held by the individual.

The present finding is particularly striking in light of Kuhl's (1984) contention that it is the "ill-defined goals" or "degenerated intentions" that fall prey to competing motivational tendencies most easily. In the present study the goal of attracting the target person was defined quite clearly, that is, subjects knew exactly what they had to do to attract the female target person. Still, self-defining needs interfered quite readily with the successful pursuit of this very clear goal. It appears, then, that self-definitional needs are quite potent in disturbing cognition-behavior relations as long as the behavioral part of this relation qualifies as a symbol of completeness.

Summary

The human we have described in this chapter is oriented towards an end state, or desired final state. Because of the individual's commitment to completeness in regard to self-defining goals, numerous facets of the person – that is, cognitions and behavior – are brought into the service of working toward completeness. To the extent that the person is lacking symbols of completeness (e. g., education, experience, membership in relevant groups or societies, personality qualities) and acknowledgement for possessing the appropriate symbols, subsequent efforts will be increasingly oriented towards completing the self-defining goal. In other words, a phenomenon of substitution will take place such that an absence of strength will lead the person to pursue indicators of strength.

This means that the relevant cognitive state of the organism in this analysis is the person's sense of falling short with regard to an aspired-to complete condition. This cognition of falling short is the source of the motivation to pursue symbols of completeness. Viewed from this perspective, the cognition-to-behavior chain in the self-completion process is quite straigthforward: Among committed individuals the cognition of weakness leads to the attempted accumulation of indicators of strength.

This process raises havoc with traditional conceptualizations of the relation between cognition and behavior, which assume a consistent interrelation. According to such a consistency position, people who sense their own weak training in an area of aspired expertise, for instance, should act in a manner that reflects this weakness. Or, if people state an aspiration and are then recognized for having this aspiration, they are expected to act as people who aspire to the goal in question. In short, people's self-related cognitions are expected to be manifested in a consistent manner.

An example may contrast this traditional consistency position with the approach presented in this chapter: A psychologist asks a group of students to indicate on a 9-point scale how much they love the arts. Then they are given an art history test. Finally, a correlation between subjects' self-reports and test perfomance is computed, and a correlation of close to zero is found.

How will researchers guided by the self-completion notion react to this finding? They should entertain the possibility that self-definitional needs may have entered the picture for some of the students. Accordingly, they should try to separate the students who are committed to the self-definition of "having extensive knowledge of the arts" from the students who are not committed to this particular self-definition. When this is achieved, new correlations between self-report and behavior are computed for both groups separately.

A significant negative correlation would be expected for committed students; these students would use the self-report for the purpose of moving towards self-definitional completeness, thereby compensating for the lack of a symbolic background for the self-definition. Subsequently, when it comes to behaving, these students – who indicated an intense appreciation of art out of incompleteness – will fail to show strong scores on the art history test. For the noncommitted students the researcher, guided by a self-completion notion, will *not* expect a negative correlation, since self-definitional needs could not affect self-reports. Accordingly, this researcher would not be surprised when the noncommitted person who indicates an appreciation of arts also tends to show a mastery of art history.

Researchers adhering to the consistency position should react differently to the finding of a zero correlation. For instance, they might refer to the Fishbein and Ajzen (1975) contention that only when attitudes towards behavior (i. e., intentions) are obtained can a positive cognition-behavior relation be expected. In other words, in the present example the cognition was perhaps too general to lead to any meaningful positive relations with behavior. In any case, researchers who abide by the consistency premise would likely renovate the study described in the present example along the following lines: The psychologist would ask for students' intentions to report on art history. From the consistency idea, one would expect that the underlying positive correlation between cognition and behavior will finally come to the fore since the cognition considered is now behaviorally specified. In other words, the cognition is made to overlap with behavior to a large degree, and a better resemblence between cognition and behavior is thereby achieved. Consequently, positive cognition-behavior relations should prevail.

Unfortunately, the above consistency approach ignores the potential workings of self-definitional needs, whose importance for cognition-behavior relations has been amply demonstrated in this chapter. Cognitions of falling short regarding a certain self-definition – despite their obvious nonspecificity – were found repeatedly to control self-symbolizing efforts whenever self-definitional needs were aroused. For instance, we found significant negative correlations between educational background and amount of attempted influence, and between educational background and positivity of self-descriptions for subjects committed to self-definitional completeness. Complementing the argument of the importance of self-definitional needs with respect to cognition-behavior relations, we also found that even when subjects held quite specific cognitions regarding a certain behavior (i. e., attracting a female), self-definitional concerns upset the consistency that was to be expected on the basis of subjects' very specific interpersonal cognitions. It appears, then, that the theme of action control cannot be discussed successfully solely on the basis of structural refinements regarding the cognitions involved in such processes.

Finally, we would like to formulate a proposal for the applied psychologist aiming at valid self-reports. In practically all facets of the connection between self-related cognitions and behavior, the individual's commitment to a self-defining goal can easily destroy the aesthetically ideal portrait of consistency between word and act. Thus, when one is searching for self-report and behavior consistency one should always entertain the idea that either the obtained self-report or the observed behavior might be in the service of self-defining needs. Then, the only "consistency" that can be expected is one of consistency with the individual's self-defining needs. That is, the quality of the individual's self-report and the form of the individual's behavior would be aimed primarily at reducing a sense of incompleteness.

References

Abelson, R. P. (1982). Three modes of attitude-behavior consistency. In M. P. Zanna, E. T. Higgins, & C. P. Herman (Eds.), *Consistency in social behavior: The Ontario symposium* (Vol. 2). Hilldsdale, N. J.: Erlbaum.

Adler, A. (1912). *Über den nervösen Charakter: Grundzüge einer vergleichenden Individual-Psychologie und Psychotherapie.* Wiesbaden: Bergmann.

Ajzen, I. (1982). On behaving in accordance with one's attitudes. In M. P. Zanna, E. T. Higgins, & C. P. Herman (Eds.), *Consistency in social behavior: The Ontario symposium* (Vol. 2). Hillsdale, N. J.: Erlbaum.

Ajzen, I., & Fishbein, M. (1980). *Understanding attitudes and predicting social behavior.* Englewood-Cliffs, N. J.: Prentice Hall.

Carver, C. S. (1975). Physical aggression as a function of objective self-awareness and attitudes toward punishment. *Journal of Experimental Social Psychology, 11,* 510–519.

Cooley, C. H. (1902). *Human nature and the social order.* New York: Scribner.

Duval, S., & Wicklund, R. A. (1972). *A theory of objective self-awareness.* New York: Academic Press.

Epstein, S. (1979). The stability of behavior: 1. On predicting most of the people much of the time. *Journal of Personality and Social Psychology. 37,* 1097–1126.

Fazio, R. H., & Zanna, M. P. (1978). On the predictive validity of attitudes: The roles of direct experience and confidence. *Journal of Personality, 46,* 228–243.

Fishbein, M., & Ajzen, I. (1975). *Belief, attitude, intention, and behavior: An introduction to theory and research.* Reading, Mass.: Addison-Wesley.

Freeman, L. C., & Aatov, T. (1960). Invalidity of indirect and direct measures of attitude toward cheating. *Journal of Personality, 38,* 443–447.

Gibbons, F. X. (1978). Sexual standards and reactions to pornography: Enhancing behavioral consistency through self-focused attention. *Journal of Personality and Social Psychology, 36,* 976–987.

Gollwitzer, P. M. (1981). The social reality of self-symbolizing: Winning completeness through others. *Dissertation Abstracts International, 42,* 3032-B. (University Microfilms No. 8128629)

Gollwitzer, P. M., & Mendez, R. (1983, April). Effects of social reality on self-symbolizing. Paper presented at the annual meeting of the Eastern Psychological Association, Philadelphia.

Gollwitzer, P. M., & Wicklund, R. A. (1985). Self-symbolizing and the neglect of others' perspectives. *Journal of Personality and Social Psychology, 3.*

Gollwitzer, P. M., Wicklund, R. A., & Hilton, J. L. (1982). Admission of failure and symbolic self-completion: Extending Lewinian theory. *Journal of Personality and Social Psychology, 43,* 358–371.

Götz-Marchand, B., Götz, J., & Irle, M. (1974). Preference of dissonance reduction modes as a function of their order, familiarity and reversibility. *European Journal of Social Psychology, 4,* 201–228.

Green, D. (1974). Dissonance and self-perception analyses of "forced compliance": When two theories make competing predictions. *Journal of Personality and Social Psychology, 29,* 819–828.

Henle, M. (1944). The influence of valence on substitution. *The Journal of Psychology, 17,* 11–19.

Hormuth, S. E. (1982). Self-awareness and drive theory: Comparing internal standards and dominant responses. *European Journal of Social Psychology, 12,* 31–45.

Jackson, D. N. (1982). Some preconditions for valid person perception. In M. P. Zanna, E. T. Higgins, & C. P. Herman (Eds.), *Consistency in social behavior: The Ontario symposium* (Vol. 2). Hillsdale, N. J.: Erlbaum.

James, W. (1910). *Psychology: The briefer course.* New York: Holt.

Kelley, H. H. (1951). Communication in experimentally created hierarchies. *Human Relations, 4,* 39–56.

Kuhl, J. (1984). Volitional aspects of achievement motivation and learned helplessness: Toward a comprehensive theory of action control. In B. A. Maher (Ed.), *Progress in experimental personality research* (Vol. 13), New York: Academic Press.

LaPierre, R. T. (1934). Attitudes and actions. *Social Forces, 13,* 230–237.

Lewin, K. (1926). Vorsatz, Wille und Bedürfnis. *Psychologische Forschung, 7,* 330–385.

Lissner, K. (1933). Die Entspannung von Bedürfnissen durch Ersatzhandlungen. *Psychologische Forschung, 18,* 218–250.

Mahler, W. (1933). Ersatzhandlungen verschiedenen Realitätsgrades. *Psychologische Forschung, 18,* 27–89.

Mead, G. H. (1934). *Mind, self, and society.* Chicago: University of Chicago Press.

Ovsiankina, M. (1928). Die Wiederaufnahme unterbrochener Handlungen. *Psychologische Forschung, 11,* 302–379.

Pallak, M. S., Sogin, S. R., & Cook, D. (1974). Dissonance and self-perception: Attitude change and belief inference for actors and observers. Unpublished manuscript, University of Iowa.

Scheier, M. F. (1976). Self-awareness, self-consciousness, and angry aggression. *Journal of Personality, 44,* 627–644.

Scheier, M. F., Buss, A. H., & Buss, D. M. (1978). Self-consciousness, self-report of aggressiveness, and aggression. *Journal of Research in Personality, 12,* 133–140.

Shibutani, T. (1961). *Society and personality: An interactionist approach to social psychology.* Englewood Cliffs, N. J.: Prentice Hall.

Snyder, M. (1979). Self-monitoring processes. In L. Berkowitz (Ed.), *Advances in experimental social psychology* (Vol. 12). New York: Academic Press.

Wicker, A. W. (1969). Attitudes versus action: The relationship of verbal and overt behavioral responses to attitude objects. *Journal of Social Issues, 25,* 41–78.

Wicklund, R. A. (1975). Objective self awareness. In Berkowitz (Ed.). *Advances in experimental social psychology,* (Vol. 8). New York: Academic Press.

Wicklund, R. A. (1980). Group contact and self-focused attention. In P. B. Paulus (Ed.), *Psychology of group influence.* Hillsdale, N. J.: Erlbaum.

Wicklund, R. A. (1982). Self-focused attention and the validity of self-reports. In M. P. Zanna, E. T. Higgins, & C. P. Herman (Eds.), *Consistency in social behavior.* Hillsdale, N. J.: Erlbaum.

Wicklund, R. A., &Frey, D. (1980). Self-awareness theory: When the self makes a difference. In D. M. Wegener & R. R. Vallacher (Eds.). *The self in social psychology.* New York: Oxford University Press.

Wicklund, R. A., & Gollwitzer, P. M. (1981). Symbolic self-completion, attempted influence, and self-deprecation. *Basic and Applied Social Psychology, 2,* 89–114.

Wicklund, R. A., & Gollwitzer, P. M. (1982). *Symbolic self-completion.* Hillsdale, N. J.: Erlbaum.

Zeigarnik, B. (1927). Das Behalten erledigter und unerledigter Handlungen. *Psychologische Forschung, 9,* 1–85.

Part II

Self-Regulatory Processes and Action Control

Chapter 5

Historical Perspectives in the Study of Action Control

Julius Kuhl and Jürgen Beckmann

In his summary of a *socio-genetic* theory of *voluntary regulation,* Leontiev (1932) describes a strategy the Chinese mailman used to avoid getting distracted while delivering an urgent telegram. "He organizes his own behavior, creating for himself additional stimuli. He hangs a number of subjects – a piece of coal, a pen, and some pepper on the end of a short rod. This he keeps before his eyes on the road. This will remind him that he must fly like a bird, run as if he was stepping on hot coals or had burnt himself with pepper" (Leontiev, 1932, p.57). This example illustrates a strategy of voluntary regulation which presumably facilitates cognition-behavior consistency. Whether or not a cognition suggesting a certain action results in its enactment depends on the actor's ability to apply appropriate self-regulatory strategies.

In this chapter, we will briefly discuss some of the early theories of volition and several reasons for the neglect of volitional processes in psychological research during the past five decades. This discussion may serve two purposes: First, it may encourage investigators to exploit the heuristic value of early theories of volition, which were considerably more comprehensive and differentiated than current theories. Second, a look at early theories may help us avoid the theoretical drawbacks that contributed to the decline of the psychology of volition. Since it is beyond the scope of this chapter to provide detailed summaries of even the most important theories of volition, we will focus on the one theory (Ach, 1910, 1935) we consider to have the greatest heuristic value. Before summarizing Ach's theory, however, we will give a brief overview of other theories.

Overview of Early Theories of Volition

The major differences between theories of volition are related to two conceptual issues. First, some authors implicitly or explicitly define volition in terms of the psychological processes mediating decisions (Atkinson, 1964; James, 1904; Lewin, 1926; Michotte & Prüm, 1910) whereas others relate volition to processes that mediate the maintenance and enactment of decisions (Ach, 1910; Hebb, 1949; Lindworsky, 1929). The former approaches (with few exceptions, e. g., Michotte & Prüm, 1910) equate volition with motivation. This definition of will as "a label for the triumph of the strongest stimulus" (Harriman, 1947) identifies volition with behavioristic stimulus-response association, or with the motivational processes that determine the relative strengths of action tendencies.

The second issue on which theories of volition differ involves the question of whether volition is a derived or a separate psychological phenomenon. Several theorists described volition in terms of some basic mental process. Ebbinghaus (1902) and Fortlage (1855) defined volition in terms of the action tendency that is suggested by the current *emotional* state of the organism. Münsterberg (1888) and Külpe (1893) described volition in terms of the anticipated *sensations* of the muscles involved in an intended movement. Ziehen (1920) proposed an *ideational* theory of volition, which held that volition consisted of a characteristic association of ideas. This concept was similar to James's (1890) theory of volition. In subsequent years, the *zeitgeist* supporting philosophical determinism further increased the reluctance to grant volition a separate psychological status. Conceiving of volition as a separate psychological process was regarded as tantamount to accepting the questionable notion of "freedom of will."

The contrary view, which defined volition as a separate mental phenomenon, was defended by Wundt (1896), various members of the Würzburg School (Ach, 1910; Lindworsky, 1929; Messer, 1934), and Hebb (1949). However, while Wundt conceived of volition as a motivational construct (i. e., underlying decisions), Ach (1910) and Hebb (1949) defined volition as a postdecisional, self-regulatory process that energized the maintenance and enactment of intended actions. Lindworsky (1929) compared volition to a switchman, a rather crude technological metaphor compared to the current information-processing metaphor (Kuhl, Chapter 6, this volume).

The question of whether volition is a *heterogenetic* (derived) or *homogenetic* (separate) phenomenon was not resolved within the German tradition of "will psychology" (Rohracher, 1932). In later years, most psychologists implicitly assumed the heterogenetic position. This view was fostered by the basic postulate of associationism, that is, that even the most complex mental events can be conceived of as derivatives of associations among simple cognitive elements. Given that the associationistic tradition is still very strong in contemporary cognitive psychology, it comes as no surprise that current cognitive psychologists still consider motivation and volition as derived phenomena in a unified *cognitive* system of associative networks (Anderson, 1983; Norman, 1980). Recently, this unitarian conception of the human mind has been challenged by a *modular* view, which distinguishes several separate "faculties" even within the cognitive system. Current research into self-regulatory processes is based on a modular conception of cognitive, motivational, and volitional processes (Kuhl, 1984; in press).

Lewin's Reduction of Volition to Motivation

The deterministic philosophical and the associationistic psychological traditions are not the only factors that have contributed to the neglect of volitional processes during the past 50 years. Within motivational psychology, Lewin's (1926) seminal paper on "Intention, Will and Need" contributed to the decline of research into self-regulatory processes. In this paper, Lewin emphasized the similarities between the concepts of intention and need. He described an intention as a *quasi-need* in order to emphasize that an intention possessed the same "dynamic qualities" that he attributed to a need. According to Lewin, an intention as well as a need is characterized by an underlying goal-directed tension system which presumably persists until the goal is reached. Also, an intention creates, just as a need does, valences *(Aufforderungscharaktere)* in the environment that direct goal-oriented behavior. By equating the concepts intention and need, Lewin reduced the problem of volition to the problem of motivation. It should be noted, however, that in his later writings (which had a considerably weaker impact on motivation theory than his earlier papers), Lewin (e.g., 1951, p.233) tended to acknowledge the special status of an intention as compared to a motivational tendency by assuming that a "decision" to perform an action has the effect of 'freezing' the current "motivational constellation for action".

The final blow to the German tradition of will psychology stemmed from Lewin's attempt to refute Ach's theory of *determining tendencies*. Lewin (1922) designed an experiment which seemed to disprove Ach's (1910) contention that associations which are experimentally invoked in a subject between two *CVC*-syllables include a tendency to reproduce the second syllable when the first is presented to the subject. Although Lewin's elegant experiment, seemingly disproving Ach's contention, did not really relate to the core of Ach's theory of volition, it probably contributed to the decline of German will psychology. Ironically, Lewin's conclusion was later rejected on the basis of several experiments conducted in Ach's laboratory (Gerdessen, 1932; Müller, 1932; Simoneit, 1926) which have not been taken into account in many evaluations of the Ach-Lewin controversy.

The Ach-Lewin controversy per se is of historical interest only. However, to the extent that it reinforced the subsequent neglect of volitional processes in many areas of psychological research (Kuhl, 1983), Lewin's equating of the concepts of intention and need created a theoretical confusion which has led motivational psychologists to ignore volitional processes almost completely. Miller, Galanter, and Pribram (1960) have made an attempt to steer clear of this confusion by emphasizing the special status that a motivational tendency (or value, in their terminology) receives when it becomes an intention. Miller et al. (1960) speculated that as soon as an action tendency assumes the status of an intention, it is transferred to a special part of working memory. This transfer, which might (according to Miller et al., 1960) even involve a transfer of information between different brain structures, presumably arouses a special mechanism which protects the current intention against competing action tendencies.

Unfortunately the distinction between intention and value (or motivation) made by Miller et al. (1960) did not have much impact on subsequent theories of human motivation, perhaps due to several inconsistencies in their book. While they stated early in the work that they would be dealing with the "will", they later criticized the concept of volition on the basis of Lewin's contention that it could be reduced to the problem of motivation. In their criticism of Lewin's dynamic interpretation of an intention, Miller et al. (1960, Chapter 4) misattributed the problem in Lewin's theory.

Although Miller et al. (1960) did not make explicit why they had decided to reduce the concept of intention to "a sequence of instructions, the execution of which has already begun" and why they denied the dynamic (i. e., intensity-related) properties of an intention, it is not difficult to infer the reason for this decision. By redefining the concept of intention in this way, they revoked Lewin's "homogenization" of the concepts of motivation (i. e., need or value) and intention. It can be shown, however, that it is possible to acknowledge the special status of an intention (as compared to a motivational tendency or a value) without denying the dynamic properties of an intention by specifying the cognitive processes that are aroused when a motivational tendency becomes in intention (Kuhl, Chapter 6, this volume).

James' Reduction of Volition to Ideation

Since William James (1890) assumed an introspective approach, he emphasized the conscious perceptions associated with the operation of "the will." As a typical illustration of an everyday act of will, he discussed his efforts to get up and leave the warm bed on chilly winter mornings. According to James, it was the conscious thought "Hollo! I must lie here no longer" which actually caused him to get up. According to his *ideo-motoric* theory, the efficiency of volition was merely a function of the relative strengths of the ideas that made up the stream of consciousness. As soon as the idea "I must lie here no longer" became sufficiently strong, the appropriate movements were initiated. Narziß Ach (1910, p. 290) criticized this theory. According to Ach, James's analysis of his introspections during his attempt to get up did not yield any valid information about volitional processes. Getting up in the morning is such a well-rehearsed act that the volitional processes mediating it do not become conscious. As we will see later in this chapter, Ach (1935) developed a set of criteria for identifying volitional processes.

Although James discussed volition in terms of the competition between conscious ideas, he mentioned several other aspects that played an important role in subsequent theories. "The essential achievement of the will, in short, when it is most 'voluntary,' is to *attend* to a difficult object and hold it fast before the mind" (James, 1890, p. 561). In this and the following quotation, a close relationship between volitional control and attentional mechanisms is postulated. "We know in the case of many beliefs how constant an effort of the intention is required to keep them in this situation and protect them from displacement by contradictory ideas" (James, 1890, p. 62). James's analysis stresses two additional aspects of volitional processes, that is, the *difficulty* of enactment and the *effort* associated with the "protective" function of volition. These two aspects of volition have been neglected in many subsequent theories of action. Modern theories of motivation predict difficulty and effort in enactment only in the rare case where two action tendencies are exactly equal in strength. In all other cases, the dominant action tendency should be performed without any difficulty provided the person has the intended behavior in her/his repertoire. (Ajzen & Fishbein, 1973; Atkinson & Birch, 1970; Heckhausen, 1977; Vroom, 1964).

James noted interesting individual differences regarding volitional control. Some people, he observed, are characterized by an "impatience of the deliberate state" and a "proneness to act or decide merely because actions and decisions are, as such, agreeable, and relieve the tension of doubt and hesitancy" (James, 1890, p. 530). Others, by contrast, are characterized by the "dread of the irrevocable." Such a factor, which was found to be crucial for intention formation by Michotte and Prüm (1910), is similar to Kruglanski and Ajzen's (1983) distinction between the "need for structure" and "fear of invalidity." According to Kuhl's theory of volition (Chapter 6, this volume), the two personality characteristics result from action orientation and state orientation. These personality factors presumably determine whether or not another phenomenon, which James called "obstructed will," occurs. James quotes Ribot's description of this phenomenon:

The patients "experience the desire to act, but they are powerless to act as they should. Their will cannot overpass certain limits: one would say that the force of action within them is blocked up: the *I will* does not transform itself into impulsive volition, into active determination" (James, 1890, p. 546). In current study, psychologists neglect this inability to enact an intention despite the desire to act as a potential cause of behavioral deficits. The two preferred explanations of behavioral deficits are: impaired cognitive or motor skills or lack of motivation (see Kuhl, 1984, for an extensive critique of the "motivation doctrine" in research on achievement motivation and learned helplessness).

Michotte and Prüm's Action-Control Perspective on Intention Formation

There are two major sources of the inability to transform a wish into active determination in spite of high motivation and the ability to perform the action in question: 1. the inability to resolve the decisional conflict; and 2. the inability to shield a current intention against competing motivational tendencies. Ach's psychology of volition primarily addressed the second problem. Michotte and Prüm (1910) dealt with the first problem in a more sophisticated way than some modern theories of motivation do. Not until very recently have theories of motivation directed attention to action control processes in this early phase of action (see Kuhl, 1984; Kruglanski, Chapter 3, this volume).

Unlike many current theories of motivation that attribute intention formation to a rational expectancy-value strategy, Michotte and Prüm (1910) describe the process of intention formation as governed by several mechanisms that prevent indecisiveness. Michotte and Prüm (1910) found that many of their experimental subjects did not base their decision on an expectancy-value rationale but simply on expectancy-related information (easiness of operation). A similar assumption was formulated recently in Kuhl's theory of action control and supported by experimental findings (Kuhl & Beckmann, 1983).

According to Michotte and Prüm (1910), the value of an alternative must exceed a certain threshold for this alternative to be chosen. Unless this threshold is surpassed by the value of some alternative, "discussion of the motives", that is, the elaboration of each alternative's value and the expectancy of being able to perform it, will carry on. But this elaboration should not continue forever. If, after long elaboration, the value of none of the alternatives approaches the threshold, "inner motives" such as the subjective values of the alternatives are replaced by "external motives", which quickly bring about a decision. This motive denotes something that James called "impatience of the deliberate state" and has effects similar to those of Kuhl's "action orientation" or Kruglanski's "need for structure." After prolonged, ineffective elaboration of the values of the competing decision alternatives, according to Michotte and Prüm (1910), an additional tendency arises that terminates elaboration and thus causes the alternative preferred at the moment to be chosen. Like James (1890), Kruglanski, and Kuhl, Michotte and Prüm describe still another strategy for reaching a firm decision. A similar strategy of "incentive escalation" has recently been incorporated into the theory of action control (Kuhl, 1984; Beckmann & Kuhl, 1984). Suppose the "motive of duration" suggests a currently preferred alternative be chosen. If its current value is not sufficient, however, the value of this preferred alternative will be increased. In opposition to Ach, Michotte and Prüm (1910) regard these processes, which avoid a loss of action control through indecisiveness, as precognitive. Michotte (1911) maintains that the experience of tension and the awareness of effort described by Ach (1910) are by no means essential characteristics of the act of volition. Despite this divergence, Ach's and Michotte and Prüm's theories are not in essence contradictory.

Ach's Psychology of Volition

In the preface to his first volume on volition, Ach (1905) emphasized the need to investigate not only the determinants of intentions, but also the processes that mediate the enactment of an intention after it has been formed. He described those processes in terms of *determining tendencies* which presumably controlled several cognitive operations ensuring the enactment of current intentions according to "the meaning of the idea of the goal" (Ach, 1935, p. 143). He developed an elaborate method of introspection, which was far more sophisticated than the self-rating methods currently used in experimental psychology. Although we have good reasons today for not accepting introspective data as unambiguous evidence for testing hypotheses (e. g., Nisbett & Wilson, 1977), there is no reason to question the heuristic value of sophisticated introspective analyses for generating testable hypotheses. An impressive account of the heuristic power of

Ach's method of introspection can be found in his analysis of the cognitive processes mediating the enactment of the intention to react to a stimulus-word with a low-probability response (similar to the color-naming task in the Stroop test). He distinguished ten stages of the process mediating the response after the stimulus had been presented, and he was able to replicate the data supporting his model in most of his subjects (Ach, 1910, pp. 256–75).

Criteria for Identifying Volitional Processes

Ach (1910) took great pains to develop a set of criteria for deciding whether or not an observed action was mediated by what he called "a primary act of the will." He called four of these criteria "phenomenological aspects of volition" because they were inferred from the introspective protocols of his subjects. The first two of these criteria – which Ach considered the most important ones – can be described today in terms of various "nodes" of a propositional network describing the memory structure that represents an intention (see Kuhl, Chapter 6, this volume).

Four Phenomenological Aspects of Volition. According to Ach (1910), an intention can be characterized by four phenomenological attributes:

1. The *objective moment (gegenständliches Moment)* refers to the intended activity in its relation to the *relational idea (Bezugsvorstellung),* which describes the condition (i.e., the context) for executing the intended activity.

2. The *ego-related moment (aktuelles Moment)* refers to the experience of the self being currently committed to the enactment of the intention.[1] It is this aspect of an intention that establishes it as a "current concern" of an actor, a concept which has been recently reintroduced into motivation theory (Klinger, 1975). The ego-related moment of an intention is the motivational basis for the attentional selectivity associated with volition; information supporting alternative action tendencies is disregarded so that the current intention is shielded against interfering sources of motivation.

3. The *state-related moment (zuständliches Moment)* is characterized by an awareness of effort required to accomplish a "narrowing of consciousness," which produces an amplification of the cognitive representation of the current intention.

4. Finally, Ach (1910, p. 244) postulated a *subjective moment* of volition *(anschauliches Moment),* which refers to sensations of tension in various parts of the body

---

[1] The accentuation of such a pledging of the self to an intention (cf. Kiesler, 1971) should indicate a complete determination to perform the intended activity with the exclusion of any alternative.

that accompany the primary act of the will. The intensity of those sensations may increase very rapidly and contribute to the abrupt and impulsive nature of that act.

The Dynamic Moment of Volition

In addition to the four phenomenological aspects of volition, Ach (1910, p. 255) emphasized the significance of an objective criterion for identifying a given behavior as being mediated by a (primary) volitional process. He identified the dynamic side of volition on the basis of difficulties inherent in the specific situation that have to be overcome before the intended activity can be performed. Ach used the term *difficulty*, however, in a sense that differs from its usual one. He did not refer to the difficulties of a task that require the possession of intellectual abilities or skills enabling a subject to perform the cognitive operations or body movements necessary to perform the intended activity. Ach's concept of difficulty refers to barriers that can be overcome by a sufficient expenditure of effort. In other words, Ach refers to aspects of difficulty that are subject to immediate personal control. Examples are stimuli in the environment arousing alternative action tendencies that are likely to derail the current intention (e. g., delicious food put in front of someone trying to lose weight), habitual response tendencies that are incompatible with, though stronger than, the current intention (e. g., the strong habitual tendency to read a word in the Stroop test though the intention is to name the incongruent color it is written in), and a state of the organism (e. g., tiredness) that makes it very hard to initiate the intended activity.

In sum, Ach's concept of difficulty refers primarily to factors which complicate the initiation and maintenance of an attempt to perform the intended action, rather than factors complicating the final execution of that action. According to Ach (1935), volitional processes are not activated unless the specific situation contains one or more aspects of control-related difficulty. He formulated a "law of difficulty," which holds that effort increases as a positive function of the perceived difficulty of implementing an intended action. In contrast to recent models of "effort calculation," which assume a similar relationship between expended effort and perceived task difficulty (Kukla, 1972; Meyer, 1973), Ach (1935, p. 346) rejected the assumption that the law of difficulty is based on a rational and reflective process. According to Ach (1935, p. 346), a rational model of effort calculation is not consistent with the spontaneous and impulsive characteristics of this process.

Cognitive Mediators of Volition

Ach (1935) discussed several cognitive processes that mediate volitional control. This part of Ach's analysis is especially useful because, compared to current theories of self-regulation, it represents a surprisingly comprehensive theory of volitional processes.

Selective Attention. Activation of the object-related and ego-related aspects of an intention presumably arouses a process of selective attention, which amplifies the cognitive representation of the current intention and helps one to ignore information related to competing action tendencies. A strong activation of the ego-related aspect of an intention presumably facilitates the protective function of volition. A cumulation of successful enactments of an intention in the past presumably strengthens the ego-related aspect of the intention and increases the awareness of the self as a causal agent (Ach, 1910, p. 265). In contrast to current theories of causal attribution (e. g., Kelley, 1972; Weiner, 1980), Ach (1910) did not regard these instances of "causal awareness" as mediated by conscious thought.

Selective Encoding. Ach (1935) described a process that presumably facilitates the protective function of volition by selectively encoding those features of a stimulus that are related to the current intention. He described several stages of a volitional process that eventually result in a restructuring of the stimulus object. Using the language of current theories of information processing, we may describe this process in terms of a successive implementation of action-related attributes in the feature list describing a given concept. A person who loves to eat apples may develop a concept of an apple in which the features "something I want to have," "something I want to taste," and so forth, rank higher in the feature list than the features "fruit," "tastes sour," etc. Ach emphasized that when the concept of an object is endowed with purposive qualities *(finale Qualitäten)*, the future enactment of the action intended toward that object is greatly facilitated because volitional control is transferred to an automatic, lower-level process. Recent research on information processing has yielded experimental evidence for the existence of such functional encoding categories (e. g., Rosch & Mervis, 1975).

Successive Attention-Adjustment. When an attempt to perform an intended action is not successful, a volitional process presumably controls successive adjustments in the focus of attention on relevant aspects of the current intention. Relevant aspects of an intention are the idea of the context specifying the conditions for performing the intended action and the idea of the elements of the intended action. The distinction between an action-oriented and a state-oriented mode of control following a failure experience relates to this volitional process (see Kuhl, Chapter 6, this volume). Ach (1935, p. 360) explicitly discussed the debilitating effect of the failure of successive attention adjustments *(sukzessive Attention)* resulting, for instance, in an inability to disregard the unpleasant feelings aroused by past failures.

Determining Feelings. Ach (1935, p. 364) ascribed a significant mediating function to the feelings that are aroused after the outcome of an action has been perceived. The more vividly the phenomenological aspects of the intention (especially the ego-related aspect) are experienced, Ach says, the stronger are the feelings aroused after a success or failure has been perceived. He was quite explicit (1910, p. 272) about the assumed functional significance of those outcome-contingent feelings: "This entire state (i. e., anger following failure) is without doubt very conducive to the instigation of an additional act of the will which actually results in a success." Because he assumed that outcome-contingent feelings facilitated the protective function of volition (e. g., disregard of task-irrelevant information) and determined the "efficiency of the will" (i. e., the ratio between the number of enacted intentions and the number of attempted actions), he called those emotions *determining feelings (determinierende Gefühle)*.

It may be noted parenthetically that, on the basis of his observations, Ach rejected theories of emotion that regarded reflective thoughts such as causal interpretations of success and failure, as primary determinants of emotion (e. g., Lindworsky, 1929). He discussed empirical results suggesting the immediacy of emotional arousal. A similar point has been made in recent criticism of cognitivistic theories of emotions (cf. Zajonc, 1980, 1984). Ach's rejection of cognitivistic theories of emotion was based on a detailed criticism of the methods used by the cognitivists which were based on what Ach called "pseudo-experiments" (e. g., asking subjects how they would respond to a verbally described hypothetical situation). Interestingly enough, the same methodological criticism applies to several current cognitivistic theories of emotion (e. g., Meyer, 1973; Weiner, 1980; Weiner, Russel & Lerman, 1979).

The Specificity of Intentions. Ach (1935, p. 244) formulated a "law of specific determination," which said that the speed and likelihood of the enactment of an intention will be a positive function of the specificity of its content. A similar assumption has been made in a recent theory of cognition (Anderson, 1983). The successful execution of an intention requires that the context containing the conditions for its enactment and the elements of the intended action be as specifically articulated as possible. Ach's student, Düker (1925), reported some experimental results supporting Ach's law of specific determination. In a recent discussion of the determinants of the intention-behavior correspondence, the specificity of an intention has been assumed to be the major determinant of the enactment of the intention (Ajzen & Fishbein, 1977).

Although the preceding summary of Ach's theory does not do full justice to the comprehensiveness and elaborateness of his approach, it may suffice as a theoretical basis for developing a general framework of volition. In the following chapter, an attempt to develop such a theoretical framework will be described.

References

Ach, N. (1905). *Über die Willenstätigkeit und das Denken. Eine experimentelle Untersuchung mit einem Anhange. Über das Hippsche Chronoskop.* Göttingen: Vandenhoeck & Ruprecht.

Ach, N. (1910). *Über den Willensakt und das Temperament.* Leipzig: Quelle und Meyer.

Ach, N. (1935). Analyse des Willens. In E. Abderhalden (Ed.), *Handbuch der biologischen Arbeitsmethoden. Bd. VI.* Berlin: Urban & Schwarzenberg.

Ajzen, I., & Fishbein, M. (1973). Attitudinal and normative variables as predictors of specific behaviors. *Journal of Personality and Social Psychology, 27,* 41–57.

Ajzen, I., & Fishbein, M. (1977). Attitude-behavior relations: A theoretical analysis and review of empirical research. *Psychological Bulletin, 84,* 888–918.

Anderson, J. R. (1983). *The architecture of cognition*. Cambridge, MA: Harvard University Press.

Atkinson, J. W. (1964). *An introduction to motivation*. Princeton, N.J.: Van Nostrand, 1958.

Atkinson, J. W., & Birch, D. (1970). *The dynamics of action*. New York: Wiley.

Beckmann, J., & Kuhl, J. (1984). Altering information to gain action control: Functional aspects of human information-processing in decision-making. *Journal of Research in Personality, 18*, 223–237.

Düker, H. (1925). Über das Gesetz der speziellen Determination. *Untersuchungen zur Psychologie, Philosophie und Pädagogik, 5*, 97–173.

Ebbinghaus, H. (1902). *Grundzüge der Psychologie. I*. Leipzig: Veit.

Fortlage, K. (1855). *System der Psychologie*. Leipzig: F. A. Brockhaus.

Gerdessen, H. (1932). Die Entwicklung der Willensbetätigung auf die Eigenschaften der Bezugsvorstellungen. *Archiv für die gesamte Psychologie*. (Suppl. 2). 1–102.

Harriman, P. L. (1947). *The new dictionary of psychology*. New York: Philosophical Library.

Hebb, D. O. (1949). *The organization of behavior*. New York: Wiley.

Heckhausen, H. (1977). Achievement motivation and its constructs: A cognitive model. *Motivation and Emotion. 1*, 283–329.

James, W. (1890). *The principles of psychology*. New York: Holt.

James, W. (1904). *The principles of psychology. Vol. II*. New York: Holt.

Kelley, H. H. (1972). *Causal schemata and the attribution process*. New York: General Learning Press.

Kiesler, C. A. (1971). *The psychology of commitment*. New York: Academic Press.

Klinger, E. (1975). Consequences of commitment to and disengagement from incentives. *Psychological Review, 82*, 1–25.

Kruglanski, A. W., & Ajzen, J. (1983). Bias and error in human judgment. *European Journal of Social Psychology, 13*, 1–44.

Kuhl, J. (1983). *Motivation, Konflikt und Handlungskontrolle*. Heidelberg: Springer.

Kuhl, J. (1984). Volitional aspects of achievement motivation and learned helplessness: Toward a comprehensive theory of action-control. In B. A. Maher (Ed.), *Progress in Experimental Personality Research* (Vol. 13). (pp. 99–171). New York: Academic Press.

Kuhl, J., & Beckmann, J. (1983). Handlungskontrolle und Umfang der Informationsverarbeitung: Wahl einer vereinfachten (nicht-optimalen) Entscheidungsregel zugunsten rascher Handlungsbereitschaft. *Zeitschrift für Sozialpsychologie, 14*, 241–250.

Kukla, A. (1972). Foundations of an attributional theory of performance. *Psychological Review, 79*, 454–470.

Külpe, O. (1893). *Grundriß der Psychologie*. Leipzig: Engelmann.

Leontiev, A. N. (1932). The development of voluntary attention in the child. *Journal of Genetic Psychology, 40*, 52–81.

Lewin, K. (1922). Das Problem der Willensmessung und das Grundgesetz der Assoziation. I und II. *Psychologische Forschung, 1*, 191–302 and *2*, 65–140.

Lewin, K. (1926). Untersuchungen zur Handlungs- und Affekt-Psychologie. II.: Vorsatz, Wille und Bedürfnis. *Psychologische Forschung, 7*, 330–385.

Lewin, K. (1951). *Field theory in the social sciences*. New York: Harper & Row.

Lindworsky, J. (1929). *The training of the will*. Milwaukee: Bruce. (Original work published in 1923).

Messer, A. (1934). *Psychologie*. Leipzig: F. Meier.

Meyer, W.-U. (1973). *Leistungsmotiv und Ursachenerklärung von Erfolg und Mißerfolg*. Stuttgart: Klett.

Michotte, A. (1911). Note complémentaire. *Archive de Psychologie, 10*, 113–320.

Michotte, A., & Prüm, E. (1910), Etude expérimentale sur le choix volontaire et ses antécédents immédiats. *Travaux du Laboratoire de Psychologie expérimentale de l'Université de Louvain, 1* (2).

Miller, G. A., Galanter, E., & Pribram, K.-H. (1960). *Plans and the structure of behavior*. New York: Holt, Rinehart & Winston.

Müller, E. (1932). Beiträge zur Lehre von der Determination. *Archiv für die gesamte Psychologie, 84*, 43–102.

Münsterberg, H. (1888). *Die Willenshandlung*. Freiburg, FRG: Mohr.

Nisbett, R. W., & Wilson, T. D. (1977). Telling more than we can know: Verbal reports on mental processes. *Psychological Review, 84,* 231–259.

Norman, D. A. (1980). Twelve issues for cognitive science. *Cognitive Science, 4,* 1–32.

Rohracher, H. (1932). Theorie des Willens. *Zeitschrift für Psychologie* (Suppl. 21).

Rosch, E., & Mervis, C. (1975). Family resemblances: Studies in the internal structure of categories. *Cognitive Psychology. 7,* 753–605.

Simoneit, M. (1926). Willenshemmung und Assoziation. *Zeitschrift für Psychologie, 100,* 161–235.

Vroom, V. H. (1964). *Work and motivation.* New York: Wiley.

Weiner, B. (1980). *Human motivation.* New York: Holt, Rinehart and Winston.

Weiner, B., Russell, D., & Lerman, D. (1979). The cognition-emotion process in achievement-related contexts. *Journal of Personality and Social Psychology, 37,* 1211–1220.

Wundt, W. (1896). *Grundriß der Psychologie.* Leipzig: Engelmann.

Zajonc, R. B. (1980). Feeling and thinking: Preferences need no inferences. *American Psychologist, 35,* 151–175.

Zajonc, R. B. (1984). On the primacy of affect. *American Psychologist, 39,* 117–123.

Ziehen, Th. (1920). *Leitfaden der physiologischen Psychologie.* Jena: Gustav Fischer.

Chapter 6

Volitional Mediators of Cognition-Behavior Consistency: Self-Regulatory Processes and Action Versus State Orientation*

Julius Kuhl

One of the most striking discrepancies between everyday experience and psychological theorizing concerns the complexity of motivational states. While most psychologists tend to focus on a single behavioral domain (e. g., achievement, affiliation, eating, learning, problemsolving, sex, etc.), we know from everyday experience that people very rarely seem to have just one behavioral inclination in a given situation. In everyday life people usually experience several motivational tendencies simultaneously and more often than not have multiple commitments to a variety of goals. At first glance our task – to explain and predict which of the competing action tendencies a person actually will implement in a given situation – seems to boil down to the objective of establishing the *dominant* (i. e., strongest) action tendency among all the competing tendencies (e. g., Atkinson & Birch, 1970).

What processes determine which action tendency becomes dominant? Most theories of motivation suggest that individuals are continuously processing motivationally relevant information, that is, information regarding the subjective value of anticipated consequences of various action alternatives and the probability of securing those consequences by performing the various actions in question (see Ajzen, Chapter 2, this volume; Ajzen & Fishbein, 1973; Atkinson, 1964; Heckhausen, 1977; Kuhl, 1982a). Presumably, the action alternative having the highest expected utility becomes dominant and is then enacted. A closer look at the problem reveals, however, that this traditional approach is not sufficient to explain goal-directed behavior. If people usually have several behavioral tendencies simultaneously and if the relative strengths of these tendencies change continuously, successful completion of an action would depend on the "pure luck" that no competing tendency becomes dominant before the goal is reached.

* This Chapter was written during the author's term as a Fellow at the Center for Advanced Study in the Behavioral Sciences, Stanford, California. It is a pleasure to acknowledge helpful comments on an earlier draft by Jürgen Beckmann and Torgrim Gjesme.

However, people are usually *persistent* in their goal-directed activities. They normally do not interrupt a goal-directed activity and switch to another one even if new information is processed that strengthens the case for an alternative activity. Despite the continuous pressure exerted by competing action tendencies, people often stick to the behavioral intention they are currently committed to until the goal is reached. This phenomenon suggests the existence of processes that prevent competing tendencies from becoming dominant before the current goal is reached. Terms such as *volition, ego-strength, will power,* and *self-regulation* have been used to describe similar control functions. Usually, these terms are used in a rather narrow sense, to refer to situations in which subjects are asked to engage in a highly aversive activity, for example, to concentrate on a boring task, or to avoid engaging in a highly pleasurable activity like eating high-caloric food, smoking, and so forth. (Mischel, 1981; Thoresen & Mahoney, 1974). The present analysis suggests a broader conception of volitional control or self-regulatory functions. It is assumed that even the enactment of seemingly simple intentions such as reading a letter, opening a window requires a certain amount of self-regulatory control. The terms *volitinal control, action control,* and *self-regulation* will be used here interchangeably to denote those processes which *protect* a current intention from being replaced should one of the competing tendencies increase in strength before the intended action is completed.

In this chapter, I will first briefly discuss the theoretical framework which has guided our research into various self-regulatory processes and strategies that help close the gap between action-related cognitions and their enactment. A deficit in one of these self-regulatory functions can cause an observed discrepancy between an action-related cognition (e. g., an attitude, a motivational tendency, or an intention) and observed behavior.

In the second part of this chapter, I will summarize our recent research on one presumably pervasive category of cognitive processes that may render enactment of an intention difficult. Cognitions related to these processes center around past, present, or future *states* of the organism rather than around *actions* that would transform a present state into a desired future state. If after dropping a valuable vase a person keeps staring at it or continues questioning how that could ever have happened, he or she may have considerable difficulty enacting any change-oriented intentions, such as picking up the pieces, trying to glue them together, or initiating some new activity unrelated to this event. Results from several experiments will be discussed which suggest that these questioning *state-oriented* cognitions are related to a *catastatic* (i. e., change-preventing) rather than a *metastatic* (i. e., change-inducing) mode of control. As long as an individual is in a catastatic mode of control, the enactment of *action-oriented* intentions seems to be more difficult than when the individual is in a metastatic mode of control. Before discussing research findings regarding distal and proximal antecedents of these two modes of control, I would like to address the more general topic of the nature of self-regulatory processes that mediate intention-behavior consistency by protecting a current intention from being replaced by a competing action tendency.

A Theoretical Framework

Recently, I proposed a theoretical framework specifying several mental processes that presumably mediate self-regulatory functions (Kuhl, 1984). According to this framework, a distinction has to be made between a *motivational tendency* and an *intention*. In contrast to a tendency, an intention is characterized by
the quality of commitment. This quality may be encoded by a characteristic *relation* component in the (propositional) memory representation of an intention
(Kuhl, 1984; Kuhl, in press). Figure 6.1 illustrates a propositional representation
of the intention "When somebody walking in front of me on the street drops something, I *will* (i.e., intend to) pick it up. The propositional structure encoding an
intention presumably is characterized in four ways: (1) the *subject* component *(S)*
specifying the agent of the intended action refers to some aspect(s) of the self; (2)
the *relation* component *(R)* encodes the quality of commitment (I *will* act), (3) the
object component *(O)* refers to some action (e.g., pick up what somebody
dropped) and points to a propositional (sub)structure which specifies the discrepancy between a present and a desired future state and enactive memory
structures encoding action alternatives that may transform the present into the
future state; and (4) the *context* component *(C)* describes the situational conditions that have to be matched in a given context for the action to be performed
(e.g., "when somebody is walking in front of me on the street").

   This model is consistent with current propositional models of human memory
(Anderson, 1983; Anderson & Bower, 1973; Kintsch, 1974; Norman & Rumelhart, 1975), and it elaborates them by specifying the content of motivationally
relevant knowledge structures.

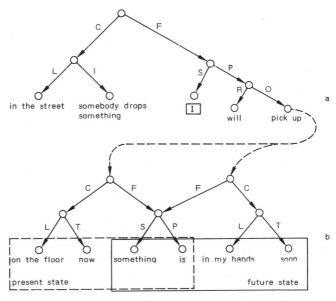

Fig. 6.1. Propositional network representation of an intention

In contrast to a purely epistemological model of human action (cf. Kruglanski & Klar, Chapter 3, this volume), I assume that humans can function perfectly well without having a declarative representation of their intentions. In fact, during the early years of life, infants are not even capable of forming such a declarative representation. In their interactions with the environment, young children progressively differentiate their enactive structures (Piaget, 1952). Even adults seem to rely in many of their habitual or routinized daily activities on enactive structures without activating declarative knowledge about these activities. Nevertheless, the development of declarative representations of one's intentions seems to be an important requirement for many self-regulatory abilities that develop later in childhood (Olson & Astington, 1983; Piaget, 1952).

According to the model of action control (Figure 2), an action-related (declarative) memory structure stored in long-term memory is activated when a match is found between the encoding of the current situation and the context component (see Figure 1) of that structure. If this action-related structure is an intentional structure (as defined by the WILL-relation), it is "admitted" to working memory. Nonintentional structures (i.e., wishes, values, norms, and expectations are transformed into an intentional structure and admitted to working memory if they conform to certain "admission rules" (Kuhl, 1984, p. 121 f.). Self-regulatory strategies are activiated if two requirements are met: First, the difficulty of enactment as defined by the strengths of competing tendencies, the amount of social pressure to engage in alternative activities, and the current degree of state orientation has to exceed a critical value (c_1). Second, the actor's perceived ability to successfully control the implementation of the intended action has to exceed a critical value (c_2). If these conditions are met, the six self-regulatory strategies discussed below may be invoked to facilitate the enactment of the current intention. These six closely related processes presumably mediate action control; they facilitate the protection and maintenance of a current intention against the pressure exerted by the competing action tendencies (Fig. 6.2).

1. Active attentional selectivity facilitates the processing of information supporting the current intention and inhibits the processing of information supporting competing tendencies. Ach (1910) reported that when his subjects were instructed to suppress an overlearned response to a stimulus word and to produce a response defined by a new instruction (e. g., "find a rhyme"), they actively focused their attention on the new instruction "rhyme" to avoid a false response (see Chapter 5, this volume). Mischel and Mischel (1983) report a series of experimental findings which show that children learn to maintain an experimentally induced intention against a seducing action alternative by avoiding visual contact with the source of distraction. Active selective attention to goal-related information can also mediate the following, more subtle, self-regulatory process.

2. Encoding control presumably facilitates the protective function of volition by selectively encoding those features of a stimulus that are related to the current intention. It is interesting to note that this assumption is in line with the conclusions derived from early introspective techniques (cf. Ach's concept of purposive

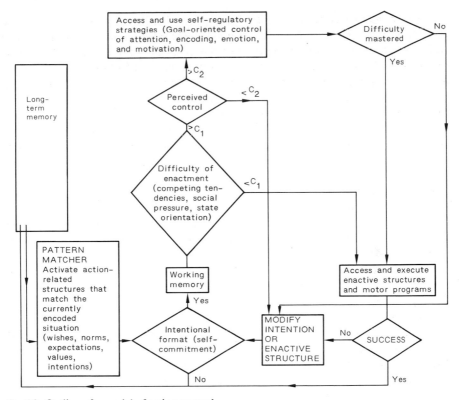

Fig. 6.2. Outline of a model of action control

qualities *(finale Qualitäten)* discussed in Kuhl and Beckmann, Chapter 5, this volume), as well as with more recent experimental evidence of a biasing of pattern-matching resources in favor of memory structures involving the current goal (Anderson, 1983; La Berge, 1973; Posner & Snyder, 1975). There is a growing body of research demonstrating that *perceptual tuning,* that is, the preconscious selection of perceptual input on the basis of higher-order conceptual structures (e. g., the current intention) occurs at very early stages of perceptual processing (Carr & Bacharach, 1976).

3. Emotion control presumably facilitates the protective function of volition by inhibiting emotional states that might undermine the efficiency of the protective function of volition. It has been suggested and found that certain emotional states (e. g., sadness, depression) render the protection and maintenance of an intention to avoid a seducing action alternative difficult (Herrmann & Wortman, Chapter 8, this volume; Izard, 1977; Mischel, Ebbesen, & Zeiss, 1972). We are currently investigating in our laboratory when, and dependent on what developmental precursors, children learn to control their emotions in order to facilitate the enactment of current intentions against powerful alternative tendencies.

4. Motivation control refers to a feedback relation from self-regulatory processes to their own motivational basis (Kuhl, 1984). This process is especially important when the current intention is supported by a weak action tendency. When a mismatch between the strength of the current intention and that of competing tendencies is detected (indicating that the current intention is insufficiently strong) a self-regulatory process is activated that increases the strength of the current intention by selectively processing information that supports it. Motivation control may involve the selective attentional process discussed above. It differs, however, from attentional control in that it explicitly aims at the strengthening of the current intention's motivational basis. While active control of attentional resources (see above: Strategy 1) presumably serves to *preserve* the current strengths of competing tendencies, motivational control aims at a change of the current hierarchy of tendency strengths. If I have the intention of mowing my lawn while feeling afraid that my motivation is too low to pull myself together and actually do it, a motivation-control strategy is to think about what will happen should I fail to perform the intended action (e. g., my neighbors will be irritated, the grass will rapidly grow to a height that will make it much more difficult to cut later, etc.).

5. Environment control describes a strategy that may develop from the more basic strategies dealing with emotion control or motivation control. Emotional and motivational states may be controlled by manipulating the environment. Making social commitments is an example. People who intend to stop smoking may inform a relevant other person about their intention because they know having informed another person will create some social pressure that may help maintain the intention. Clinically oriented theories of self-control (e. g., Thoresen & Mahoney, 1974) place particular emphasis on the process of environment control, probably because deficits in this process are more easily removed than deficits in the other, less easily observable processes.

6. Parsimony of information-processing is an aspect of volitinal control that relates to the definition of stop rules for information-processing. Theoretically, an actor could go on forever processing new information about various consequences of potential action alternatives without ever performing any of them. Efficient action control requires optimalizing the *length* of the decision-making process. Whenever the actor believes that further processing of information bearing on potential action alternatives may jeopardize the execution of the current intention, the process of appraising action alternatives should be brought to a halt, especially if further processing may reveal information that undermines the motivational power of the current intention.

Active vs. Passive Maintenance of the Current Intention. It should be noted that attentional selectivity, control of encoding, emotion, motivation, environmental circumstances, and rules for terminating the decision-making process may be initiated by *active* or by *passive* (automated) control processes. In the foregoing discussion, I have focused upon active strategies mediated by self-related (declarative) knowledge about one's current intention. These strategies are presumably based on metacognitive knowledge about the facilitating function of (1) selectively attending to information supporting the current intention, (2) selectively encoding goal-related features of incoming information, (3) activating positive emotions and enhancing the motivational basis of the current intention by manipulating (4) the internal or (5) external incentive structure, and (6) avoiding overlong decision making. Similar control processes may occur on an automatic level. Even children who have not developed active self-regulatory strategies can nonetheless maintain and protect their intention for some time, although these passive control functions seem to be less persistent and less flexible (Mischel, 1981).

The superiority of active control functions may be attributable to the long-lasting support they receive from the declarative record of the enactive representation of the current intention. The (declarative) knowledge about the intention enables the actor to maintain some reference to the intention in an active state even if the intention itself does not yet have (or no longer has) the status of the enactive intentional structure that is currently active in working memory. In addition, the superiority of active control functions may be enhanced by the greater priming of goal-related memory structures that active attentional processes can produce, as compared to passive attentional mechanisms (Neely, 1977). However, passive maintenance of a current intention does have the advantage of requiring less processing capacity. Whenever sufficient to protect a current intention until it is enacted, passive processes are more useful than active processes because the former are faster, more reliable, and less demanding of the processing system. It is interesting that automation of active self-regulatory strategies has been regarded as an adaptive aspect of volitional control in many theories of volition (e. g., Ach, 1935; Lindworsky, 1923; Düker, 1983).

Action and State Orientations

Until now I have related the difficulty of enacting a current intention to the amount of pressure exerted by competing action tendencies. The demand placed on self-regulatory processes should be a positive function of the strengths of competing action tendencies, and it should be inversely related to the strength of the current intention. However, the *difficulty of enactment* is assumed to be affected by another factor. Difficulty of enactment should increase with the strength of factors that support a catastatic (change-preventing) as opposed to a metastatic (change-promoting) mode of control. As mentioned earlier, the metastatic mode of control is considered closely related to but not identical with the action-oriented state of the organism. Action orientation seems to be necessary but not sufficient for the development of the metastatic mode of control. For example, a person may be action-oriented in the sense that she or he focuses on a fully developed action structure without being in the metastatic control model if that structure is being processed on a fantasy-related level. According to this conception, wishful thinking is considered an instance of the catastatic mode of control even if it is based on fantasizing about fully developed plans of action. Further determinants of the metastatic and catastatic modes of control need to be investigated in the future. In our research we have investigated the effect of action and state orientations as one pair of factors presumably mediating meta- and catastatic modes of control.

An organism is said to be action-oriented if attention is focused on a fully developed action structure. If attention is focused on some internal or external state, the organism is said to be state-oriented. This state may be characterized by perseverating cognitions related to some present, past, or future state of the organism, or even by the absence of any coherent conscious thought (e. g., absent-mindedness).

Antecedents of Action and State Orientations. I assume that whether an individual becomes action- or state-oriented, depends on at least two factors: (1) the degree of perceived incongruence between any two pieces of information processed; and (2) the extent to which an individual has developed *degenerated intentions,* that is cognitive representations in which one or more elements of an intention are ill-defined, weakly activated, or not specified at all (Kuhl, 1984). The first factor, incongruence, may provoke action *or* state orientation, depending on its degree. Mild degrees of incongruence between an individual's expectations and new information, between conflicting expectations, or between unconscious and conscious representations may stimulate an action-oriented attentional focus and a strong involvement in the current activity (Berlyne, 1960; Csikszentmihalyi, 1975; Eckblad, 1981; Kuhl & Wassiljew, 1983). If incongruence exceeds a critical level, a state-oriented response is expected, that is attention focuses on the incongruence-producing information and the experienced cognitive-emotional state resulting from it.

The second assumed antecedent of action and state orientations (i. e., the numger of degenerated intentions developed), may be considered a special case of incongruence deriving from a disturbed balance between the elements of a cognitive structure representing an intention. Depending on which element is degenerated, various forms of state orientation may develop. If that part of the object component of an intention related to the intended action alternative is not specified (Figure 1), as may be the case after many futile attempts to reach the intended goal by the various actions available, failure-oriented state orientation may develop with attention focused on past failures. Failure-oriented state orientation may interfere with efficient cognitive functioning at tasks that place heavy demands on attentional capacity (Kuhl, 1981; Kuhl & Weiß, 1983). It also frequently happens that the cognitive representation of the current intention specifies an action alternative, but that an excessive amount of attention is focused on the desired goal state. As we will see later, this goal-centered or extrinsic form of state orientation may have disruptive effects on the intrinsic task involvement that seems essential to the efficient performance at complex tasks (Kuhl & Wassiljew, 1982). Finally, if the relation component of the propositional structure encoding the current intention (Fig. 6.1) is degenerated, a vacillating type of state orientation is expected. Since the relation component encodes the quality of commitment (I *WILL*) of an intention, vacillating state orientation may result in indecisiveness and in an inability to terminate the decision-making process and actually perform the intended action (Kuhl & Beckmann, 1983). Since this type of action vs. state orientation is expected to be especially important during the decision-making process, that is prior to the initiation of an action, it may be called *decision-related* action vs. state orientation.

Empirical Evidence

The theory outlined in the preceding section focuses on *proximal* determinants of action and state orientations, which specify the immediate antecedent processes of those two orientations. An experimental test of the theory requires some hypotheses regarding more *distal* determinants that can be more easily manipulated and assessed than process variables such as perceived incongruence or degenerated intentions. How can one experimentally manipulate the likelihood that subjects become action or state oriented? How can we assess individual differences in the probability of becoming action or state oriented?

To answer the latter question I have developed a questionnaire that assesses the three types of action and state orientations, that is the disposition toward failure-related, performance-related, and decision-related action and state orientation. Psychometric analyses of the first version of this *Action-Control Scale* have yielded encouraging results: internal consistency coefficients ranging between 0.71 and 0.82 (Cronbach's alpha) and discriminant validity coefficients ranging between 0.01 and 0.36 (Kuhl, 1984).

Several experimental methods for inducing action or state orientation have been developed. State orientation has been induced by confronting subjects with a series of unexpected and uncontrollable failures (Kuhl, 1981; Kuhl & Wassiljew, 1982; Kuhl & Weiß, 1983), by asking subjects to write an essay describing their present feelings about and the causes of their past performance (Kuhl, 1981, Experiment 1), and by an attempt to induce a degenerated intention (Kuhl & Helle, 1984). Attempts to induce action orientation have been made by instructing subjects to keep verbalizing their hypotheses while working on a concept formation problem (Kuhl, 1981; Kuhl & Weiß, 1983), and by having hospitalized patients record various events on the ward (Kuhl, 1983b). In the remainder of this chapter, I will summarize the results obtained in these studies.

Cognition-Behavior Consistency

Our first empirical studies were designed to test the assumption that action and state orientations are associated with meta- vs. catastatic modes of control. According to this assumption, action-related mental structures such as intentions, motives, or attitudes are more likely to be expressed in behavior when an individual is action-oriented than when state-oriented.

Motive-Behavior Congruence. In one study (Kuhl & Geiger, in press), elementary school children (ages 8 to 11) were administered the Action-Control Scale and subsequently given the opportunity to engage in three activities that were designed to satisfy three different needs, that is, an achievement-related activity (throwing balls at a target), a curiosity-related activity (exposing hidden drawings in a "fun-pad" game), and a helping-related activity (helping sort the mixed-up pieces from three different puzzles so that the children from a kindergarten class would be able to use them). Since these activities did not involve uncontrollable failure or long-term goals, we selected the decision-related subscale as the predictor of motive-behavior consistency. The children were administered a multimotive test constructed to assess individual differences in the three motives (i.e., achievement, curiosity, and altruism). The results were rather striking. Although the correlations between the three motive scores and latency of or proportion of time spent in motive-congruent activities were close to zero when computed across the entire sample (i.e., ignoring the distinction between action- and state-oriented subjects), motive-congruent behavior did show up in the action-oriented subgroups. As can be seen from Figure 6.3, the latencies of motive-congruent behavior were smaller in the action-oriented groups than in state-oriented groups. The differences between action- and state-oriented groups reached statistical significance ($p < 0.05$) in two of the three motive groups, that is, in the group consisting of subjects whose dominant motive was achievement-oriented and in the group of subjects having a dominant helping motive. A comparison of indices of motive-congruent *and* motive-incongruent behavior between motive groups revealed that within the action-oriented group, achievement-oriented subjects (i.e., subjects for whom the achievement scores were

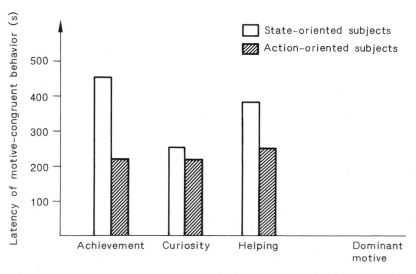

Fig. 6.3. Latency of motive-congruent behavior in a free-choice situation as a function of individual differences in state and action orientation

highest among the three motive scores) had a latency of the achievement-oriented activity that was significantly lower than the latencies of helping-oriented or curiosity-oriented activities. This configuration of motive-behavior congruence was *not* found within the group of state-oriented subjects. A similar indication of motive-behavior consistency was found (in the action-oriented group only) regarding latency of helping-oriented behavior (Kuhl & Geiger, in press).

Intention-Behavior Congruence. Another study was designed to examine whether action- and state-oriented subjects differed in the extent to which they carried out their intentions (Kuhl, 1982 c). Forty-eight sixth-grade German students were given a list of 22 after-school activities, and were asked to rate the strength of their intention to perform each of the activities. The next day, subjects were asked to report how much time they had actually spent doing each activity. The results showed that the correlations between the strength of intentions to engage in a free-choice activity and the amount of time spent performing it were considerably higher (i.e., $0.40 < r < 0.80$) in action-oriented subjects than in state-oriented subjects (i.e., $0.30 < r < 0.40$). Interestingly enough, this pattern of results was reversed for some highly routinized activities that the subjects were not free to choose because they were socially required to engage in them (e.g., brushing their teeth, shining their shoes). For some of these activities, intention-behavior correlations were significantly higher in state-oriented than in action-oriented subjects. This result may be attributable to an increase in the tendency to perform activities that require little self-regulatory support (i.e., routinized and socially required activities) while an individual is in the catastatic mode of control. In fact, state-oriented subjects may develop a tendency to perform routinized and externally controlled behaviors to obviate heavy demands on their relatively low self-regulatory capacities.

The Alienation Effect. The assumption made in the preceding paragraph has an interesting implication. If the catastatic mode of control presumably associated with state orientation facilitates routinized activities and impairs the enactment of intentions that require some support of self-regulatory strategies, state-oriented subjects may frequently fail to perform the most preferred action alternative. While state-oriented persons are still in the process of making up their minds whether or not they should perform a preferred action (i.e., while they are trying to mobilize the self-regulatory support needed to initiate that action), they may suddenly find themselves performing some well-learned or socially expected behavior that does not require a similar amount of self-regulatory support.

The results of a recent study are consistent with this line of thought (Kuhl & Eisenbeiser, in press). Subjects were asked to engage in a rather unattractive sorting activity. After some time the experimenter said that the sorting part of the experiment was finished, and that he needed some time to prepare the final phase of the experiment. During the waiting period the subject was invited to read one of the comics or magazines lying on a table or to continue the sorting if she or he preferred. Although the reading activity was rated as far more attractive than the sorting task by all subjects in a parallel sample, about 83% of the subjects did not switch to the reading activity. This result was attributed to a cumulative increase in motivation to perform the ongoing activity (cf. Atkinson & Birch, 1970). Among the 12 subjects who did switch to the more attractive activity, 10 subjects had been classified as having a *low* disposition toward (vacillating) state orientation. This result nicely illustrates one aspect of the theory according to which the seemingly "easy" enactment of a preferred and fully available action alternative places considerable demands on self-regulatory functions that shield the preferred action tendency against competing alternative tendencies. Self-regulatory support of the preferred tendency seems especially necessary when a less preferred action tendency is rather easy to enact because it is overlearned, externally controlled (e.g., by social norms), or currently ongoing.

Failure-Oriented State Orientation and Helplessness. Several experiments have been conducted to investigate a more subtle aspect of intention-behavior consistency than the one examined in the studies reported so far. In studies using the so-called learned-helplessness paradigm, subjects are confronted with a series of uncontrollable failures on a training task and are subsequently administered a test task. The frequently replicated finding is that subjects exposed to uncontrollable failure show an impaired performance on the test task compared to control groups that were allowed to succeed on the training task or that did not receive any pretreatment (e.g., Hiroto & Seligman, 1975; see Abramson, Seligman, & Teasdale, 1978; Roth, 1980, for reviews). Performance decrements have been found even when the test task (e.g., an anagram problem) differed considerably from the training task (e.g., a discrimination problem).

Although quite different theoretical interpretations have been proposed to account for generalized performance deficits following a failure pretreatment (see Carver & Scheier, Chapter 11, this volume), an assumption common to virtually all these theories is that subjects lose confidence in their ability to control success at the training task and transfer this expectancy deficit to the test task (but see Frankel & Snyder, 1978, for a different hypothesis). As a result, motivation to invest effort in the test problem decreases. This theory has been criticized because it is inconsistent with theory and research on achievement motivation, in particular, and personality functioning in general (Kuhl, 1984). It has been theoretically postulated (Mischel, 1973) and empirically demonstrated (Endler & Magnusson, 1976; Mischel, 1968) that humans discriminate between various situations rather than being "born generalizers" (Seligman, 1975). A direct test of the generality of ability concepts and control expectations regarding a variety of tasks revealed that subjects develop highly task-specific concepts even after very brief experience with a new task (Kuhl, 1977). Also, theory and research on achievement motivation suggest that following repeated failure on one task, many subjects show an increase rather than a decrement in their motivation to succeed on a different achievement-related task because success on the latter task may substitute for the failure on the first (Atkinson, 1958; Atkinson & Feather, 1966; Blankenship, 1982).

We have conducted several experiments to test a different theoretical explanation of generalized performance deficits following failure. According to this

view, subjects who show performance decrements following uncontrollable failure may be quite confident when confronted with a test task that seems to have
nothing in common with the training task they failed on and may strongly intend
to concentrate on the new task, and still have a deficit in those self-regulatory
functions that help put that task-oriented intention into effect. Specifically, the
consistency between their task-oriented intentions and their attentional behavior
while working on the test task may be disrupted to the extent that they have developed a catastatic mode of control. The catastatic mode may be activated by
the incongruence between subjects' (success) expectations and their failures at
the training task or by attentional demands resulting from a perseverating *degenerated intention* to solve the training problem.

The latter factor explains experimental results showing that instructions that
help subjects give up any perseverating intention to solve the training task (e. g.,
telling them that the task is insoluble or unimportant) erase performance deficits
on the test task (Koller & Kaplan, 1978; Roth & Kubal, 1975). The mediating
role of the former factor (i. e., the unexpectedness of uncontrollable failure) was
demonstrated by a recent finding in our laboratory which showed that subjects
could be completely immunized against the debilitating effect of exposure to uncontrollable failure by making them expect to fail on the training task because
"you had a low score on a pretest which correlates highly with performance on
this problem" (Kuhl & Weiß, 1983). It should be noted, parenthetically, that the
assumed relationship between state orientation and experienced incongruence
also accounts for the less often confirmed claim of traditional helplessness theory (Seligman, 1975) that exposure to noncontingent success should produce the
same kind of performance deficits as exposure to noncontingent failure. Since
most people generally seem to have success-related goals, failure should produce stronger expectancy-outcome discrepancies and, as a result, stronger performance deficits than noncontingent success by the present theory.

The results of our experiments confirmed both our skepticism regarding the
mediating role of generalized deficits in motivation and in expectation of control
and our hypothesis regarding the mediating role of functional deficits in some
self-regulatory processes. Expectancy of control measures obtained between the
pretreatment and the test phase of the experiments were – compared to those for
control-groups – not lower in those subjects that showed performance decrements at the test task following a failure pretreatment. Similarly, self-reported
motivation regarding the *test task* was not lower in experimental groups receiving failure pretreatment than in the control groups (Kuhl, 1981; Kuhl & Weiß,
1983). In light of these negative findings, it is indeed surprising that "after
10 years of intensive research on human helplessness and its applications, the basic postulates of the learned helplessness theory have not been tested adequately.
In particular, the pivotal hypothesis of the theory that it is the expectation of act-
outcome noncontingency which causes the behavioral and emotional characteristic of helplessness has gone largely unexamined" (Alloy, 1982, p. 444). Recently
an attempt was made to infer from this failure to obtain information regarding
expectation of control that expectancy variables or any other cognitive variable
do *not* play a role in the causation of performance decrements following failure

(Oakes & Curtis, 1982). This conclusion is not consistent with our and others' positive evidence regarding the mediating role of cognitive variables.

What ist the positive evidence in our experiments for the mediating role of action and state orientations? In one experiment (Kuhl, 1981), we attempted to induce state orientation by asking subjects after the failure training to reflect on the present situation, their current emotional state, and the causes of their failure, and to write a short essay about their thoughts. Performance decrements were higher in this group than in a group that was exposed to uncontrollable failure without being asked to reflect upon state-related aspects of the situation. In a second experiment (Kuhl, 1981), we subdivided the subjects according to the median of their scores on the subscale for failure-related action vs. state orientation. The results showed that significant performance decrements on the test task following exposure to uncontrollable failure (compared to a control group) occurred only in the state-oriented group. Interestingly enough, the performance decrements found in this group were reversed in a condition in which an attempt was made to induce action orientation by instructing subjects to verbalize their hypotheses about the correct solution ("Explicit Hypothesis Testing") while working on the training problem. In this condition, state-oriented subjects showed enhanced performance compared to control subjects. This result is consistent with our theory that many subjects may be quite motivated to perform well on the test task, but that the facilitating effect this heightened motivation should have fails to occur when state orientation prevents subjects from disregarding interfering state-related cognitions.

In a recent study, these results were replicated and extended by some interesting findings obtained by several thought-sampling measures (Kuhl & Weiß, 1983). Again, performance decrements were observed in the state-oriented failure group only. These deficits were reversed in state-oriented subjects from an explicit-hypothesis-testing group and in state-oriented subjects in an expected-failure group who were informed at the beginning of the training phase that they had scored poorly on a pretest that was allegedly a strong predictor of performance on the training task. Postexperimental thought-sampling measures revealed that state-oriented subjects in the no-intervention failure condition reported a significantly greater frequency of thoughts about (1) the causes of their performance (*while* working on the test problem!), (2) the shape they were in, (3) how long the experiment might last, and (4) how good their own abilities to solve tasks like these might be.

A rather dramatic confirmation of the hypothesis that performance decrements in state-oriented subjects may be mediated by perseverating state-related cognitions was found in another recent study (Grosse, 1982; see Kuhl, 1983a, p. 290f., or Kuhl, 1984, p. 153 for more accessible summaries of this study). In this study, state-oriented subjects exposed to uncontrollable failure reported considerably more perseverating thoughts and negative feelings about failing at the training task than action-oriented subjects did, even after working for up to 60 minutes on a variety of other tasks they had succeeded at.

These results are consistent with our assumption that performance decrements following failure are mediated by an inability to postpone task-irrelevant state-oriented cognitions until one has finished a task one is trying to master. The results are not compatible with a purely motivational interpretation of generalized performance decrements following exposure to uncontrollable failure. The traditional motivational explanation of those performance deficits seems to be based on an implicit assumption of a universal cognition-motivation-behavior consistency. If one presupposes that behavioral outcomes always reflect subject's cognitions and motivations, one is forced to conclude that performance deficits are attributable to expectancy and motivational deficits. Within the present theoretical framework, this reasoning is not plausible. If some subjects sometimes fail to bring about the enactment of their current intention owing to a deficit in certain self-regulatory functions, performance decrements may occur even if the subjects are perfectly confident and fully desirous of performing the task.

Anxiety-Free State Orientation. A final word should be said about how the results reported here relate to the test-anxiety hypothesis, which attributes the functional deficits following failure to task-irrelevant cognitions frequently found in highly test-anxious subjects (Easterbrook, 1959; Humphreys & Revelle, 1984; Lavelle, Metalsky, & Coyne, 1978; Mandler & Sarason, 1952; Morris & Liebert, 1970; Wine, 1982). Theoretically, there should be some overlap between test anxiety and state orientation. It is hard to conceive of a subject's being anxious without focusing on some (future) state (i.e., the one she or he is afraid of). Anxiety research supports the view that as long as subjects having a disposition to develop anxiety focus on action alternatives that may help them prevent the event they fear from occurring, the actual level of anxiety aroused remains minimal (Lazarus, 1966; Lazarus & Averill, 1972). On the other hand, there may be many instances where high levels of state orientation are aroused that are *not* accompanied by a state of anxiety (Kuhl, 1984, Chapter 5). Accordingly, there should be a moderate rather than a substantial correlation between state orientation and test anxiety. In fact, in a recent study, the correlation between state orientation and test anxiety was 0.36 (Grosse, 1982). The theory of state orientation (Kuhl, 1984) goes beyond traditional accounts of task-irrelevant cognitions based on theories of test anxiety in two respects. First, it provides a more specific and more detailed theoretical account of the cognitive processes mediating the occurrence of task-irrelevant thoughts in test-anxious subjects. Second, it encompasses cases in which task-irrelevant cognitions occur even though anxiety level is rather low (Kuhl, 1984). The latter case is exemplified by our results showing that the role of state orientation as a factor moderating the debilitating effect of exposure to uncontrollable failure remained unattenuated in a subsample of subjects low in test anxiety.

Goal-Centered State Orientation. Thus far I have summarized research illustrating various instances in which cognition behavior consistency may be disrupted as a result of a vacillating or failure-related type of state orientation. It should be noted that, with very few exceptions, the results reported thus far could not be replicated with any other subscale of the *Action Control Scale* which speaks to the discriminant validity of each subscale. This general conclusion also applies to the subscale assessing extrinsic/intrinsic state and action orientations. How could we design an experimental situation in which this type of action vs. state orientation would make a difference? Recent theory distinguishes between two modes of processing, that is, a *sequential-analytic* mode and a *holistic-intuitive* mode, with high action-oriented task involvement being associated with an attentional and emotional state that facilitates a switch to the holistic-intuitive mode (Kuhl, 1983 c). This mode of processing presumably facilitates performance on problems that are considerably more complex than the tasks usually employed in research on problem-solving and achievement motivation (e. g., Atkinson & Feather, 1966; Newell & Simon, 1972).

In a recent study, an attempt was made to construct a problem that was more similar than the typical laboratory tasks to the very complex and ill-defined problems that confront human beings in their everyday lives (Dörner, Kreuzig, Reither, & Stäudel, 1983). In a computer simulation, subjects played the role of mayor of a town without knowing much about the complex and dynamic interdependencies between the numerous variables in the computer program simulating aspects of the town's economic and social development. The result that is most interesting in the present context was that "good" mayors (as defined by a compound measure of the town's economic and social development) did not differ from "poor" mayors in any of the numerous indicators of various cognitive and information-processing abilities included in the study. The only significant predictor of successful mayoralty was the subjects' score on a questionnaire measuring emotional reactions to highly complex and ambiguous situations. The emotional state assessed in that questionnaire bears a striking resemblance to the emotional state that is presumably associated with high intrinsic task-involvement and the holistic-intuitive mode of processing (Kuhl, 1983 c). This resemblance suggests that intrinsic task involvement may in fact facilitate performance in extremely complex, ambiguous situations in which an exclusive reliance on the sequential-analytic mode of processing – involving step-by-step focusing of attention on specific bits of information – may be entirely inadequate.

The results of a recent study are consistent with this suggestion (Kuhl & Wassiljew, 1983). Subjects were asked to learn the rather complex oriental game *Kalah*. After one hour's training, they were asked to play three games against a computer and to verbalize the reason for each decision they made while playing the game. It was expected that subjects who tended to pay too much attention to the ultimate goal (i. e., winning the game) while playing would have more problems in developing the complex strategies necessary to win the game than subjects having a high disposition toward intrinsic task involvement. The results confirmed this expectation. Subjects who scored high in intrinsic task involvement were more confident of winning and more motivated to play the game than goal-centered subjects. Subjects high in task involvement verbalized more complex strategies and performed better than goal-centered subjects. Interestingly enough, subjects high in task involvement lost their superiority in an experimental condition in which they were exposed to uncontrollable failure at a completely different (concept formation) task administered prior to the experiment, in an allegedly separate experiment. This effect was attributed to an impairment of the holistic-intuitive mode of processing presumably resulting from the emotional state aroused by the exposure to uncontrollable failure. Since there was no indication for an expectancy or motivational deficit regarding the Kalah game in goal-centered subjects following failure pretreatment, the results suggest that subjects fully intended to perform well but failed to enact their task-oriented intentions. In other words, a switch from the holistic-intuitive mode of processing to a sequential-analytic mode may be another mediator of intention-behavior inconsistency, especially in rather complex situations.

The "Overjustification" Effect Reconsidered. The finding that intrinsic task involvement was associated with a higher degree of motivation to solve a complex problem than goal-centered state orientation may shed some light on the still insufficiently explained finding that rewarding a child for performing a pleasurable activity may undermine her or his motivation regarding that activity (Deci, 1975; Lepper, Greene, & Nisbett, 1973; Greene & Lepper, 1974). Recent developmental studies suggest that the prevailing attributional explanation of this "overjustification" effect does not seem adequate (Morgan, 1981; 1983). Since we felt that none of the various alternative explanations of the effect yielded a satisfactory explanation of the existing data (e.g., Calder & Staw, 1975; Deci, 1975; Kruglanski, 1975; Reiss & Sushinsky, 1975), we designed an experiment to test our hypothesis that the overjustification effect is attributable to an imbalance in the cognitive representation of the various elements of an action plan. Specifically, we assumed that the reward offered for performing an activity may over-activate the goal-centered element of an action plan (see Figure 1). This goal-centered type of state orientation may then result in a loss of interest in the activity itself and a gradual restructuring of the action elements of that plan toward the attainment of the goal (see Eckblad, 1981, for a similar theoretical position). Action elements that may have been a source of the original (intrinsic) motivation but that appear to be of little instrumental value for obtaining the reward (e.g., making very elaborate and colorful drawings) may be removed from the action plan in favor of more instrumental elements.

The results of a recent study were consistent with this theoretical analysis. Children who scored high on goal-centered state orientation lost considerably more of their interest in a pleasurable activity after being rewarded for doing it for 24 minutes than children high in action-oriented task-involvement (Kuhl & Melendez, 1984). In addition, this study revealed an interesting interaction between the duration of the period the subject had to perform the activity to obtain the reward and the degree of the overjustification effect. The nature of this interaction was more compatible with the present theoretical interpretation than with the traditional attributional interpretation (Kuhl & Melendez, 1984).

Self-Regulatory Processes

The experimental evidence reported thus far is consistent with the theoretical claim that the three types of state orientation are associated with a catastatic mode of control, which renders the enactment of the current intention difficult. However, this research does not tell us whether the action – vs. state-orientation distinction is associated with differences in those self-regulatory functions mentioned earlier (e.g., parsimony of information processing, motivation control, etc.) Although one might expect that those functions are less developed in state-oriented subjects, it is also conceivable that state-oriented and action-oriented subjects do not differ in the quality of their self-regulatory strategies, and that intention-behavior inconsistencies found in state-oriented subjects occur solely because enacting intentions is just more difficult for them than for action-oriented subjects. In several experiments we have investigated the extent to which differences in the efficiency of self-regulatory strategies contributes to the differential intention-behavior consistency of the two types of subjects. A second aim of this research was to obtain information about the functional significance of various self-regulatory processes *irrespective* of whether or not they were related to the state- vs. action-orientation distinction.

Selective Attention. In one study we found direct evidence of individual differences in selective attention (Kuhl, 1983a, p.268). Subjects were shown a series of cards and asked to memorize three words set in a square. Each card also contained three words that were not to be memorized. These words were set in a circle. After the subject had inspected each card for fifteen seconds, the experimenter presented a list of 100 words, which contained all the words shown to the subjects plus additional neutral words. The subjects were asked to indicate which words they recognized, irrespective of whether these had been set in the circle or in the square. The results showed that action-oriented subjects recognized significantly fewer "irrelevant" words than state-oriented subjects. This effect was reversed, however, in an experimental condition in which the number of cards was doubled (from 7 to 14). This reversal might indicate that action-oriented subjects change their attentional focus when the task at hand becomes uncontrollable (Kuhl, 1982b).

Parsimony of Information-Processing. According to the theory of state orientation, the assumed degeneration of the relation node of an intention (i.e., the WILL-relation) results in the vacillating type of state orientation. State-oriented subjects are expected to process more information before arriving at a decision than action-oriented subjects. Excessive prolongation of the decision-making process jeopardizes the enactment of an intended action because the longer a person hesitates to initiate an intended action, the greater the likelihood that a competing action tendency will take over.

In one experiment (Kuhl & Beckmann, 1983), subjects were asked to choose games of dice they were to play later in the experiment. Each game was described in terms of the chances of winning and the number of points to be won.

The games were defined in such a way that the pay-off could easily be maximized simply by choosing a game for which the chances of winning exceeded a critical value. The number of points to be won hardly affected expected utility because the number of points subracted when the subject lost a game was negligible compared to the number of points gained when the subject won a game. The results showed that most action-oriented subjects ignored the information about the number of points to be won when making their choices, whereas state-oriented subjects used a more complex (conjunctive) decision rule based on both expectancy-related and value-related information. These results support the assumption that action orientation is associated with more parsimonious processing of decision-related information than state orientation.

Motivation Control. According to the theory of action control (Fig. 6.2), volitional control of the enactment of a current intention may be mediated by increasing its motivational basis. This strengthening of the motivation to act on a current intention presumably increases the likelihood that it will win the race against competing action tendencies. This assumption was tested in a recent study (Beckmann & Kuhl, 1984). Twenty-two students who were looking for an apartment were given a list of apartments ranging from very low in attractiveness to very high. Subjects were asked to inspect the list of alternatives twice before they made a final decision. Each time the subjects had inspected the list, they were asked to rate the attractiveness of each apartment on the list. In addition, subjects were asked to indicate tentatively which apartments they might later choose. The results showed that for action-oriented subjects, the difference in mean attractiveness between the tentatively chosen and the rejected alternatives increased significantly from the first to the second point of measurement, whereas subjects high in (vacillating) state orientation showed no indication of an increase in the attractiveness of the tentatively chosen alternatives as compared to the rejected ones. This result suggests that action-oriented subjects selectively attend to incentive-related information that increases the subjective value of the tentatively chosen alternative. This evidence regarding a volitional process of *incentive escalation* suggests a reinterpretation of many of the data traditionally attributed to dissonance reduction with respect to the process of motivation control as described in the action-control model (Beckmann & Irle, Chapter 5, this volume).

Emotion Control. In a field study we investigated to what extent individual differences in action and state orientation as assessed by the Action Control Scale are associated with emotions that may impair or facilitate the enactment of change-oriented intentions (Kuhl, 1983 b). Subjects were hospitalized patients who had just undergone a hernia operation. Two and seven days after their operation, they were asked to rate the intensity of the pain they currently had. On the seventh day after the operation, subjects also responded to several questions regarding their daily activities on the ward. Ratings regarding the intensity of pain were significantly higher in state-oriented than in action-oriented subjects. State-oriented subjects also requested significantly more analgesics than action-oriented subjects. The questionnaire responses indicated that state-oriented subjects tended to be more passive and contemplative than action-oriented subjects (e. g., looking at their incision, thinking about the operation, worrying, trying to be careful), whereas action-oriented subjects engaged significantly more often in change-oriented activities than did state-oriented subjects (e. g., practicing moving their legs, making plans for the time after they leave the hospital, walking around the ward, etc.)

Working Memory Capacity and Depression. The action-control model (Fig. 6.2) contains another assumption regarding the cognitive processes mediating individual differences in intention-behavior consistency. If degenerated intention(s) have – like any other intention – a high priority for access to working memory, subjects having a degenerated intention should have a reduced working memory capacity. The catastatic mode of control may, in fact, be mediated by a degenerated intention blocking that part of working memory which is reserved for the current intention. If that part (or "slot") is occupied by a degenerated intention (which cannot be enacted), the special functional characteristics provided for that part of working memory may backfire. That is, access to the protective volitional strategies and high maintenance of activation of the current-goal part of working memeory would protect and maintain an unexecutable intention, rendering the initiation of a new (executable) intention difficult. Depression may be an extreme case of the frequent occupation of working memory by degenerated intentions.

In a recent experiment (Kuhl & Helle, 1984), we tested this implication of the action-control model using a subgroup of hospitalized depressive patients and three control groups (i. e., hospitalized schizophrenics, hospitalized alcoholics, and college students). The four groups of subjects were subdivided into an experimental and a control group. In the experimental condition, we attempted to induce a degenerated intention by instructing the subject that she or he should clear up a table that had many articles (i. e., papers, pencils, computer cards, files, etc.) spread out on it. The context of execution was ill-defined, however, in that we told the subject that she or he would not have a chance to start right away because several tests had to be done first. It would be up to her or him to find an appropriate time during the course of the experiment to start clearing up the table. Subjects assigned to the control condition were also confronted with the messed-up table. They were not, however, instructed to clear it up. Subsequently, all subjects were administered several memory tasks intended to assess short-term memory capacity. Since the scores on the Beck Depression Inventory (BDI) were available for all subjects, the total sample of subjects was subdivided into a highly depressive group and a weakly depressive one (median-split). A significant interaction between experimental conditions and BDI groups was found. As can be seen from (Table 6.1, this interaction effect is attributable to a significant reduction in short-term memory capacity in subjects who had a high (i. e., above median) BDI score and in whom a degenerated intention had been induced. Similar effects were found when other methods of estimating short-term memory capacity were used. Table 1 contains additional results confirming our assumption about the cause of the reduced memory capacity observed in depressive subjects in whom a degenerated intention was induced. Depressive subjects assigned to the experimental condition reported a significantly greater frequency of thinking about the messed-up table. Interestingly, the results reported in Table 6.1 were replicated when the BDI Factor was replaced by a state- orientation factor (based on a median-split of scores involving failure-related state orientation).

Table 6.1. Mean short-term memory span (as defined by Woodworth & Schlosberg, 1954) and reported frequency of thinking about the "messed-up table" as a function of the degree of depression (BDI-score) and experimental conditions

Depression (BDI-Score)	low		high	
Experimental condition (clear-up-table instruction):	with	without	with	without
Memory span	4,84	4,83	4,42	4,83
Frequency of thoughts about messed-up table	2,3	2,2	3,0	1,3
n	13	14	14	13

Conclusion

The main purpose of this chapter has been to summarize my own theoretical and experimental work regarding volitional mediators of intention-behavior consistency. It has been proposed that it is theoretically naive and empirically untenable to assume that one can unequivocally derive an individual's behavior from the outcome of the decision-making process, that is, from the current intention. Whether or not the current intention will be enacted depends on the *difficulty of enactment* in relation to the efficiency of *self-regulatory processes*. Difficulty of enactment is a function of (1) the amount of external pressure against the performance of the current intention (e.g., social norms or instructions that support alternative actions), (2) internal pressure against the enactment of the current intention (i.e., number and strength of competing action tendencies), and (3) the factors producing a catastatic (vs. a metastatic) mode of control (e.g., high incongruence and degeneration of intentions). According to our data, efficiency of volitional control seems to be facilitated by: (1) selective attentional mechanisms; (2) parsimony of information-processing; (3) motivation control; and (4) emotion control. We are currently investigating the role of the two remaining factors assumed in the action-control model (Fig. 6.2), that is, (5) environment control and (6) encoding control. Our results suggest that individual differences in and experimental manipulations of action and state orientations affect intention-behavior consistency both by changing the difficulty of enactment (i.e., by affecting the catastatic vs. metastatic mode of control) and by affecting the efficiency of various self-regulatory functions.

I believe that the most exciting challenge for future research is to develop methods of assessing proximal rather than distal determinants of action-control processes. In our current research we are trying to obtain a more detailed picture of the cognitive processes mediating self-regulatory functions. This research, which focuses on a more molecular level of analysis than the research reported in this chapter, explores the applicability of various methods borrowed from information-processing research to the study of volitional control (e. g., backward masking, the dichotic-listening task, delayed auditory feedback,and the Stroop task). Eventually, we hope this line of research will enable us to write a new chapter on volitional control that bridges the present gap between cognitive and motivational research on self-regulation.

References

Abramson, L. Y., Seligman, M. E. P., & Teasdale, J. D. (1978). Learned helplessness in humans: Critique and reformulation. *Journal of Abnormal Psychology, 87,* 49–74.

Ach, N. (1910). *Über den Willensakt und das Temperament.* Leipzig, East Germany: Quelle & Meyer.

Ach, N. (1935). Analyse des Willens. In E. Abderhalden (Ed.), *Handbuch der biologischen Arbeitsmethoden. Bd. VI.* Berlin: Urban & Schwarzenberg.

Ajzen, I., & Fishbein, M. (1973). Attitudinal and normative variables as predictors of specific behaviors. *Journal of Personality and Social Psychology, 27,* 41–57.

Alloy, L. B. (1982). The role of perceptions and attributions for response-outcome contingency in learned helplessness: A commentary and discussion. *Journal of Personality, 50,* 443–479.

Anderson, J. R. (1983). *The architecture of cognition.* Cambridge, MA: Harvard University Press.

Anderson, J. R., & Bower, G. H. (1973). *Human associative memory.* Washington: Hemisphere.

Atkinson, J. W. (1958). *Motives in fantasy, action, and society.* Princeton, NJ: Van Nostrand.

Atkinson, J. W. (1964). *An introduction to motivation.* Princeton, NJ: Van Nostrand.

Atkinson, J. W., & Birch, D. (1970). *The dynamics of action.* New York: Wiley.

Atkinson, J. W., & Feather, N. T. (1966). *A theory of achievement motivation.* New York: Wiley.

Beckmann, J., & Kuhl, J. (1984). Altering information to gain action control: Functional aspects of human information-processing in decision-making. *Journal of Research in Personality, 18,* 224–237.

Berlyne, D. E. (1960). *Conflict, arousal, and curiosity.* New York: McGraw-Hill.

Blankenship, V. (1982). The relationship between consummatory value of success and achievement-task difficulty. *Journal of Personality and Social Psychology, 42,* 901–914.

Calder, B. J., & Staw, B. M. (1975). The interaction of intrinsic and extrinsic motivation: Some methodological notes. *Journal of Personality and Social Psychology, 31,* 76–80

Carr, T. H., & Bacharach, V. R. (1976). Perceptual tuning and conscious attention: Systems of input regulation in visual information processing. *Cognition, 4,* 281–302.

Csikszentmihalyi, M. (1975). *Beyond boredom and anxiety.* San Francisco: Jossey-Bass.

Deci, E. L. (1975). *Intrinsic motivation.* New York: Plenum.

Dörner, D., Kreuzig, H. W., Reither, F., & Stäudel, T. (Eds.). (1983). *Vom Umgang mit Unbestimmtheit und Komplexität.* Bern: Huber.

Düker, H. (1983). *Über unterschwelliges Wollen.* Göttingen, FRG: Hogrefe.

Easterbrook, J. A. (1959). The effect of emotion on the utilization and the organization of behavior. *Psychological Review, 66,* 183–210.

Eckblad, G. (1981). *Scheme theory.* New York: Academic Press.

Endler, N. S., & Magnusson, D. (Eds.). (1976). *Interactional psychology and personality.* Toronto: Hemisphere.

Frankel, A., & Snyder, M. L. (1978). Poor performance following unsolvable problems: Learned helplessness or egotism? *Journal of Personality and Social Psychology, 36,* 1415–1423.

Greene, D., & Lepper, M. R. (1974). Effects of extrinsic rewards on children's subsequent intrinsic interest. *Child Development, 45,* 1141–1145.

Grosse, B. (1982). Verhaltenskorrelate von Handlungs- und Lageorientierung: Reliabilität und Validität der Handlungskontroll-Skala (unpublished thesis). Ruhr-University Bochum, FRG.

Heckhausen, H. (1977). Achievement motivation and its constructs: A cognitive model. *Motivation and Emotion, 1,* 283–329.

Hiroto, D. W., & Seligman, M. E. P. (1975). Generality of learned helplessness in man. *Journal of Personality and Social Psychology, 31,* 311–327.

Humphreys, M. S., & Revelle, W. (1984). Personality, motivation, and performance: A theory of the relationship between individual differences and information processing. *Psychological Review 91,* 153–184.

Izard, C. (1977). *Human emotions.* New York: Plenum.

Kintsch, W. (1974). *The representation of meaning in memory.* Hillsdale, NJ: Erlbaum.

Koller, P. S., & Kaplan, R. M. (1978). A two-process theory of learned helplessness. *Journal of Personality and Social Psychology, 36,* 1177–1183.

Kruglanski, A. (1975). The endogenous-exogenous partition in attribution theory. *Psychological Review, 82,* 387–406.

Kuhl, J. (1977). *Meß- und prozeßtheoretische Analysen einiger Person- und Situationsparameter der Leistungsmotivation.* Bonn: Bouvier.

Kuhl, J. (1981). Motivational and functional helplessness: The moderating effect of state versus action orientation. *Journal of Personality and Social Psychology, 40,* 155–170.

Kuhl, J. (1982a). The expectancy-value approach in the theory of social motivation: Elaborations, extensions, critique. In N. T. Feather (Ed.), *Expectations and actions: Expectancy-value models in psychology.* Hillsdale, NJ: Erlbaum.

Kuhl, J. (1982b). Action vs. state orientation as a mediator between motivation and action. In W. Hacker, W. Volpert, & M. von Cranach (Eds.), *Cognitive and motivational aspects of action.* Amsterdam: North-Holland-Publishing Co.

Kuhl, J. (1982c). Handlungskontrolle als metakognitiver Vermittler zwischen Intention und Handeln: Freizeitaktivitäten bei Hauptschülern. *Zeitschrift für Entwicklungspsychologie und Pädagogische Psychologie, 14,* 141–148.

Kuhl, J. (1983a). *Motivation, Konflikt und Handlungskontrolle.* Heidelberg, FRG: Springer.

Kuhl, J. (1983b). Motivationstheoretische Aspekte der Depressionsgenese: Der Einfluß von Lageorientierung auf Schmerzempfinden, Medikamentenkonsum und Handlungskontrolle. In M. Wolfersdorf, R. Straub, & G. Hole (Eds.), *Der depressiv Kranke in der psychiatrischen Klinik: Theorie und Praxis der Diagnostik und Therapie.* Regensburg: Roderer.

Kuhl, J. (1983c). Emotion, Kognition und Motivation: II. Die funktionale Bedeutung der Emotionen für das problemlösende Denken und für das konkrete Handeln. *Sprache und Kognition, 4,* 228–253.

Kuhl, J. (1984). Volitional aspects of achievement motivation and learned helplessness: Toward a comprehensive theory of action control. In B. A. Maher (Ed.), *Progress in Experimental Personality Research* (Vol. 13). (pp. 99–171). New York: Academic Press.

Kuhl, J. (in press). Motivation and information processing: A new look at decision-making, dynamic conflict, and action control. In R. M. Sorrentio & E. T. Higgins (Eds.). *The handbook of motivation and cognition: Foundations of social behavior.* New York: Guilford Press.

Kuhl, J., & Beckmann, J. (1983). Handlungskontrolle und Umfang der Informationsverarbeitung: Wahl einer vereinfachten (nicht optimalen) Entscheidungsregel zugunsten rascher Handlungsbereitschaft. *Zeitschrift für Sozialpsychologie, 14,* 241–250.

Kuhl, J., & Eisenbeiser, T. (in press). Mediating versus meditating cognitions in human motivation: Action control, inertial motivation, and the alienation effect. In J. Kuhl & J. W. Atkinson (Eds.), *Motivation, thought, and action.* New York: Praeger.

Kuhl, J., & Geiger, E. (in press). The dynamic theory of the anxiety-behavior relation: A study on resistance and time allocation. In J. Kuhl & J. W. Atkinson (Eds.), *Motivation, thought, and action.* New York: Praeger.

Kuhl, J., & Helle, P. (1984). Motivational and volitional determinants of depression: The degenerated-intention hypothesis. (Submitted). Max Planck Institute for Psychological Research, Munich, FRG.

Kuhl, J., & Melendez, J. (1984). Undermining intrinsic motivation: Individual differences and duration of exposure to extrinsic rewards (Manuscript in preparation). Max Planck Institute for Psychological Research, Munich, FRG.

Kuhl, J., & Wassiljew, I. (1983). *Intrinsic task-involvement coping with failure, and problem-solv-*

ing: Motivational and emotional determinants of the complexity of action plans (Manuscript submitted for publication). Max Planck Institute for Psychological Research, Munich, FRG.

Kuhl, J., & Weiß, M. (1983). *Performance deficits following uncontrollable failure: Impaired action control or global attributions and generalized expectancy deficits?* (Manuscript submitted for publication.) Max Planck Institute for Psychological Research, Munich, FRG.

La Berge, D. (1973). Attention and the measurement of perceptual learning. *Memory and Cognition, 1,* 268–276.

Lavelle, T. L., Metalsky, G. I., & Coyne, J. C. (1978). Learned helplessness, test anxiety, and acknowledgment of contingencies. *Journal of Personality and Social Psychology, 88,* 381–387.

Lazarus, R. S. (1966). *Psychological stress and the coping process.* New York: McGraw Hill.

Lazarus, R. S., & Averill, J. R. (1972). Emotion and cognition: With special reference to anxiety. In Ch. D. Spielberger (Ed.), *Anxiety: Current trends in theory and research* (Vol. 2). New York: Academic Press.

Lepper, M. R., Greene, D., & Nisbett, R. E. (1973). Undermining children's intrinsic interest with extrinsic rewards: A test of the overjustification hypothesis. *Journal of Personality and Social Psychology, 28,* 129–137.

Lindworsky, J. (1923). *Der Wille. Seine Erscheinung und Beherrschung nach den Ergebnissen der experimentellen Forschung.* Leipzig: Barth.

Mandler, G., & Sarason, S. B. (1952). A study of anxiety and learning. *Journal of Abnormal and Social Psychology, 47,* 166–173.

Mischel, W. (1968). *Personality and assessment.* New York: Wiley.

Mischel, W. (1973). Toward a cognitive social learning reconceptualization of personality. *Psychological Review, 80,* 252–283.

Mischel, W. (1981). Metacognition and the rules of delay. In J. Flavell & C. Ross (Eds.), *Cognitive and social development: Frontier and possible futures.* New York: Cambridge University Press.

Mischel, W., Ebbesen, E. G., & Zeiss, A. R. (1972). Cognitive and attentional mechanisms in delay of gratification. *Journal of Personality and Social Psychology, 21,* 204–218.

Mischel, H. N., & Mischel, W. (1983). The development of children's knowledge of self-control strategies. *Child Development, 54,* 226–254.

Morgan, M. (1981). The overjustification effect: A developmental test of self-perception interpretations. *Journal of Personality and Social Psychology, 40,* 809–821.

Morgan, M. (1983). Decrements in intrinsic motivation among rewarded and observer subjects. *Child Development, 54,* 636–644.

Morris, L. W., & Liebert, R. M. (1970). Relationship of cognitive and emotional components of test anxiety to physiological arousal and academic performance. *Journal of Consulting and Clinical Psychology, 35,* 332–337.

Neely, H. N. (1977). Semantic priming and retrieval from lexical memory: Roles of inhibitionless spreading activation and limited-capacity attention. *Journal of Experimental Psychology: General, 106,* 226–254.

Newell, A., & Simon, H. A. (1972). *Human problem-solving.* Englewood Cliffs, NJ: Prentice-Hall.

Norman, D., & Rumelhart A. D (1975). *Explorations in cognition.* San Francisco: Freeman.

Oakes, W. F., & Curtis, N. (1982). Learned helplessness: Not dependent upon cognitions, attributions, or other such phenomenal experiences. *Journal of Personality, 50,* 387–408.

Olson, D. R., & Astington, J. W. (1983). Children's acquisition of metalinguistic and metacognitive verbs. (Manuscript submitted for publication.) Ontario Institute for Studies in Education, Toronto: Canada.

Piaget, J. (1952). *The origins of intelligence in the child.* New York: International Universities Press.

Posner, M. I., & Snyder, C. R. R. (1975). Attention and cognitive control. In R. L. Solso (Ed.), *Information processing and cognition.* Hillsdale, NJ: Erlbaum.

Reiss, S., & Sushinsky, L. W. (1975). Overjustification, competing responses, and the acquisition of intrinsic interest. *Journal of Personality and Social Psychology, 31,* 1116–1125.

Roth, S. (1980). A revised model of learned helplessness in humans. *Journal of Personality, 48,* 103–133.

Roth, S., & Kubal, L. (1975). The effect of noncontingent reinforcement on tasks of differing importance: Facilitation and learned helplessness. *Journal of Personality and Social Psychology, 32*, 680–691.

Seligmann, M. E. P. (1975). *Helplessness: On depression, development, and death.* San Francisco: Freeman.

Thoresen, C. E., & Mahoney, M. J. (1974). *Behavioral self-control.* New York: Holt, Rinehart, & Winston.

Wine, J. D. (1982). Evaluation anxiety: A cognitive-attentional construct. In H. W. Krohne & L. C. Laux (Eds.), *Achievement, stress, and anxiety.* Washington, D.C.: Hemisphere.

Woodworth, R. S., Schlosberg, H. (1954). *Experimental psychology.* London: Methuen.

Appendix: The Action-Control Scale

The Action-Control Scale consists of three subscales:
(1) Performance-related action vs. state orientation *(AOP)* (i.e., activity vs. goal orientation).
(2) Failure-related action vs. state orientation *(AOF).*
(3) Decision-related action vs. state orientation *(AOD).*

Each subscale contains 20 items. Each item specifies a situation and two response alternatives, one indicating an action-oriented and the other one indicating a state-oriented response. Each subscale contains ten items assessing *behavioral* manifestations of action and state orientations and ten items assessing *cognitive* manifestations.

For most experimental purposes, it has proven useful to combine the cognitive and the behavioral items. The three scores for *AOP, AOF,* and *AOD* are computed by summing up all action-oriented answer alternatives endorsed by the subject, separately for each scale. In a recent study (n = 115), the following estimates of internal consistency (Cronbach's alpha) have been obtained: .74 *(AOP);* .79 *(AOF)* and .79 *(AOD).*

The action-oriented response alternatives are:
AOP (behavioral): 1 a, 2 a, 3 a, 4 b, 5 b, 6 b, 7 a, 8 b, 9 a, 10 b,
      (cognitive): 11 b, 12 a, 13 b, 14 b, 15 a, 16 b, 17 a, 18 b, 19 b, 20 b.
AOF (behavioral): 1 a, 2 a, 3 b, 4 b, 5 b, 6 b, 7 b, 8 b, 9 b, 10 a,
      (cognitive): 11 b, 12 b, 13 a, 14 a, 15 b, 16 a, 17 b, 18 b, 19 b, 20 a.
AOD (behavioral): 1 b, 2 a, 3 b, 4 a, 5 a, 6 b, 7 a, 8 b, 9 a, 10 b,
      (cognitive): 11 b, 12 a, 13 b, 14 b, 15 b, 16 a, 17 b, 18 a, 19 b, 20 b.

Correlations between *AO*-scores and several personality variables indicated the theoretically expected overlap with test anxiety, extraversion, self-consciousness, achievement motivation, future orientation, and cognitive complexity (see Kuhl, 1984). The moderate size of these correlations (< |.36|), however, indicates that a substantial proportion of variance in action-orientation scores cannot be accounted for by any of the personality variables mentioned earlier.

AOP
1. When I've had good ideas while playing a difficult game
   ___ I soon look for something else to do
   ___ I could play on for hours
2. When I've done extremely well in an important contest
   ___ I'd like best to continue
   ___ I then like to do completely different things
3. When I receive an award for excellent achievement
   ___ I like to continue practicing in the same area immediately
   ___ I like to do things that have nothing to do with this area
4. When I've finished an excellent piece of work
   ___ I like to do something else for a while
   ___ it makes me want to do some more in the same area
5. When I've won an interesting game often
   ___ I like a change and do something else
   ___ I could play on and on

6. When I've made a decisive win in a game
____ I have a longing to stop the game after a while
____ I'd just as soon keep right on playing
7. When the food really tastes good and I've already had enough
____ I continue eating because the food looks so good
____ I stop eating.
8. When I'm reading something interesting
____ I busy myself with other things sometimes for a change
____ I often stick with it for a long time
9. When the TV schedule seems interesting to me
____ I watch one program after the other
____ I soon need a change anyway
10. When I do something interesting with friends
____ I soon get interested in something else anyway
____ I'd rather not stop with what we're doing
11. When I've constructed something complicated that didn't take me very long
____ I soon turn my attention to something else
____ I congratulate myself again and again about how well it turned out
12. If I get lucky in a situation where my chances were poor
____ I play it back in my mind over and over again
____ it's not long before I think about other things
13. When I've accomplished something really important
____ I think about other things relatively soon
____ I can't think about anything else at first
14. When my effort is graded unexpectedly well
____ I don't think about it long
____ I think about how well I did
15. When I try something new and I'm successful with it
____ I keep thinking of it for a while
____ I think about something else after a little while
16. If I were to win a lot of money (e. g., in a lottery)
____ I would immediately think about how to spend the money
____ I would keep thinking about how I could have been so lucky
17. When somebody surprises me with a gift that really pleases me
____ I think about the nice surprise for a long time
____ I soon busy myself with other things after the initial surprise is over
18. When the doctor tells me that I don't have any internal injuries after I've taken a bad fall
____ it's a closed case for me
____ my relief lasts a long time
19. When I really liked a vacation
____ I busy myself with other things soon after I return
____ after my return, I think a lot about the vacation
20. If someone has irritated me and I really told him off
____ then the matter is finished for me
____ the feeling of satisfaction stays with me for quite a while

AOF

1. When I've made several futile attempts to start an assignment
____ I start something else relatively soon
____ I don't feel like doing anything at all
2. When my work is labeled "unsatisfactory"
____ then I really dig in
____ at first I am stunned
3. When I notice that I'm not getting anywhere with something important
____ it kind of cripples me
____ I lay it aside for a while and do something else
4. When something breaks down unexpectedly

___ it takes a while before I can get myself to do something about it
___ I undertake the necessary steps immediately

5. When I can't memorize something even though I've tried and tried
___ I find it hard to start anything else
___ I do something else for a while

6. When grades do not match the effort I put into a task
___ it takes a while before I get over the disappointment
___ I then work extra hard

7. When something important to me just keeps going wrong
___ I gradually get discouraged
___ I forget about it for a while and do something else

8. When something makes me sad
___ I lose all desire to do anything
___ I try to divert my attention to other things

9. When several things go wrong for me on the same day
___ I really don't know what to do with myself
___ I can still do things as though nothing had happened

10. When my whole ambition is to finish something successfully and it doesn't work out
___ I would like to start the whole thing over again from the beginning
___ it's hard for me to do anything at all

11. If I lost something of value and all efforts to find it proved futile
___ I would have a hard time getting over it
___ I wouldn't think about it very long

12. If I've worked on a project for four weeks and everything turns out wrong
___ Its a long time before I get over it
___ I don't let it bother me for very long

13. When I'm lagging far behind in a contest of some sort
___ I think about how I can make the best of the situation
___ I think about whether or not I might make a fool of myself

14. When a new appliance falls on the floor by accident
___ I concentrate fully on what should be done
___ I can't stop thinking about how this could happen

15. If somebody is unfriendly to me
___ it can put me in a bad mood for quite a while
___ it doesn't bother me for long

16. When I'm in pain
___ I am able to concentrate on other things
___ I can hardly think about anything else

17. When I'm taking an important test and I notice that I'm not doing too well
___ it gets harder and harder for me to concentrate on the questions
___ I don't think much about it until the test is over

18. When I have to write a letter and can't think of anything to say
___ I think about whether or not there's something else I can do
___ I can't think about anything else

19. When I notice I've been used
___ I can't stop thinking about it for a long time
___ I soon forget about it

20. When a friend suddenly behaves in a way that shows the friend has withdrawn from me
___ I immediately consider how I should behave towards her/him
___ I try hard to figure out what's the matter

AOD

1. If I had to work at home
___ I would often have problems getting started
___ I would usually start immediately

2. When I want to see someone again
___ I try to set a date for the visit right away

_____ I plan to do it some day
3. When I have a lot of important things to take care of
_____ I often don't know where to start
_____ It is easy for me to make a plan and then stick to it
4. When I have two things that I would like to do and can do only one
_____ I decide between them pretty quickly
_____ I wouldn't know right away which was most important to me
5. When I have to do something important that's unpleasant
_____ I'd rather do it right away
_____ I avoid doing it until it's absolutely necesary
6. When I really want to finish an extensive assignment in an afternoon
_____ it often happens that something distracts me
_____ I can really concentrate on the assignment
7. When I have to complete a difficult assignment
_____ I can concentrate on the individual parts of the assignment
_____ I easily lose my concentration on the assignment
8. When I fear that I'll lose interest during a tedious assignment
_____ I complete the unpleasant things first
_____ I start with the easier parts first
9. When it's absolutely necessary that I perform an unpleasant duty
_____ I finish it as soon as possible
_____ it takes a while before I start on it
10. When I've planned to do something unfamiliar in the following week
_____ it can happen that I change my plans at the last moment
_____ I stick with what I've planned
11. When I know that something has to be done soon
_____ I often think about how nice it would be if I were already finished with it
_____ I just think about how I can finish it the fastest
12. When I'm sitting at home and feel like doing something
_____ I decide on one thing relatively fast and don't think much about other possibilities
_____ I like to consider several possibilities before I decide on something
13. When I don't have anything special to do and am bored
_____ I sometimes contemplate what I can do
_____ it usually occurs to me soon what I can do
14. When I have a hard time getting started on a difficult problem
_____ the problem seems huge to me
_____ I think about how I can get through the problem in a fairly pleasant way
15. When I have to solve a difficult problem
_____ I think about a lot of different things before I really start on the problem
_____ I think about which way would be best to try first
16. When I'm trying to solve a difficult problem and there are two solutions that seem equally good to me
_____ I make a spontaneous decision for one of the two without thinking much about it
_____ I try to figure out whether or not one of the solutions is really better than the other
17. When I have to study for a test
_____ I think a lot about where I should start
_____ I don't think about it too much; I just start with what I think is most important
18. When I've made a plan to learn how to master something difficult
_____ I first try it out before I think about other possibilities
_____ before I start, I first consider whether or not there's a better plan
19. When I'm faced with the problem of what to do with an hour of free time
_____ sometimes I think about it for a long time
_____ I come up with something appropriate relatively soon.
20. When I've planned to buy just one piece of clothing but then see several things that I like
_____ I think a lot about which piece I should buy
_____ I usually don't think about it very long and decide relatively soon

Chapter 7

Dissonance and Action Control*

Jürgen Beckmann and Martin Irle

Imagine that you have bought a car which you expect to drive faster than 120 miles per hour. After breaking it in you try to reach the car's top speed and find that it is less than 120 miles per hour. Selling the car would cause financial loss, so you have to keep it. Will you be dissatisfied with the car from now on and think you made a wrong decision each time you drive it? Fortunately, there are certain cognitive strategies which allow you to enjoy driving the car after this experience of inconsistency between your expectation about the car's top speed and its actual performance. You may search for positive information, for example that 110 miles per hour is still much faster than most other cars can do; that your car is safer and more comfortable than other cars, etc. Such cognitive operations will reduce the discontent you felt when experiencing the inconsistency between the car's actual speed and your expectation about it.

The theory of cognitive dissonance (Festinger, 1957) deals with such experiences of cognitive inconsistency and allows predictions about the modes of inconsistency reduction. The theory has influenced the study of numerous social psychological issues and instigated a great number of empirical research studies since its formulation by Leon Festinger more than 25 years ago (for an overview of theory and research see: Wicklund & Brehm, 1976). In these works, processes of dissonance reduction have been considered primarily as isolated processes that serve the function of generating consistency in the cognitive representation of the environment and one's own person.

The analysis given in this chapter will seek to determine whether the findings accumulated in this research tradition can serve to explain the existing theoretical gap between cognition and behavior. From everyday experience we know that solely the intention to perform a specific action may not be sufficient for the actual performance of that action. For example, often despite an early morning

* Preparation of this chapter was supported by the Deutsche Forschungsgemeinschaft. The authors are grateful to Dieter Frey, Julius Kuhl, and Gün Semin for helpful comments on an earlier draft.

resolution to clean one's shoes one finds in the evening that one has again failed to do so. Kuhl (1982) has pointed out that this problem is neglected in contemporary action theory, where the assumption seems to be that the action tendency, dominant at one point in time, will find its direct correspondence in behavior. To bridge the gap between motivation and behavior, as demonstrated in the example given above, a theory of action control was formulated (Kuhl, 1984; see Chapter 6, in this volume). It is assumed that whenever an actor perceives that the enactment of a current intention is threatened by the occurrence of competing motivational tendencies he will mobilize additional effort to screen the current intention. Kuhl (1984) postulates several cognitive processes which may be employed by the actor to screen the current intention. Two such instances are first, selective processing of information and second, parsimonious use of information once it is selected. Furthermore, the effectiveness of such action-control processes is assumed to depend on an actor's *action* vs *state orientation*. In the case of state orientation a person tends to ruminate about some past, present or future state rather than concentrating on the enactment of his/her current intention. In opposition to this, a state of action orientation facilitates the enactment of intentions (see Kuhl, Chapter 6, in this volume for more detail).

In the present chapter the theory of cognitive dissonance is analyzed from this action control point of view. The question examined is to what extent the cognitive processes postulated by dissonance theory to restore consistency between cognitions can contribute to bridge the gap between cognition and behavior. To be sure, this interpretation of the theory of cognitive dissonance is not entirely new. But the present level of analysis is extended from an isolated examination of processes between cognitions to a consideration of their function within the process of action. This broader perspective yields some suggestions for specifications of dissonance theory itself. Furthermore, it points to the relevance of the cognitive processes described in dissonance theory for a theory of action control.

The Relevance of the Theory of Cognitive Dissonance for Processes of Action Control

How is cognitive dissonance related to action control? An examination of the large body of research existing on dissonance theory reveals that nearly all of them concentrate on dissonance reduction as an independent process. Action and action control is never taken into consideration.

But, consider the following example:

I want to buy a new car, and have already put a particular model on my short list. To make the final decision as to whether I should actually buy this model or another one that I like as well I will try to obtain information concerning the advantages and disadvantages of the two cars. From test reports in motor journals I find out that the car I like second best came out better in some items than the one I prefer most. On the other hand, friends who have already bought the car I prefer tell me that they are completely satisfied with it and that it is really fun to drive it. Even after I have gathered additional information I may still be facing the following dilemma: There are as many items that speak in favour of the preferred car as against it. At the same time the less preferred car has some advantages over the other but has some drawbacks as well.

According to the theory of cognitive dissonance[1] this is a situation where I should begin selective information processing about the alternatives open to decision that support my intention to buy car A. For example, I could search only for information which I know will report advantages of the car that I tentatively decided to buy (advertisements for this car). The end envisaged by these processes is to reconcile the state of dissonance that is generated by the inconsistency of the different cognitions (information about advantages and disadvantages of the cars).

As can be seen from this example, the process of restoring cognitive consistency by means of preference for decision-supporting information may not be an end in itself but serve another function, namely, to secure the ability to decide and act. According to this action-theory interpretation therefore, dissonance reduction is considered as not primarily having the function of establishing a non-contradictory cognitive system but to promote control for the transformation of some intention (to buy car A) into the appropriate behavior.

Given the complex environments that we usually live in and which contain more information than we are able to deal with (cf., March & Simon, 1958), situations like the one described above seem inevitable. These situations are comparable to the one of Buridan's ass who starved to death between two stacks of hay because he was not able to decide which of the two stacks of hay he should eat. We experience a similar dilemma whenever two incompatible wishes are instigated at the same time or, when we find in the pursuit of only one goal – as in the example of the car purchase – that as much information speaks for an alternative as speaks against it. Nevertheless, only very few of us have to experience a sad fate similar to that of Buridan's ass who remained in a state of indecisiveness because he found no means to solve the conflict. In everyday life we usually make use of more or less effective control strategies which prevent us from failing to transform our cognitions (intentions) into behavior when faced with such a difficult decisional conflict (Kuhl, 1984).

The cognitive processes described in the theory of cognitive dissonance (Festinger, 1957) can be considered strategies that help to solve decisional conflicts and thereby promote the transformation of intentions into (effective) behavior. We do not assume that dissonance reduction always serves this end. In numerous studies consistency may have been the one and only end of dissonance reduction. Nevertheless, a number of processes of dissonance reduction will be understood only if we take account of their mediating function within the context of action, that is, of their function to stabilize decision situations in order to secure the ability to act. If information processing were objective and impartial

---

[1] Our example deals with a predecision situation. Festinger (1964) has restricted the theory of cognitive dissonance to the postdecision situation. Nevertheless, it was shown theoretically and empirically that the theory can be employed in predecision situations (see Irle, 1982 for more detail).

we would only very seldom be able to transform an intention into behavior, faced as we are with the contradictory information contained in our environment. Hence, we assume that processes of dissonance reduction have a regulatory function within a process of action that counteracts the danger of inability to act and that helps to ensure action control. Considered this way, processes of dissonance reduction serve the (more or less) undisturbed execution of an actual purpose and not only the establishment of consistency within the system of a person's cognitions. We shall now present some preliminary evidence for this assumption by describing a number of studies selected as examples from the existing body of research. This evidence will be supplemented by some new research findings.

Dissonance Reduction by Means of Discrepancy Induction

At first sight the title of this section may seem paradoxical since dissonance is a kind of discrepancy. But, we will argue that two different kinds of discrepancies are bound up with processes of dissonance reduction within the context of an action. One type of these discrepancies may be described as being a (psycho-)logical discrepancy between cognitions and can be called a *qualitative discrepancy*. It is the kind of discrepancy usually associated with cognitive dissonance. Yet underlying the qualitative discrepancy there is a *quantitative discrepancy* between the given decision or action alternatives. Reducing the qualitative discrepancy (dissonance) is achieved by increasing the quantitative discrepancy between the given action alternatives (respectively, the attraction attached to them). This can be demonstrated in the case of the car purchase example:

I have made up my mind to buy a car. Then I receive unfavorable information about car A and favorable information about the competing model B. Hence, the prior quantitative discrepancy between the two alternatives in terms of their attractiveness, on the basis of which I made my decision, decreases. The decrease of the quantitative discrepancy makes the qualitative discrepancy salient in that choosing A implies renunciation of the positive aspects of B and acceptance of the negative aspects of A. This dissonance can be reduced by information processing being influenced so as to again increase the quantitative discrepancy between the two alternatives. As soon as one alternative becomes sufficiently preferred to the other by means of such cognitive manipulation (i.e., that the quantitative discrepancy surpasses a certain threshold), the qualitative discrepancy (dissonance) becomes insignificant in that action control is no longer endangered. To summarize, when the quantitative discrepancy between the decision or action alternative is increased the qualitative discrepancy or dissonance is reduced.

This process may be described as being a feedback loop like Miller, Galanter & Pribram's (1960) TOTE unit. Such a feedback loop is activated only when a qualitative discrepancy (dissonance) becomes salient. This warning signal of some threat to action control comes into the focus of attention whenever the quantitative discrepancy between competing decision or action alternatives falls short of a certain threshold. A small quantitative discrepancy signals the danger of a wrong decision since new information may in this case easily change the preference order. The smaller the quantitative discrepancy, the greater the danger that by means of new information the previously most preferred alternative will be superseded by a competing alternative. Thus, if such a dissonance becomes salient, cognitive processes are put into operation that act on the (attractiveness ratings of the) decision alternatives until the quantitative discrepancy between them exceeds the given threshold. When the quantitative discrepancy has been increased sufficiently, dissonance will vanish from the focus of attention. This may equally be achieved by screening out competing decision alternatives. For example, if I sit at my desk to write an important paper when friends of mine start to play soccer outside, the strength of the action alternative *playing soccer* will increase and compete with the alternative *writing the paper*. I will become aware that both alternatives are incompatible and a decision for one of them implies the rejection of the positive experiences implied by the other one.

Should I fail to succeed in eliminating the awareness of this dissonance I will not be able to enjoy whatever alternative I decide to perform. Both alternatives require high amounts of concentration. Dissonance, or thinking about the renunciation of the positive aspects of the rejected alternative would consume a part of my processing capacity. Since I am really enthusiastic about playing soccer, this cognition would be quite resistant to change though it may have lesser resistance to change than the cognition of writing the paper. Nevertheless, given the high degree of resistance to change playing soccer possesses, it would be very difficult for me to change this cognition. But, I may try to decrease the intensity of its representation, for example by turning to especially interesting problems concerning the issue I am writing about. This might hold my attention completely and I will withdraw my attention from my soccer-playing friends. Thereby the dissonance between the two alternatives open to choice will no longer impair the writing of the paper. In this example, dissonance is not reduced by reconciling the cognitions involved, but through a change in the intensity of their representation. It may be said that in this case the qualitative discrepancy (dissonance) between the cognitions involved is reduced by means of increasing the quantitative discrepancy between the competing action tendencies.

This is exactly the way Kelly (1962, p. 81) conceived of dissonance reduction: "The tendency postulated is not toward bringing the organism to rest, or restoring balance, but toward maximizing imbalance between forces in order to reconcile force and action." "It seems to me the cognitive dissonance theory has it that he (Buridan's ass – the authors) will do anything to avoid getting caught midway between two stacks of hay."

Classical conflict theories (Lewin, 1935; Miller, 1944, 1951) addressing the same problem held that a solution to the conflict between the competing decision alternatives was due to a change in the situation. Thereby these theories deny the individual an active part in action control. Only in his late works did Lewin (1951, 1952) introduc the individual's capacity to "freeze" a constellation of action tendencies once a decision has been made. Festinger (1957) went one step further; not only do individuals have the capacity to freeze some constellation of decision alternatives but furthermore they may act on these alternatives to solve the decisional conflict – for example by increasing the discrepancy between them. The individual changes the cognitive representation of the conflict situation in a way that allows the performance of an action. It may be said that Festinger postulates an additional motivation which occurs whenever cognitive inconsistencies threaten an individual's action control. This motivation, for example in the case of the car purchase, adds to the motivation to buy a new car when inconsistent information blocks decision-making. Therefore, this may be considered "a very different motivation from what psychologists are used to dealing with" (Festinger, 1957, p. 3), a motivation serving the end to secure action control within an action process. In this respect, the underlying model coincides with Kuhl's (1984) two-stage-motivation model. In this model one type of motivation *(choice motivation)* selects a goal for an action. This motivation is, whenever execution of this intention is threatened by some obstacle, supported by a second type of motivation *(control motivation)* which puts into operation cognitive processes to eliminate these obstacles. Processes of dissonance reduction may be considered a subset of these processes, although establishing consistency may not always – as discussed here – be subordinate to some other goal. From the traditional perspective, restoring consistency within a person has often been considered the end of dissonance reduction. Concerning this issue, the distinction of choice motivation and control motivation is less useful. In this case, the motivation to reduce cognitive dissonance can be compared to White's (1959) competence motivation. Finally, the consistent order of cognitions aspired to guarantees competence to act since, only if the relevant cognitions form noncontradictory relations, is an unequivocal deduction of what action to take possible.

Empirical Evidence for the Divergency Effect

There are a number of experiments furnishing evidence for dissonance reduction as a means for decision or action control as discussed in the preceding section. Yet, the authors of these experiments did not reflect upon the functional importance of dissonance reduction for a successful performance of a present intention. The experiments discussed as follows are restricted to the phase following a decision to act. So, from our point of view dissonance reduction serves the function of controlling the intention within the phase of its actual performance. If I have finally bought car A, driving should be fun and should not be spoiled by thinking of the advantages of car B, which I did not buy.

Brehm (1956) conducted the classical experiment in which such intention control by means of a divergency or "spreading apart of alternatives" effect occurred. In this experiment subjects had to rank order nine consumer goods. Subsequently three different experimental conditions were set up. Subjects of the control group were simply given a product chosen by the experimenter. So, in this condition subjects did not make a decision themselves. In the two dissonance conditions the subjects had to choose between two products. In the group with high dissonance, these products were the ones for which initial attractiveness ratings differed 1.5 points. In the group with low dissonance the choice was between the products for which the initial ratings differed 2 or 3 points. Subsequently, the subjects read reports about some of the products. They then rated their attractiveness again. As hypothesized, according to the theory of cognitive dissonance, Brehm (1956) found that subjects in both dissonance groups enlarged the divergence between the decision alternatives by increasing the attractiveness of the chosen product and decreasing that of the rejected product. The smaller the initial discrepancy was, the greater the changes of attractiveness ratings.

Such a spreading-apart effect may be produced by seeking information selectively in favour of the decision made (Ehrlich, Guttman, Schönbach, & Mills, 1957). Another possibility in the absence of further information is to concentrate on the positive aspects of the chosen alternative (Younger, Walker, & Arrowood, 1977). Within the dissonance literature findings concerning selective exposure to decision consonant information are far from being unequivocal (for an overview see Freedman & Sears, 1965, Wicklund & Brehm, 1976; Frey, 1981). Some experiments have shown results that seem to contradict the original theory (e.g., Rosen, 1961; Feather, 1963; Sears, 1965; Freedman, 1965). In each of the experiments subjects showed a propensity to expose themselves to seemingly dissonant information. Apart from a number of factors that have to be controlled in order to test the selective exposure hypothesis adequately (see Wicklund & Brehm, 1976, p. 175), we assume that one of the drawbacks of these studies is that they do not take into account that dissonance reduction processes are embedded in a course of action. Unless this aspect is paid regard to, no strict test of the hypothesis is possible since it is not known whether some information is actually dissonant for the actor or only from the point of view of the experimenter. As will be shown later, the perspective taken on dissonance reduction in this paper may help to remedy these problems.

Forced Compliance

Evidence from research on forced compliance may also be interpreted from the perspective taken here. In experiments with this research paradigm subjects usually commit themselves to a behavior that is dissonant with their attitudes. Since there is little external justification for engaging in such behavior, they are forced to screen the intention to perform this attitude-discrepant behavior against tendencies to engage in behavior that would follow their attitudes. This is achieved in the same way (i.e., by means of control) as described in the preceding section. The strength of the tendency to show the attitude-discrepant behavior is increased whereas the strengths of the competing tendencies are decreased. In fact, this may be considered another example of the divergency effect or of discrepancy enlargement in order to control for the performance of the intended behavior.

In forced compliance experiments different incentives are usually offered to subjects for engaging in the attitude-discrepant behavior. These incentives will affect the strength of the tendency. If the incentive is high, individuals do not need to employ cognitive operations in order to augment the discrepancy between the competing tendencies. But, if the incentive given cannot sufficiently increase the strength of the action tendency to which the individual is committed in order to generate a discrepancy, which allows to perform the intended behavior adequately, the individual has to resort to cognitive processes described in the theory of cognitive dissonance.

The most consistent findings regarding forced compliance have been produced within the *forbidden toy* paradigm that was employed for the first time by Aronson and Carlsmith (1963). In this experiment nursery school children had to first rank a number of toys according to their attractiveness. Subsequently, they were told not to play with the toy they had rated second in attractiveness. For a violation of this prohibition they were threatened in one condition with mild and in another condition with rather severe punishment. Afterwards, the children were left alone to play with all toys except the one they were forbidden to play with. Subsequently, the children had to rate the attrativeness of the toys once again. It was found, as hypothesized, that the threat of mild punishment led to more devaluation of the forbidden toy than the threat of severe punishment. From our point of view, to maintain action control, that is in order not to have their playing spoiled by the wish to play with the forbidden toy (that would surely intrude if the discrepancy in attractiveness between the alternatives was small), the children that were threatened with merely mild punishment had to decrease the attractiveness of the forbidden toy.

It is interesting to note that these results are found only if the conditions of commitment and volition are met (Frey & Irle, 1972), since one of the originators of action-control theory, Narziß Ach, included similar notions in his model. According to Ach (1910) volitional processes, that is, action control processes must contain the element of "I really will" thereby connecting the intention to the self (cf. Kuhl & Beckmann, Capter 5, in this volume). Connecting a behavioral intention to the self is really the essence of the notion "commitment" (cf. Kiesler, 1971).

Control of Motivation by Means of Dissonance Reduction

In a number of studies, dissonance reduction has been related directly to action tendencies. These were the experiments that induced Kelly (1962) to consider the theory of cognitive dissonance as a model of discrepancy induction. Strangely enough, the researchers conducting these experiments did not consider the matter from Kelly's view point (i. e., as a solution to a conflict between competing action tendencies). Therefore, this experimental paradigm appears quite often as a separate issue although, as we will show, it has the same structure as that of the experiments with regard to spreading apart of alternatives or forced compliance discussed in the preceding sections.

Brehm and Crocker (1962) conducted an experiment in which dissonance reduction was employed to control for the incompatible action between fasting and food intake. The subjects in these experiments had committed themselves to go without breakfast and lunch on the day of the experiment. When they arrived at the experimental room in the afternoon they were shown some food which they expected to eat later in the session. The experimenter asked them to rate their hunger first. Subsequently, the subjects had to work on an irrelevant task. Following that task the experimenter asked the subjects to commit themselves to several more hours of fasting. In the condition with high dissonance he told them that no additional credit points could be given for this additional participation. In the condition with low dissonance they were offered 5 Dollars for the additional participation. Brehm and Crocker (1962) maintain that dissonance is aroused in proportion to the absence of justification for additional fasting. Thus, they do not consider a dissonance between two incompatible action tendencies, which would allow this research to be integrated within the framework of research concerning the divergency effect. Instead, they proceed from the questionable assumption that for all subjects the cognition "I get credit points" follows from the cognition "I do not take in food". It is questionable whether people are so strictly utilitarian as this assumption expects them to be. They most likely do not expect a (quasi) materialistic reward for each action that is performed.

We assume, therefore, that dissonance was not aroused primarily because of the lack of a reward, but that dissonance existed between the cognition "I do not take in food" and "I am very hungry" in the condition without reward since in this condition the discrepancy between the competing tendencies was small. Therefore, this situation corresponds to that of a choice between decision alternatives. In this case the alternatives consist of cognitive representations of motivational tendencies. Dissonance in this situation will be higher the smaller the discrepancy of strengths between the competing motivational tendencies. Therefore, in the experiment by Brehm and Crocker (1962) dissonance should be stronger the less the strengths of the two competing motivational tendencies differ. Consequently, the strength of the motivation to reduce dissonance should increase in proportion to the reduction of discrepancy between the two motivational tendencies. It becomes evident that the motivation to reduce dissonance is in fact a special kind of motivation. While processing information for an action the motivation to reduce dissonance actually occurs in addition to the underlying motivation that selects the goal of the action. This supplementary motivation intervenes in order to avoid indecisiveness or instability of decisions, or as we would say, in order to avoid a loss of action control.

An additional incentive for fasting (here: a 5 Dollar reward) reduces the dissonance without the individual having to initiate cognitive activities toward that end. Using the terminology of the theory of cognitive dissonance, the reward is a consonant cognition that is added to the cognition of fasting. Thereby, the attractiveness of this alternative, or expressed in motivation theoretical terms – the strength of this motivational tendency is increased. If we assume that the motivation to fast is stronger than the motivation to eat because of the instruction to fast, the discrepancy between the alternatives is increased. As a result, dissonance between the two motivational tendencies is no longer salient. Therefore,

employment of cognitive operations for the reduction of dissonance is not required. If no consonant cognition, like a 5 Dollar reward, is externally offered
for one of the competing tendencies, the individual has to change the cognitive
representation of the strengths of the two motivational tendencies by means of
cognitive operations. This can be achieved through a change of the cognition
which is least resistant to change in the given situation. In the experiment by
Brehm and Crocker (1962) the subjects had committed themselves to the cognition "I will not take in food". This means that this cognition was bound to the
subject's self and thereby rendered highly resistant to change (cf. Kiesler, 1971).
Therefore, the cognition about hunger is the least resistant to change and, consequently, the cognition that will be changed. By reducing the cognitively represented strength of hunger, the discrepancy between the two competing alternatives is increased and the salient dissonance is reduced. In the experiment by
Brehm and Crocker (1962) the individuals who had not been externally offered a
consonant cognition for fasting (reward) and therefore could not reduce dissonance this way, showed lower estimates of hunger in the postmeasure than in the
premeasure. In contrast to that, individuals who had been offered a reward (consonant cognition) for fasting showed an increase in their hunger ratings from the
pre- to the postmeasure. Additional support for this major proposition was
found: when subjects were asked how many sandwiches, cookies, and pints of
milk they would like to have after the evening session, subjects with external possibility to reduce dissonance (reward condition) desired more food items than
subjects without an external possibility to reduce dissonance.

Our alternative interpretation of the dissonance processes in Brehm and
Crocker's (1962) experiment can be applied to a number of other experiments in
the paradigm of cognitive control of motivation. Such an application provides
further support to our action control assumption. In an experiment by Brehm
(1962), subjects were asked to abstain from all liquids from bedtime until the
experimental session the following afternoon. On their arrival at the laboratory
in the afternoon half of the subjects were confronted with a salient cue for thirst.
In this high stimulation condition a pitcher of water with paper cups sat on the
testing desk throughout the whole session. In the low stimulation condition the
pitcher of water was absent. On their arrival a premeasure of thirst was taken
from all subjects. Then the subjects were asked to continue refraining from taking liquids until the following afternoon. One-half of the subjects was offered a
1 Dollar reward for their abstention (called high dissonance condition), the other
half was offered a 5 Dollar reward (called low dissonance condition).

In the condition with low reward and high stimulation a significant decrease
of thirst ratings was found from the pre- to the postmeasure (at least for male
subjects). In the other three conditions only changes of thirst ratings were found.
Low stimulation resulted in a small decrease of thirst ratings in the case of a
small reward and a small increase in the case of a high reward. In the case of
high stimulation and high reward the thirst rating increased from the pre- to the
postmeasure. Nevertheless, this increase does not significantly deviate from both
low stimulation conditions.

According to our interpretation of this experiment, dissonance should not be produced by a reward that is too small, but instead by the conflict between the two action alternatives open to choice. The pitcher of water present in the high stimulation condition is likely to instigate the motivation to drink. This should make salient the dissonance between the cognition about the thirstiness and the commitment not to drink. The greater the instigated thirst the smaller becomes the discrepancy between the competing alternatives and hence, the more cognitive activity is necessary to reduce the dissonance if no consonant cognitions are offered externally. Therefore, people in the high stimulation condition should show stronger attempts to reduce dissonance than people in the low stimulation condition. Nevertheless, because of the previous deprivation, people in the low stimulation condition should experience some amount of dissonance. Since different possibilities to reduce the dissonance are given according to the reward manipulation, it is to be expected that dissonance reduction is strongest in the condition without possibility of external reduction (high reward). It was found that in the low stimulation condition the small reward is apparently sufficient to make the tendency to stay without liquids strong enough. Therefore, the discrepancy between the competing tendencies does not fall short of the critical threshold. Accordingly, an increase in thirst rating is found in this condition. On the contrary, if the small reward is combined with high stimulation the discrepancy between the competing tendencies risks falling short of the critical threshold. Consequently, the cognitive representation of the competing tendencies has to be influenced so as to increase the discrepancy between them. A significant decrease of the thirst rating is thus found in this condition. In the high stimulation condition, the discrepancy seems to become so small that even the high reward offered is not entirely sufficient. Therefore, in this condition cognitive operations have to be employed in addition to the external reward to increase the discrepancy between the competing tendencies and to reduce dissonance. This is manifested by the finding that thirst measures in the high reward condition, in spite of the high stimulation, differ little from thirst measures in the high reward – low stimulation condition.

The experiments just described illustrate our thesis that the cognitive processes specified within the theory of cognitive dissonance are apt to maintain action control. Furthermore, the dynamic aspect of an action process is made salient. There is no unequivocal point of decision which changes the character of the cognitive processes fundamentally. The realization of an intention is constantly threatened by competing action tendencies in the phase of actual performance as well in the phase of orientation prior to the initiation of behavior. Without motivation – or intention – control, by means of dissonance reduction processes for example, there would always be the risk of the current intention being replaced by a competing tendency with no execution taking place. One would not be able to fast in order to lose weight, nor have a meal and enjoy it.

If the preceding assumption holds true, dissonance reduction processes should become increasingly necessary the more the actual performance of an intention (that is perhaps made dominant only by cognitive operations) is delayed. The longer this period of time, the more likely the prospect that competing motivational tendencies increase in strength and threaten to replace this intention before it is performed. Therefore, in the case of delayed transformation of an intention into behavior, the amount of dissonance reduction processes should increase with the duration of delay, at least as far as there is high commitment to this intention.

A study by Fellner and Marshall (1970) supports this assumption. Fellner and Marshall investigated the decision behavior of people who had been contacted as potential kidney donors for a member of their family. It was found that two types of decision behavior could be differentiated: One group of persons made their decision immediately after they had been contacted. These people did not wait until they received further information, although this decision would have serious consequences for them and although the physicians of the transplantation team urged them to postpone their decision until they had received complete information. In contrast to this group a second group avoided searching for a decision by themselves. They preferred to leave the decision up to the selection process which selected the most appropriate donor.

In the context of our discussion it is of special interest that for the fast deciders dissonance and dissonance reduction – created by increasing the attractiveness of the chosen action alternative (to donate) and decreasing that of the rejected alternative (not to donate) through selective information processing – increased with the duration of delay between their decision and the actual transplantation. Apparently the quick deciders were people who were "action-oriented" in the sense that they tried to screen their intention against competing action tendencies in order to be able to decide quickly and stick to this decision without conflict. This study shows clearly how processes of dissonance reduction are employed as a means for motivation (or intention) control. Furthermore, it demonstrates that the motivation to employ dissonance reduction may be instigated by the perceived threat to the formed intention.

Implications for the Theory of Cognitive Dissonance

Up to this point our paper has concentrated essentially on demonstrating that dissonance reduction can be interpreted as being cognitive operations by which individuals try to screen the execution of an intention once formed against the pressure of competing action alternatives. The theory of cognitive dissonance thus contains propositions about a specific class of action-control processes. Accordingly, if the large body of research conducted to test this theory is considered this theory seems to be helpful in closing the gap between intention and action.

We will now show that a molar view of dissonance reduction implied by the action-control perspective may help to make the theory of cognitive dissonance itself more precise. Commonly, processes of dissonance reduction are analyzed as isolated phenomena, that is on the molecular level of some processing of information. From this level of analysis the activity of the individual in question seems solely to be directed towards generating consistency among his/her cognitions. But, Festinger (1957) already postulated that a striving to obtain consistency will only occur for cognitions that are relevant to each other in a given situation. One of the theory's unsolved problems concerns the specification of when two cognitions are relevant to each other. Since this antecedence condition has not been further specified by Festinger (1957), the critical question, whether dissonance occurs at all for the subjects in these experiments was left to the individual experimenters (see, the criticism by Chapanis & Chapanis, 1964; Irle, 1975).

In view of this problem, strict tests of the theory are hardly feasible because contradictory results (e.g., subjects search for information which is dissonant from the experimenter's point of view: see Rosen, 1961; Freedman & Sears, 1965), may in all cases be explained away by declaring that no dissonance occurred for the subjects. If this molecular level of isolated consideration of information processing is maintained and it is at the same time taken into account that these processes are usually embedded in some course of action, obviously there is a solution to this problem. The goals that an individual is pursuing in a given situation determine which cognitions come into the focus of attention and thereby form relevant relations. These current concerns direct attention, memory, and thought context (Klinger, 1971; Klinger, et al., 1976). Irle (1975) took similar considerations as a basis for his qualification of the antecedence conditions of the theory of cognitive dissonance. He assumes that only those cognitions will come into the focus of attention in a given situation and establish relevant relations which have a certain value instrumentality and value importance in Rosenberg's (1956) sense. Value instrumentality and value importance should not be considered as fixed quantities. It is obvious that the value instrumentality of cognitions depends on the values or goals pursued in a given situation. Furthermore, as everyday-life experience and the findings of Klinger, et al. (1976) suggest, these current concerns may also influence the value importance of cognitions, that is, an individual's rank order of values. If motivation for a specific action is strong in a given situation, even values that seem to be of high importance for an individual over rather long periods of time and therefore easily become salient, may not get into the focus of attention. Songer-Nocks (1976), for example, found that people seem to ignore norms when there are strong motivational pressures.

Novels have been written about the fates of people who, in a state of high motivation, committed an action that they later found to be incompatible with values which had always been extremely important for them. Furthermore, there are reports about people in extreme life situations (e.g., concentration camp) who in a state of high motivation brought about by high deprivation (e.g., hunger) seemed to be able to act in contradiction to the most deep-rooted taboos (e.g., cannibalism), or at least in contradiction to cultural norms which they had adhered to their whole lives long (Cohen, 1953). But there were also some people in these situations who were not able to screen such intentions necessary for survival against the norms they contradicted and who died as a consequence of this (Bidermann, 1967).

The study by Darley and Batson (1973) may be taken as further evidence. In this study students of theology had to prepare a short talk which they were asked to give in an adjoining building. One half of the subjects was asked to talk about the good Samaritan parable, the other half was asked to talk about nonpastoral vocational facilities of clergymen. In addition, the subjects were put under different grades of time pressure (high, medium, no time pressure; see the operationalization of "need for structure" by Kruglanski & Klar, Chapter 3, in this volume). On the way to the adjoining building the subjects had to pass by a person obviously requiring help. Now, since clergymen are trained to internalize the role of a good Samaritan, one would expect them to help in any case. But this was not so. Help varied as a function of the condition subjects were in. The results can be interpreted according to the dissonance-theoretical position advocated here, although Darley and Batson (1973) did not consider dissonance or dissonance reduction processes. When the student's attention focused on the current concern "talk about nonpastoral vocational facilities of clergymen" and in addition they were under time pressure, help given was 87% less than when a talk about the good Samaritan parable had to be given and there was no time pressure. Since a clergyman's education focuses on charity as a central value of the Christian religion, this value should have been highly important and highly resistant to change for the students. Therefore, it has to be presumed that this value did not become salient in spite of its high importance because the current concern completely occupied the focus of attention. Consequently, the subjects did not experience dissonance. Otherwise, these students of theology should have been expected to help, since according to their assumed cognitive system, the cognition about their current behavior should have been less resistant to change than the norm of charity. Hence, the assumption seems justified that the relevance of cognitions is defined first by the goal currently pursued, and second by the importance of certain values. The salience of an individual's values will be inversely related to the strength of the motivation to reach a certain goal. If the strength of this motivation is low, an individual's values without direct reference to the actual goal may get into the focus of attention and form relevant relations with cognitions connected directly to the currently pursued goal.

As can be further concluded from Darley and Batson's (1973) findings, the definition of relevance, respectively, the selection of those cognitions that will get into the focus of attention, depends on additional factors as well. In this experiment the perceived time pressure was such an additional determinant. Time pressure is a factor that sets limits on successful execution of an action. This factor increases the difficulty of purpose realization. As Ach (1910) postulated, a perceived difficulty of purpose realization should release additional energy, which promotes the execution of the action against the difficulty. Kuhl (in press; in this volume) has specified a number of cognitive mediators in his theory of action control which are employed in this case.

This theory differentiates between two states of the organism. One of these states, called *action orientation,* is characterized by the metacognitive directive to do everything possible to secure the actual realization of a current purpose. On the other side, *state-orientation* is defined as repetitive focusing on past, present or future states. Several mediating processes are specified in the theory, which are postulated to intervene in a state of action orientation to promote the execution of the action. In the present discussion, the postulated processes of selectively attending to the intended action, and parsimonious information processing in connection with the action are of special interest.

In an experiment Kuhl (1981) found evidence that in a state of action orientation people attend selectively only to information relevant to the execution of the intended action. In contrast to this, in the case of state orientation people also attended to information irrelevant to the execution of the action. In a study by Kuhl and Beckmann (1983) it was demonstrated that in a state of action orientation only part of the available action-relevant information was used by the subjects to decide between the given action alternatives – probably to proceed quickly to a realization of their purpose. In the case of state orientation the whole body of available information was processed.

With regard to the concept of relevance of cognitions in the theory of cognitive dissonance, the definition presented above, including value instrumentality and value importance, has to be extended because of these findings. It has to be assumed that the relative importance of these two factors is affected by action vs. state orientation. In the case of state orientation, cognitions which are not instrumental for the pursued goal will more likely get into the focus of attention than in the case of action orientation. There is a greater probability, therefore, that inconsistencies with competing goals, values, etc. that are involved in the course of an action will become salient in the case of state orientation. An action-oriented person may not be bothered by dissonance caused by inconsistency with values that usually guide action when selecting the means to obtain scarce goods. This is exemplified, for instance, in the case of people who survived in extreme life situations (Cohen, 1953; Biderman, 1967). For people whose mode of control is state orientation, on the other hand, this dissonance may become salient and because of this their range of permissible means to reach the action goal will be narrower. This assumption is supported by the finding of a recent experiment by Beckmann (1984-a, Chapter 4). In this experiment subjects who rated fairness as an important value had to choose between an action alternative that promised high profit but was unfair to other subjects and a fair alternative for which high profit was unlikely. It was found that state-oriented subjects chose the fair alternative most of the time whereas action-oriented subjects tended to choose the unfair, high profit alternative.

The resistance to change of cognitions in a dissonant relationship may also be affected by the processes described in the theory of action control. According to Irle's (1975, 1982) revision of the theory of cognitive dissonance, the resistance to change of cognitions is solely dependent on the number of cognitions the cognition in question is connected to in a consonant relationship and the number of cognitions it is connected to in a dissonant relationship. As a study by Möntmann and Schönborn (1982) shows, the resistance to change is not dependent on the absolute number of such connections. Since human information-processing capacity is limited, attentional processes play a crucial role in determining factual resistance to change. It seems a plausible assumption that processes like selective or parsimonious information processing postulated for the state of action orientation should affect the resistance to change of cognitions in a given situation. If selective processing of action relevant cognitions is considered, it may be assumed that those cognitions related most directly to the action will possess the highest resistance to change in the case of action orientation.

In opposition to this, in the case of state orientation, resistance to change is more likely to depend on an individual's overall cognitive system. This implies that cognitions highly resistant to change on the basis of an individual's overall cognitive system – like charity in the case of the clergymen – will be relatively easy to change if they are dissonant to cognitions bound closely to the pursued action goal and if the actor is action-oriented because of, for example, time pressure. Contrary to this, in the case of state orientation where a larger amount of action-irrelevant cognitions is considered as well, the high resistance to change of the respective cognition (e. g., charity) should become salient.

These assumptions help to explain a number of findings which are contradictory to the theory of cognitive dissonance. Although, as described above, numerous experiments furnished evidence for a decision-stabilizing divergency effect by means of dissonance reduction, in a number of experiments a convergency effect was found. In spite of increasing the discrepancy of attractiveness ratings of chosen and rejected alternatives to secure a maximum of decision stability and by way of this, action control, subjects in these studies actually decreased the discrepancy (Brehm & Wicklund, 1970; Walster, 1964). It may be assumed that the subjects in these experiments turned their attention to the rejected alternatives after the decision. Therefore, the rejected alternatives were not only not decreased in attractiveness, but were even increased in attractiveness.

A study (Beckmann, 1984-b) was conducted in order to test whether dissonance reduction is determined by an action vs. a state orientation. The design chosen for this study was similar to that of Festinger and Carlsmith's (1959) study: In this study it was found that performing a boring task led to increased attractiveness ratings of the experiment for the condition with low incentive (insufficient justification) as compared to the high incentive condition. In the study by Beckmann (1984-b) action vs. state orientation was introduced as an additional factor. Orthogonal to the factor action vs. state orientation which was controlled experimentally, an incentive manipulation was introduced. The first part of the experiment which was described to subjects as a word-learning study consisted of learning ten foreign words. Following this learning phase the experimenter told half the subjects in the insufficient justification condition that because of budget cuts he would not be able to pay them the DM 10.– promised for participation in the experiment. In spite of this, all subjects agreed to continue the experiment when asked to by the experimenter. According to our theoretical considerations, as presented above, it was expected that action-oriented subjects would be able to reduce the dissonance due to the denial of the reward through increasing the attractiveness of the experiment like in Festinger and Carlsmith's (1959) study. Increasing the attractiveness of the experiment would increase the motivation to perform the task. On the contrary, state-oriented subjects should not be able to increase the attractiveness of the experiment because of relatively high perceived resistance to change of this cognition. The results of the experiment supported these assumptions (see Fig. 7.1).

Differences in dissonance reduction have also been found in experiments about cognitive deformation of alternatives prior to a final decision. A number of these studies offer evidence for the assumption that, contrary to Festinger's

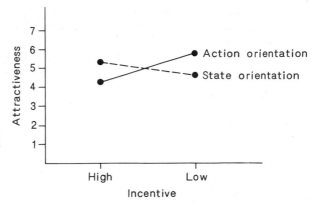

Fig. 7.1. Rated attractiveness of the experiment in high and low (insufficient) incentive condition for action- and state-oriented subjects (Beckmann, 1984-b)

(1964) conviction, information distortion (e. g. Grabitz, 1971 a, 1971 b), respectively, a divergency effect (Mills & O'Neal, 1971; O'Neal, 1971; O'Neal & Mills, 1969) occurs not only following but also prior to a decision and in favour of the intention that has been formed. There are some experiments that seem to support the opposite.

Linder and Crane (1970) and Linder, Wortman, and Brehm (1971) found a convergence of the attractiveness ratings of the alternatives prior to the decision. It can be assumed that because of the concrete situations given in the different experiments, action orientation was instigated in those that showed a divergency effect, whereas state orientation was instigated in those that showed a convergency effect. In the experiments by O'Neal and Mills the subjects had to choose from a number of photographs the ones they liked most. Such a decision situation without consequences for the subject is likely to instigate action orientation (cf., Kuhl & Beckmann, 1983). In the experiments by Linder, Wortman, and Brehm (1971), state orientation was essentially required in order to accomplish the task. The subjects in these experiments were told that they would have to answer a number of intimate questions. Before they chose the interviewer they were told that exact answering of the questions was very important. Mann, Janis and Chaplin (1969) report a divergency effect as a result of their experiment. But to obtain this effect they had to exclude a number of subjects from the data analysis because these people showed decision revisions. It seems reasonable to assume that the attractiveness ratings of these people changed in a convergent direction prior to the decision revision.

Beckmann and Kuhl (1984) conducted an experiment to test if in the case of action orientation a decision-bolstering divergency effect occurs prior to a decision whereas people in a state of state orientation cannot achieve this because they are focussing on the whole body of available information. Action vs. state orientation was controlled before the experimental task. This task consisted of selecting from 16 presented apartments those that one would like to rent. The subjects had to rate the attractiveness of each of the apartments twice. First, a premeasure unrelated to a decision was taken. Before the postmeasure was taken subjects were informed that they would soon be asked to choose the apartment they wanted to rent. It was assumed that action-oriented subjects would change their attractiveness ratings in such a way that a divergency effect would be produced. State-oriented subjects, on the contrary, should rather show a convergence of their attractiveness ratings from the pre- to the postmeasure. These assumptions were supported. The action-oriented subjects increased their attractiveness ratings of those alternatives they chose later and decreased the attractiveness ratings of those alternatives they rejected. State-oriented subjects showed only slight decreases of attractiveness ratings of the chosen ones as well as of those alternatives later rejected. Since the attractiveness of the chosen alternatives was decreased slightly, this may be interpreted as a tendency toward a convergency effect in the case of state orientation (see Fig. 7.2).

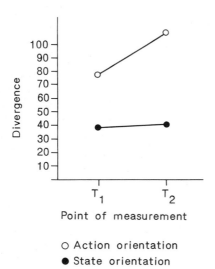

Point of measurement

○ Action orientation
● State orientation

Fig. 7.2. Change of divergence of attractiveness ratings from first to second point of evaluation for action- and state-oriented subjects (Beckmann & Kuhl, 1984)

Conclusion

The way dissonance reduction is discussed in this chapter – as processes that enable an individual to maintain action control when this is endangered by inconsistency – seems to be in contradiction to the established understanding of the theory of cognitive dissonance. Usually the theory is considered a theory about establishing consistency among one's cognitions. As was already stressed, we see no contradiction here in saying that in particular situations a general consistency among one's cognitions may be aimed at. Wicklund and Frey (1981) hold a more radical position here. These authors maintain that dissonance reduction is always employed to support an action goal. An external observer who does not take into account that the selection of the cognitions coming into the focus of attention is determined by a person's current concerns, may wrongly interpret the processes occuring as an increase instead of as a reduction of dissonance since he/she may take the whole body of an individual's cognitions as the basis for this conclusion.

A number of experiments came to be regarded as refutations of the theory of cognitive dissonance (Mills, Aronson & Robinson, 1959; Rosen, 1961). But, there is a specific reason for this: the individual's subjective situation was not taken into account. In the actual experiments the subjects were searching for information that was inconsistent with other cognitions within their cognitive system. However, within the given perspective this inconsistency would become salient only for the experimenter and not for the subjects for whom reaching their action goal was the focus of attention. Canon (1964) was able to show that the inclusion of the given action goal is of crucial importance for the processes involved. As Festinger (1957) postulated, dissonance can only occur between cognitions which are relevant to one another in a given situation. This means that dissonance can only occur for cognitions simultaneously present in the focus of attention. But, as we stated in the present paper, the selection of what gets into the focus of attention is dependent on two factors: a person's current concerns and his/her action or state orientation in the given situation. This consideration seems to be in line with Festinger's initial assumptions, since he states that a "behavioral" element may make two cognitions relevant to each other (1957, p. 12). This aspect seems to have gotten lost in the development of the theory. One of the reasons for this may be that relevance has been increasingly identified as a connection to the self. Thereby, the theory was reduced to a cognitivistic self-theory (Aronson, 1968, 1969; Bramel, 1968). These reductions of the theory apparently arise from the insufficiently exact definition of the theory's antecedent conditons (e.g., the relevance notion). An integration of dissonance reduction into an action perspective implying a specification of the relevance concept will contribute to the solution of this problem without necessarily reducing the contents of the theory. Dissonance reduction may therefore on one hand serve to establish consistency between cognitions within an individual's cognitive system. On the other hand, however, the enactment of an intention may in some cases be facilitated by cognitions which are basically inconsistent with other cognitions within the individual's cognitive system but which still determine his/her behavior. The question then arises: Which of these two tendencies will become dominant in a given situation – the motivational tendency to reach the action goal or the motivational tendency towards consistency of the self? The findings by Mills, Aronson and Robinson (1959), Rosen (1961), and Canon (1964) suggest that the relationship between these two tendencies is similar to the relationship proposed by White (1959) for competence and action motivation. According to White (1959) the action motivation always has priority over the competence motivation (the motivation to establish a consistent self). The latter motivation could only become dominant if no other (action) motivation was strong. As has been stated already, a state-oriented mode of information processing may generate differences here. In the case of state-oriented thinking, the tendency to establish a consistent self may enter into competition with an existing action motivation. It may occupy part of the individual's processing capacity and thereby impair the performance of the action intention.

The discussion of the theory of cognitive dissonance as a theory describing processes of intention control, as presented here, may be of value for theories of action control. First, within the theory of cognitive dissonance, the control strategies are presented in an elaborated model. In contrast to that, the mediating processes in most of the theories of action control are comparably undifferentiated. Second, the assumptions stated in the theory of cognitive dissonance have been confirmed in numerous empirical studies. If researchers concerned with topics within the framework of action control took advantage of this in their investigations they would find that a number of experiments addressing these questions need not to be undertaken since the answers have already been given within dissonance theory research.

References

Ach, N. (1910). *Über den Willensakt und das Temperament*. Leipzig: Quelle und Meyer.

Aronson, E. (1968). Dissonance theory: Progress and problems. In R. P. Abelson, E. Aronson, W. J. McGuire, T. H. Newcomb, M. J. Rosenberg & P. H. Tannenbaum (Eds.) *Theories of cognitive consistency: A sourcebook* (pp. 5–27). Chicago: Rand McNally.

Aronson, E. (1969). The theory of cognitive dissonance: A current perspective. In L. Berkowitz (Ed.), *Advances in Experimental Social Psychology*, Vol. 4, (pp. 115–133). New York: Academic Press.

Aronson, E., & Carlsmith, J. M. (1963). Effect of severity of threat on the devaluation of forbidden behavior. *Journal of Abnormal and Social Psychology, 66,* 584–588.

Beckmann, J. (1984). *Kognitive Dissonanz: Eine handlungstheoretische Perspektive*. Heidelberg: Springer-Verlag. (a)

Beckmann, J. (1984). *Extrinsic vs. intrinsic reduction of cognitive dissonance*. (Manuscript submitted for publication.) Max-Planck-Institute for Psychological Research, Muinch, FRG. (b)

Beckmann, J., & Kuhl, J. (1984) Altering information to gain action-control: Functional aspects of human information processing in decision-making. *Journal of Research in Personality, 18,* 224–237.

Biderman, A. D. (1967). Life and death in extreme captivity situations. In M. H. Appley & R. Trumbull (Eds.) *Psychological stress: Issue in research*. (pp. 242–264). New York: Appleton-Century-Crofts.

Bramel, D. (1968). Dissonance, expectation, and the self. In R. P. Abelson, E. Aronson, W. J. McGuire, T. M. Newcomb, H. M. Rosenberg & P. H. Tannenbaum (Eds.), *Theories of cognitive consistency: A sourcebook* (pp. 355–365). Chicago: Rand McNally.

Brehm, J. W. (1956). Postdecision changes in the desirability of alternatives. *Journal of Abnormal and Social Psychology, 52,* 384–389.

Brehm, J. W. (1962) An experiment on thirst. In J. W. Brehm & A. R. Cohen (Eds.) *Explorations in cognitive dissonance*. New York: Wiley.

Brehm, J. W., & Crocker, J. C. (1962). An experiment on hunger. In J. W. Brehm & A. R. Cohen (Eds.) *Explorations in cognitive dissonance*. (pp. 137–143). New York: Wiley.

Brehm, J. W., & Wicklund, R. A. (1970). Regret and dissonance reduction as a function of postdecision salience of dissonant information. *Journal of Personality and Social Psychology, 14,* 1–7.

Canon, L. K. (1964). Self-confidence and selective exposure to information. In L. Festinger (Ed.) *Conflict, decision and dissonance* (pp. 83–95). Stanford: Stanford University Press.

Chapanis, N. P. & Chapanis, A. (1964). Cognitive dissonance. Five years later. Psychological Bulletin, *61,* 1–22.

Cohen, E. A. (1953). *Human behavior in the concentration camp.* New York: Norton.

Darley, J. H., & Batson, C. D. (1973). From Jerusalem to Jericho: A study of situational and dispositional variables in helping behavior. *Journal of Personality and Social Psychology, 27,* 100–108.

Ehrlich, D., Guttman, J., Schönbach, P, & Mills, J. (1957). Postdecision exposure to relevant information. *Journal of Abnormal and Social Psychology, 54,* 98–102.

Feather, N. T. (1963). Cigarette smoking and lung cancer: A study of cognitive dissonance. *Australian Journal of Psychology, 14,* 55–64.

Fellner, C. H., & Marshall, J. R. (1970). Kidney donors. In J. Macauley & L. Berkowitz (Eds.), *Altruism and helping behavior.* (pp. 269–281). New York: Academic Press.

Festinger, L. (1957). *A theory of cognitive dissonance.* Evanston Ill.: Row, Peterson.

Festinger, L. (Ed.). (1964). *Conflict, decision, and dissonance.* Standord: Stanford University Press.

Festinger, L., & Carlsmith, J. M. (1959). Cognitive consequences of forced compliance. *Journal of Abnormal and Social Psychology, 58,* 203–210.

Freedman, J. L. (1965). Preference for dissonant information. *Journal of Personality and Social Psychology, 2,* 287–289.

Freedman, J. L., & Sears, D. O. (1965). Selective exposure. In L. Berkowitz (Ed.), *Advances in Experimental Social Psychology,* Vol. 2 (pp. 57–97). New York: Academic Press.

Frey, D. (1981). *Informationssuche und Informationsbewertung bei Entscheidungen.* Bern: Huber.

Frey, D., & Irle, M. (1972). Some conditions to produce a dissonance and an incentive effect in a "Forced-compliance" situation. *European Journal of Social Psychology, 2,* 45–54.

Grabitz, H. J. (1971). Zur Beziehung von Inertia-Effekt und sequentieller Position widersprechender Ereignisse bei der Revision subjektiver Wahrscheinlichkeiten. *Psychologische Forschung, 35,* 35–45. (a)

Grabitz, H.-J. (1971). Die Bewertung von Information vor Entscheidungen in Abhängigkeit von der verfolgten Alternative und der Verläßlichkeit der Information. *Zeitschrift für Sozialpsychologie, 2,* 383–388. (b)

Irle, M. (1975). *Lehrbuch der Sozialpsychologie.* Göttingen: Hogrefe.

Irle, M. (Ed.). (1982). *Decision making: Social psychological and socioeconomic analyses.* Berlin/New York: De Gruyter.

Kelly, G. (1962). Comments on J. Brehm, Motivational effects of cognitive dissonance. In M. D. Jones (Ed.), *Nebraska Symposium on Motivation 1962* (pp. 78–81). Lincoln, NE: University of Nebraska Press.

Kiesler, C. A. (1971). *The Psychology of Commitment.* New York: Academic Press.

Klinger, E. (1971). *Structure and functions of fantasy.* New York: Wiley.

Klinger, E., Barta, S. G., Mahoney, T. W., et al. (1976). Motivation, mood, and mental events. Patterns and implications for adaptive processes. In G. Serban (Ed.), *Psychopathology of human adaptation* (pp. 95–112). New York: Plenum.

Kuhl, J. (1981). *Aufmerksamkeitslenkung und Handlungskontrolle.* (Unpublished manuscript.) Ruhr-Universität Bochum, FRG.

Kuhl, J. (1982). Action vs. state-orientation as a mediator between motivation and action. In W. Hacker, W. Volpert & M. von Cranach, M. (Eds.), *Cognitive and motivational aspects of action.* Amsterdam: North-Holland-Publishing Co.

Kuhl, J. (1984). Volitional aspects of achievement motivation and learned helplessness: Toward a comprehensive theory of action-control. In B. A. Maher (Ed.), *Progress in Experimental Personality Research.* Vol. 13, New York: Academic Press.

Kuhl, J., & Beckmann, J. (1983). Handlungskontrolle und Umfang der Informationsverarbeitung: Wahl einer einfachen (nicht optimalen) Entscheidungsregel zugunsten rascher Handlungsbereitschaft. *Zeitschrift für Sozialpsychologie, 14,* 241–250.

Lewin, K. (1935). *A dynamic theory of personality: Selected papers.* New York: McGraw-Hill.

Lewin, K. (1951). *Field theory in social science.* New York: Harper and Row.

Lewin, K. (1952). Group decision and social change. In G. Swanson, T. Newcomb & E. Hartley (Eds.), *Readings in Social Psychology* (pp. 197–211). New York: Holt, Rinehart & Winston.

Linder, D. E., & Crane, K. A. (1970). A reactance theory analysis of predecisional cognitive processes. *Journal of Personality and Social Psychology, 15,* 258–264.

Linder, D. E., Wortman, C. B., & Brehm, J. W. (1971). Temporal changes in predecision preferences among choice alternatives. *Journal of Personality and Social Psychology, 19,* 282–284.

Mann, L., Janis, I. L., & Chaplin, R. (1969). Effects of anticipation of forthcoming information on predecisional processes. *Journal of Personality and Social Psychology, 11,* 10–16.

March, J. G., & Simon, H. A. (1958). *Organizations.* New York: Wiley.

Miller, G. A., Galanter, E., & Pribram, K.-H. (1960). *Plans and the structure of behavior.* New York: Holt, Rinehart & Winston.

Miller, N. E. (1944). Experimental studies of conflict. In J. M. V. Hunt (Ed.), *Personality and the behavioral disorders* (Vol. I) (pp. 431–465). New York: Ronald Press.

Miller, N. E. (1951). Learnable drives and rewards. In S. S. Stevens (Ed.), *Handbook of experimental psychology* (pp. 435–472). New York: Wiley.

Mills, J., Aronson, E., & Robinson, H. (1959). Selectivity in exposure to information. *Journal of Abnormal and Social Psychology, 59,* 250–253.

Mills, J., & O'Neal, E. (1971). Anticipated choice, attention, and the halo effect. *Psychonomic Science, 22,* 2331–233.

Möntmann, V., & Schönborn, C. (1982). *Operationale und theoretische Probleme zum Verlauf der Änderungsresistenz von Kognitionen.* Manuscript, University of Mannheim, FRG.

O'Neal, E. (1971). Influence of future choice importance and arousal upon the halo effect. *Journal of Personality and Social Psychology, 19,* 334–340.

O'Neal, E., & Mills, J. (1969). The influence of anticipated choice on the halo effect. *Journal of Experimental Social Psychology, 5,* 347–351.

Rosen, S. (1961). Postdecision affinity for incompatible information. *Journal of Abnormal and Social Psychology, 63,* 188–190.

Rosenberg, G. J. (1956). Cognitive structure and attitudinal affect. *Journal of Abnormal and Social Psychology, 53,* 367–372.

Sears, D. O. (1965). Biased indoctrination and selectivity of exposure to new information. *Sociometry, 28,* 363–376.

Songer-Nocks, E. (1976). Situational factors affecting the weighting of predictor components in the Fishbein model. *Journal of Experimental Social Psychology, 12,* 59–69.

Walster, E. (1964). The temporal sequence of post-decision processes. In L. Festinger (Ed.), *Conflict, decision and dissonance* (pp. 112–127). Stanford: Stanford University Press.

White, R. W. (1959). Motivation reconsidered: The concept of competence. *Psychological Review, 66,* 297–333.

Wicklund, R. A., & Brehm, J. W. (1976). *Perspectives on cognitive dissonance.* Hillsdale, NJ: Erlbaum, 1976.

Wicklund, R. A., & Frey, D. (1981) Cognitive consistency: Motivational vs. non-motivational perspectives. *Social Cognition, 1,* 141–163.

Younger, J. C., Walker, L., & Arrowood, A. J. (1977). Postdecision dissonance at the fair. *Personality and Social Psychology Bulletin, 3,* 284–287.

Chapter 8
Action Control and the Coping Process

Claudia Herrmann and Camille B. Wortman *

For the past several years, Wortman and her associates have been studying how people react to undesirable life events, such as permanent paralysis, chronic illness, criminal victimization and loss of a loved one (e.g. Bulman & Wortman, 1977; Wortmann & Dunkel-Schetter, 1979; Silver & Wortman, 1980). We have examined such variables as victims' emotional reactions to the crisis, their attributions of blame for what has happened, their ability to find meaning in the crisis, and the support they receive from others. Our research has focused on how these variables influence one another and change over time as the coping process unfolds. We have also studied the impact of these variables on the victim's long-term adaptation to the event. In this chapter, an attempt is made to broaden the focus of this past work. On the basis of the theory of action control that has been introduced recently by Kuhl (1981, 1983 a, 1983 b, 1984), we examine the interplay between people's reactions to an undesirable life event and the performance of behaviors or actions.

In the past, researchers have shown a great deal of interest in the process of how people cope with undesirable life events (see Silver & Wortman, 1980, for a review). Some of this work has focused primarily on peoples' *emotional reactions* to life crises. Investigators have attempted to determine what types of emotions are experienced by victims of life crises, and whether people go through particular stages or sequences of emotional reactions, such as shock, anger, depression, and acceptance, as they attempt to cope with the crisis (see Silver & Wortman, 1980, for a review and critique). More recently, investigators have become interested in the relationship between specific emotional reactions and successful long-term adjustment, and in whether those who express their emotions show better long-term adjustment than those who inhibit emotional expression (see Wortman, 1983, for a discussion).

Other researchers have focused primarily on the *cognitions* that are experi-

* The authors want to thank Julius Kuhl for his helpful comments on an earlier draft of this paper.

enced by victims of life crisis, especially their attributions of causality and blame for what has happened (see, e.g., Abramson, Seligman, & Teasdale, 1978; Herrmann, 1981, 1984; Silver, Wortman, & Klos, 1982). This focus has been especially prominent in the discipline of social psychology. Investigators have attempted to explore why victims of seemingly uncontrollable events react with feelings of self-blame and whether feelings of blame are adaptive or maladaptive in coming to terms with the crisis (see, e.g., Bulman & Wortman, 1977; Janoff-Bulman, 1979; Miller & Porter, 1983; Wortman, 1976; Wortman, 1983). In recent years, investigators have begun to broaden this focus to include other cognitions such as attempts to find meaning in the victimizing experience (e.g., Bulman & Wortman, 1977; Silver, Boon, & Stones, 1983), and attempts to minimize one's distress by making comparisons with less fortunate others (e.g., Taylor, Wood, & Lichtman, 1983).

Although considerable research has been conducted on victims' emotional reactions and cognitions, surprisingly little attention has been paid to their overt behavior. It has been noted that a victimizing experience, such as being raped or losing one's spouse, often interferes with peoples' performance of valued activities in work and family roles (see, e.g., Silver & Wortman, 1980; Wortman, 1983). It has also been suggested that victimization can shatter one's assumptions about the world, and thereby interfere with subsequent goal-directed behavior (Wortman, 1983). In general, however, investigators of the coping process have focused primarily on victims' *reactions* to the crisis rather than on goal-directed *actions*.

Some theoretical formulations dealing with the initiation and maintenance of goal-directed behavior have appeared in the literature (see, e.g., Miller, Galanter, & Pribram, 1960; Hacker, 1973; Volpert, 1974, 1982; von Cranach, Kalbermatten, Indermühle, & Gugler, 1980). However, the issues raised by these theories have not been incorporated into psychological research on coping with undesirable life events.

In recent years, a theory of action control has been proposed by Kuhl (1981, 1983a, 1984) that may have important implications for the process of coping with undesirable life events. This theory discusses the conditions under which cognitions and emotional reactions may inhibit the performance of intended actions. This model has been applied to behavior in both laboratory (Kuhl, 1981; Heckhausen & Kuhl, in press) and field (Kuhl, in press) settings, and respresents a promising theoretical approach for examining the process of coping with life crises.

Coping with Life Crises: The Gap Between Theory and Practice

The focus of this chapter is to bring two areas of research together by examining the process of coping with life crises and Kuhl's theory of action control (1981, 1983a, 1984) in relation to one another. First, we consider whether people who have experienced undesirable life events may have difficulties in carrying out goal-directed activities. We then review several theoretical explanations that may

account for the failure to enact one's intentions and engage in goal-directed behavior. Only Kuhl's (1981, 1983 a, 1984) theory can account for instances where
people fail to perform important actions despite the ability and motivation to do
so. Next, we provide a brief summary of the theory of action control as it relates
to victims of undesirable life events. We then explore how life crises are likely to
effect the performance of goal-directed actions, and consider whether action orientation is adaptive or maladaptive for successful coping with an undesirable
life event.

In the concluding section of the chapter, we explore ways in which a joint consideration of work on action control and the coping process might enrich both of
these domains. By applying the action control model to those who have experienced life crisis, it might be possible to determine the model's boundaries of applicability, and to delineate new areas where further theoretical elaboration
might be helpful. Similarly, a consideration of coping research in light of the action control model highlights new areas of potential importance, and suggests
some intriguing issues for subsequent research.

Undesirable Life Events and the Enactment of Intentions

When an individual experiences an undesirable life event, such as bereavement,
life-threatening illness, criminal victimization, or physical disability, the psychological impact can be overwhelming. An environment that was previously tolerable has now become unpredictable and threatening. One's hopes, plans, and
dreams for the future may be fundamentally altered. The enactment of many former behaviors or roles may no longer be appropriate or possible. The profound
changes that frequently accompany life crises may trigger intense anxiety and
distress. Against the backdrop of this distress, the victim may be faced with a variety of coping tasks which demand immediate action. In the case of widowhood, for example, such matters as funeral arrangements, insurance settlements,
taxes, concerns about future livelihood and care for children must be handled
along with a piercing sense of loss. In short, most undesirable life events are accompanied by profound disruption, required behavioral changes, and competing demands for action. In these cases, the enactment of goal-directed behaviors
might create extreme difficulties for victims. Let us illustrate with two examples.

Jim has recently become paralyzed in a car accident. He realizes that many
tasks that were previously routine, such as getting dressed or moving around the
house, require physical capacities and skills that he no longer possesses. He is
forced to admit that he has become quite dependent on other people's help. This
presents a major problem for him, because he has always viewed himself as a
very independent person. He wants to overcome this problem and plans to ask
people for assistance more often. However, he finds himself in many situations
where he needs help, but does not ask for it.

Sandra is told by her doctor that she is terminally ill with cancer and might not
live for more than a few months. After she has accepted this diagnosis as valid,
she decides to spend her remaining time doing more of the things that she en-

joys. She makes a list of activities that she intends to pursue. Yet she finds herself thinking about how unfortunate it is that she does not have more time left to enjoy these activities. The days pass and Sandra becomes increasingly depressed. She does not do any of the activities that she has planned.

Jim and Sandra do not succeed in doing what they consider to be important for coping successfully with their current situation. Their cases are not unique. The question of why people do not do what they plan to do may be of extreme importance when they are facing a life crisis. In fact, failure to engage in those behaviors which one intends to accomplish may be even more detrimental during a life crisis than during normal periods of life. Clearly, it is important to learn more about the conditions under which those facing life crises may have difficulties in sustaining goal-directed activities. In addition to its theoretical relevance, this knowledge might be useful in the design of interventions that would help victims to enact their intentions, and thereby enhance the likelihood of successful coping. It is surprising that this issue has not received more attention from coping researchers.

There are a number of theories that have focused on the performance of goal-directed actions. In the following section we briefly describe some of them and examine how they explain the failure to perform intended actions. We then explore their possible relevance to the situation faced by those who have experienced life crises.

Theoretical Explanations for the Incongruence Between Intention and Action

According to expectancy-value theories of motivation (e.g., Atkinson & Birch, 1978; Heckhausen, 1980), the performance of an action depends upon the strength of the motivational tendency to perform it. Thus, if a rape victim decides that she wants to start going out again at night, but ends up staying at home, these theories would ascribe it to a lack of motivation to go out. Alternatively, they would assume that a competing motivation, such as the need for security, is stronger than the motivation for social contact.

Behavior theorists (e.g., Kanfer, 1970) maintain that the performance of an action is contingent on the presence of the relevant stimulus and reinforcement conditions. According to this theoretical approach, Sandra in our example above might fail to pursue her plans because her intended activities are under insufficient stimulus and reinforcement control. Another behavioral model explains the failure to perform intended actions as stemming from a lack of the necessary skills (e.g., Lewinsohn, 1974). For instance, Jim's paralysis requires him to engage in help-seeking behaviors which might involve skills that he has not developed. A similar rationale is provided by the theory of self-efficacy (Bandura, 1977). According to this theory, the likelihood of performing an intended action is dependent on the perception of one's own ability (self-efficacy). Thus, Sandra and Jim might not perform their intended actions because they believe that they might fail to accomplish them successfully.

Each of these theoretical approaches provides a possible explanation of the incongruence between behavioral intentions and the performance of actions. However, none of them can account for those instances in which people do not perform the task in question, although both high motivation and high ability are present. This specific case is addressed by the theory of action control (Kuhl, 1981, 1983 a, 1984).

The Theory of Action Control

Kuhl (1981, 1983 a, 1984, in this volume) has proposed a theory of action control that specifies the processes which intervene between intentions and the performance of the intended behavior. The likelihood of the performance of an action is said to be determined by the presence of two tendencies: action orientation and state orientation, which are conceptualized as two extremes on a continuum. A person is called action oriented when his or her attention is simultaneously or successively focused on (a) some aspect of the present state, (b) some aspect of a future state, (c) a discrepancy between the present and future state, and (d) at least one action alternative that can eliminate the discrepancy. If one of the four cognitive elements is missing, the person is said to be state oriented. Thus, state orientation exists if a person ruminates about the present state without taking account of the future. Even if people focus on both the present and the future, they are state oriented unless they attend to the discrepancy between the two states and focus on action alternatives that might overcome the discrepancy.

According to the theory, a high degree of action orientation directs and controls cognitive and affective processes that facilitate the enactment of intentions. Because it involves repetitive and dysfunctional focusing on fixed aspects of the situation, state orientation impedes the performance of intentions. Kuhl (1981, 1983 a, 1984) maintains that the degree of a person's action control is jointly determined by dispositional and situational parameters. He maintains that some people have a dispositional tendency to ruminate about fixed aspects of their situation, while others characteristically focus on actions that will facilitate their desired state. The situational determinants of action control that have received the most attention are the unpredictability and the uncontrollability of the outcome (Kuhl, 1983 a, 1984).

Action Control and the Coping Process

In the following paragraphs we will examine the interface between action control and coping with undesirable life events.

Effects of Undesirable Life Events on Action Control

Can undesirable life events be characterized as situations that influence the likelihood of the enactment of intentions? When an undesirable life event is experienced, is action orientation or state orientation more likely to be induced in the
victim? In the following section we will attempt to illustrate why stressful life experiences often bring about a period of state orientation.

Surprise

An unexpected event, or surprise, is likely to induce state orientation. Kuhl
(1984) maintains that the confrontation with unexpected information leads to
the inhibition of physiological and cognitive processes that are necessary for the
initiation of actions. A characteristic feature of many undesirable life events is
that they are sudden and unexpected. According to Kuhl's (1981, 1983a, 1984)
theory, such events are likely to induce state orientation.

   There is some evidence that when life events are sudden and unexpected, they
are particularly likely to be associated with the repetitive and dysfunctional
thought patterns that are characteristic of state orientation. In their study of bereavement, Parkes and Weiss (1983) have found that those who had "short forewarning" (less than two weeks notice their loved one was going to die) seemed
less able to put the death behind them, and to move on to the next stage of their
lives. At an interview conducted 2–4 years after the loss, 61% still asked themselves why the death had happened, 61% occasionally or always sensed the presence of the dead person, and 44% agreed with the statement, "It's not real; I feel
that I'll wake up and it won't be true." Such thoughts were much less prevalent
among respondents who had greater forewarning about their spouse's death. Interestingly, only one of the respondents in the short forewarning group was
judged by coders to be coping well or very well with the loss; 63% of respondents
in the long forewarning group were judged as doing well or very well. According
to Parkes and Weiss (1983), unexpected and untimely losses have this effect because they transform the world into "a frigthening place ... Where there has
been security, there is now anxiety ... (by repeatedly thinking about the deceased spouse), the widow or widower can maintain a sense of an ongoing relationship with him or her and from this sense of continuity derive feelings of security and continued meaning." (pp. 245–246).

Degenerated Intentions

After an initial disorientation due to the unexpectedness of the undesirable life
event, longer phases of state orientation might be activated by another aspect of
the life crisis. The changes faced by the victim might create so-called "degenerated intentions" Kuhl (1984) refers to intentions that persist in spite of the fact
that they cannot be carried out or are no longer appropriate as degenerated intentions. These degenerated intentions are said to stem from the inability to relinquish unattainable goals. These old plans take up storage capacity which

would be needed for the enactment of new intentions. If they persevere, they interfere with the proper execution of the new intentions.

There is considerable evidence that victims of undesirable life events often engage in behaviors directed toward recovering the lost object, even when the loss is permanent and such goal-directed actions would seem to serve little purpose. This phenomenon has been frequently noted in the bereavement literature, where the urge to recover the lost object is expressed in a variety of "searching behaviors," including:

"restless scanning of the environment, in the strong perceptual set which the bereaved person develops for the lost person, in the tendency to misperceive and thus to identify strangers as being the one who ist lost, in the tendency to return to places associated with the lost person and to treasure objects associated with him . . ." (Parkes, 1972b, p. 345).

As one widow expressed, "I can't help looking for him everywhere . . . I walk around searching for him . . . I go to the grave . . . but he's not there. It's as if I was drawn towards him . . ." (Parkes, 1972a, p. 44). According to Parkes (1972a), the bereaved person recognizes that searching is irrational, but nonetheless experiences a strong desire to engage in this type of behavior. Cornwell, Nurcombe, and Stevens (1977) have found searching behavior to be quite common among parents who experienced the sudden death of their infant. Alertness for the sight of babies was an enduring phenomenon. According to the authors, "parents remained unconsciously primed to search for their lost baby for well over a year after its death" (p. 657). Even those who have lost a limb seem to experience a strong urge to recover the lost object. Parkes (1972b) has reported that one third of the amputees studied admitted that they sometimes looked to see if their limb was, after all, present. Two thirds indicated that they had given some thought to the question of what had been done with their limb after it had been amputated.

Although this phenomon has not been investigated in detail, there is some reason to suspect that searching behaviors may interfere with goal-directed activities. As Parkes (1972a) has suggested.

"when the mind is occupied with so important a task it cannot easily concentrate on another. The total preoccupation with the business of searching leaves no room for other interests, and the newly bereaved people I have met showed little concern for food, sleep, personal appearance, work, or family." (p. 50).

Another example of degenerated intentions among victims of life crises is the tendency to persevere in the performance of role behaviors that are no longer appropriate. For example, Parkes (1972a) has noted that widows frequently continue to set the table for two, and to shop for foods that their husbands would have liked. Similarly, Cornwell et al. (1977) report that many mothers momentarily forget the baby's death, and experience a strong desire to go about their mothering tasks. Some mothers go about such activities as heating bottles in response to an illusion that they had heard the baby crying. Parkes (1972b) has suggested that these behaviors are residues of old working models which need to be replaced or modified. He maintains that change is brought about through repeated and painful disappointments as the victim recognizes that his or her world is fundamentally altered. According to Parkes (1972b), until this change is achieved, „his ability to control, predict and plan his life in a meaningful and satisfying way is reduced" (p. 348).

State Oriented Cognitions

Earlier, we noted that there has been considerable interest in the cognitions and ruminations that are experienced by victims of undesirable life events. Are the cognitions that typically occur in response to life crises indicative of state or action orientation? Cognitions are action oriented only when they focus on alternatives for reducing the discrepancy between a present and intended future state. In many cases, the cognitions experienced by victims of life crises involve the repetitive and possibly dysfunctional focusing on fixed elements of their situation.

Some theorists (e. g., Klinger, 1975, 1977) stress that it is very common for victims of undesirable life events to engage in thoughts about the event. Klinger has noted that immediately after a goal has been blocked, it completely dominates a person's thought content. If the victim keeps focusing on the event without at the same time focusing on action possibilities which would help to achieve the desired future state, these action-irrelevant thoughts may hinder the performance of intended actions.

There is clear evidence from the literature that victims of undesirable life events experience intrusive and repetitive thoughts about what has happened (see, e. g., Horowitz, 1976, 1979; Janis 1971). For example, Parkes and Weiss (1983) have reported that following the loss of a spouse, approximately two-thirds of the respondents experienced a strong yearning for the deceased and a preoccupation with thoughts about him or her. These thoughts may focus on the lost person, or may constitute a review of the events leading up to the loss. As one widow expressed it, "I go through that last week again and again" (Parkes, 1972a, p. 48). Similarly, among parents who experienced the sudden loss of their infant, for example, it was common to ask such questions as, "What did I do during the pregnancy to cause this? What didn't I do? What should I have done differently?" (Helmrath & Steinitz, 1978). These thoughts are often accompanied by visual memories of what has happened. Rees (1971) has reported that a large majority of the bereaved experience visual and auditory hallucinations of the deceased. Even many years after the bereavement, a large proportion of widows and widowers still experience a strong sense of the presence of the lost person (see also, Parkes, 1972b). In general, the memories are striking in their clarity and vividness. As one widow expresses, "I still see him, quite vividly, coming in the door." or "I can almost feel his skin or touch his hands" (Parkes, 1972a, p. 48).

There is also evidence to suggest that these recurrent thoughts may persist for a considerable length of time. Parkes and Weiss (1983) report that approximately half of their respondents continued to experience a frequent preoccupation of thoughts about the deceased 13 months following the loss of their spouse. A survey conducted by Harris revealed that 61% of the 1168 Vietnam veterans in their sample still think about the death and destruction that went on in Vietnam (Williams, 1983). Finally, Silver, Boon, and Stones (1983) report that among women who were victims of father-daugther incest an average of 20 years earlier frequent ruminations about the incest experiences are still commonly experienced.

In short, victims seem to spend a substantial amount of time ruminating about what has happened to them. These ruminations can be characterized as state-oriented coping attempts. As we will discuss below, repeated contemplations of the event may or may not have adaptive value for long-term coping. In the short run, however, the victim who engages in such cognitions is unlikely to perform intended actions, except for well-learned routine tasks and actions that are externally controlled (Kuhl, 1981, 1983a, 1984).

Vacillating State Cognitions

Kuhl (1984) refers to the excessive appraisal of action alternatives as "vacillating state cognitions." This type of state orientation is generated when the individual cannot decide what he or she should do or wants to do in order to reach a desired goal.

Undesirable life events often confront the victims with difficult decisions concerning their future course of life. In many cases, a life crisis may render former goals inappropriate, and may require a person to reevaluate and to redefine these goals. A man who planned a career in professional sports, for example, would have to decide on a different vocation if a serious accident made his original career plans impossible to achieve. Such a process of redefinition may be challenging and time-consuming. Once new goals are identified, the victim may have little information or prior experience to guide him in determining which action alternatives will be functional for reaching these goals. In other cases, an undesirable life event may thrust the victim into a new life role where the immediate goals and the plans for reaching them are unclear. For example, a widow may be required to make complex decisions about her husband's business affairs. In still other cases, people might not know which goal-directed actions will be most likely to reduce their distress. For instance, a woman who has a bad marriage may be uncertain whether to initiate divorce proceedings or to work toward improving the relationship, and she may engage in thoughts about the different possibilities. In all these cases, state orientation will result if the victim is uncertain about which actions should be undertaken and keeps on dwelling on the possible choices.

Emotions

The degree of action control is said to be determined by the presence of action-facilitating affect (Kuhl, 1981, 1983a, 1984). However, the theory of action control does not make any statements about *which* affects lead to action orientation and which lead to state orientation.

Anxiety, depression, and anger are described in the literature as the most prevalent reactions to undesirable life events (c.f., Silver & Wortman, 1980). We assume that all of these different emotional reactions to undesirable life events can induce state orientation by mediating cognitions which are centered around the past (e. g., anger that is focused on the injustice of an accident caused by a drunk driver), the present (e. g., grief about the loss of a loved one), or the future (e. g., fear that one could be raped or attacked again) and not focused on the action alternatives for reducing the discrepancy between the present and the intended future state. Moreover, emotional distress and state-oriented cognitions may well escalate one another. In discussing victims of incest, Silver et al. (1983) have noted that thoughts and painful memories about the past often produce emotional pain. Similarly, increased distress is likely to trigger a review of what has happened in order to gain some perspective or find some meaning in what has happened (see also Singer, 1978).

Conclusion

We can conclude that the majority of undesirable life events induce cognitions and emotions that are likely to reduce the victims' ability to enact the intentions that they consider to be adaptive. However, is the assumption that life events are likely to make their victims state oriented compatible with the models of coping which are described in the literature?

The theoretical formulations about how people react to undesirable life events are extremely diverse (c. f. Silver & Wortman, 1980). Some models suggest that victims first go through a stage of increased activity which is then followed by passivity as they attempt to cope with the aversive outcome. For instance, Klinger (1975, 1977) describes "an incentive-disengagement cycle" in which a person initially reacts with greater effort when a particular goal is removed or blocked. If these coping reactions are unsuccessful, the person finally gives up and becomes depressed. The individual is then said to become disengaged from his incentives. A similar sequence, one of reactance-depression, is described by Wortman and Brehm (1975).

On a superficial level, these stages of invigoration and increased attempts to alter one's fate seem to be indicative of action orientation. However, none of the studies on coping have investigated whether the conditions for action orientation were present. Therefore, we do not know if actions performed by victims are the result of high action control. It is possible that the activities pursued in a state of invigoration are not so much guided by rational cognitive processes, but rather by external factors, such as situational demands and constraints and other people's reactions and expectations. For instance, a bereaved wife might perform all the socially desirable and necessary actions for the arrangement of the funeral, because the execution of these actions is facilitated by external sources of control. However, she might not enact her intention to take a walk in order to relax unless this behavior is encouraged by a friend or under some other kind of external control.

In short, an examination of the coping literature suggests that undesirable life events are likely to make victims state oriented and that phases of increased activities after the exposure to an undesirable life event are not necessarily an indication of high action control.

Significance of Action Control for Successful Coping

As we have discussed above, victims of undesirable life events are, at least during some stages of the coping process likely to be state oriented. In this condition, cognitive activities do not facilitate goal-directed actions. On the contrary, the enactment of intentions will be impeded. What impact is this likely to have on the victim's ability to cope successfully with the crisis? Because the optimization of actions may be very important in a crisis situation, it might be argued that state orientation will have detrimental consequences. On the other hand, it is possible to suggest that when a life crisis occurs, action orientation may not always be desirable, appropriate, or even possible. Below, we examine this issue in more detail, and attend to the possible negative and positive consequences of state orientation.

Negative Effects of State Orientation

If a person is in a prolonged phase of state orientation, many actions that are important to him or her and which might be crucial for good adjustment will not be performed. This can have several consequences that may impede the successful resolution of the crisis.

First, the experience of repeated failure to enact one's intentions can lead to a "secondary" helplessness effect (cf. Abramson et al. 1978). It is important to keep in mind that any life crisis can be conceptualized as a series of discrete coping tasks. In trying to cope with the situation as it unfolds, the victim may develop intentions regarding a wide range of behaviors and activities. For example, a widow may intend to get rid of her husband's clothes, to seek out a consultant for adivce regarding financial matters, and to invite a friend over for companionship. If they find themselves in a protracted phase of state orientation, victims may be exposed to repeated failures to enact their intentions. Particularly if victims attribute this failure to personal factors, such as incompetence, they may experience the cognitive and motivational deficits that are characteristic of learned helplessness (Abramson et al., 1978). Because such failures to do what one had planned to do may occur repeatedly, attributions to personal inadequacy (i. e. internal, stable and global attributions according to the terminology of Abramson et al., 1978) may be especially likely to occur. For this reason, these failures may gradually erode the victim's self-esteem, and diminish feelings of personal effectiveness and mastery.

State orientation might also result in the reduction of behaviors that elicit positive reinforcement. Several theorists (e. g., Ferster, 1973, 1974; Lewinsohn, 1974) have maintained that lack of behavior-contingent positive reinforcement is a major factor in causing and maintaining depression. Moreover, by not engaging in activities that are enjoyed or valued, the victims may become trapped in a vicious circle in which their failure to perform valued activities aggravates depressive symptoms and undermines the motivation to enact subsequent intentions (Kuhl 1983b).

Earlier, we presented evidence to suggest that victims frequently experience intrusive thoughts or ruminations about the traumatic event. In many cases, the repeated ruminations that are characteristic of state orientation may be deeply troubling to the victim. For example, a widow may focus repeatedly on some unkind words that she said to her husband shortly before his death (Parkes & Weiss, 1983). Vietnam veterans may have intrusive thoughts about horrific battle scenes involving the death and mutilation of close friends or civilians (Williams, 1983). In order to ease the pain of these thoughts, the victim may turn to coping mechanisms that impede long-term resolution of the crisis. For example, Williams (1983) has explained how Vietnam veterans deal with their distress by becoming numb or "emotionally dead" (p. 13). Some men supposedly believe that if they allow themselves to experience feelings, "they may never stop crying or may completely lose control of themselves" (p. 13). According to Williams, this emotional distancing can lead to the failure to achieve the intimacy necessary to sustain close relationships. A similar emotional numbing has been noted among the bereaved (cf. Parkes & Weiss, 1983), who frequently withdraw from all social contact following the loss of their spouse. Clearly, this type of emotional numbing, which may occur as a result of state orientation, can lead victims to "go through life with an impaired capacity to care for or even love others – and suffer true estrangement and alienation" (Williams, 1983, p. 13). Other strategies used to buffer the pain and anxiety associated with unpleasant ruminations about the past, such as increased alcohol and drug use, may also impair long-term coping. As Lacoursiere, Godfrey, and Ruby (1980) have noted, symptom reduction by means of alcohol or drugs may lead to tolerance and an increased consumption of these substances, which may then lead to problems caused by the alcohol and drugs themselves.

Another potential problem with state orientation is that while in this state, victims may fail to perform an important action at a crucial point in time. In many cases, the availability of an action alternative depends on the performance of a former action. For instance, consider the case of a woman who has lost her job. If she puts sufficient effort into searching and applying for a new position, she might be successful in obtaining one. However, if she is in a phase of state orientation (for example, if she dwells on the injustice of having been fired), she might fail to make the efforts that are necessary for getting another job offered. The longer she remains unemployed, the more difficult it may be for her to make a favorable impression on prospective employers. The continuing unemployment might undermine her self-esteem, confront her with new problems (for example, how to pay the rent), and exacerbate her situation in numerous other ways.

Prolonged periods of state orientation might also impair long-term adjustment by reducing the victim's options for obtaining subsequent resources. For example, state orientation may decrease the likelihood that the victim will engage in behaviors that initiate or sustain social relationships. Earlier, we pointed out how the bereaved may show reduced interest in social activities. This may gradually undermine the availability of the victim's social resources, and minimize the victim's chances of receiving effective social support from others.

Thus, continuing phases of state orientation can easily create a vicious circle for the victim: The failure to perform certain actions will reduce the probability or limit the possibility of performing later ones, because their existence depends on the performance of the former ones.

Is there any experimental evidence that when confronted with an undesirable life event, victims who are state oriented are less likely to show effective long-term adjustment than those who are action oriented? To date, we have only been able to locate a single study that provides a test of this hypothesis (Kuhl, 1983 b). A scale designed to measure disposition toward action control was administered to patients who had undergone a hernia operation. Patients who were classified as state oriented reported significantly more pain and requested twice the amount of analgesics as those who were classified as action oriented. In addition, action oriented patients engaged significantly more often in change oriented activities than state oriented patients. For example, the action oriented individuals practiced moving their limbs, made plans for the time after their discharge from the hospital and engaged in such activities as reading and talking. State oriented patients were reported to be more passive and contemplative. They engaged in such activities as looking at their wounds and thinking about the operation.

We know of no other research on reactions to life crises in which the respondent's degree of action control has been assessed. However, a number of studies provide suggestive evidence that ruminating about the loss, or the victimizing experience, is associated with poor long-term coping. In their longitudinal study of bereavement, Parkes and Weiss (1983) found that preoccupation with thoughts about the deceased evidenced during their initial interview (conducted three weeks after the loss) was associated with an overall assessment of poor adjustment at 13 months after the loss, and at the final follow-up, which was conducted two to four years later. At the final interview, more than half of the respondents who were preoccupied with thoughts of their spouse at the initial interview expressed agreement with the statement, "I wouldn't care if I died tomorrow," and 56% showed moderate to severe anxiety during the interview. (Among these who had not been preoccupied with thoughts of the deceased at the interview, percentages were 17% in each case). Similarly, Silver et al. (1983) found that many of the incest victims they studied continued to think about what had happened, and to search for some meaning in the experience years after its termination. The more frequent the respondent searched for meaning, perhaps indicative of a state orientation, the more psychological distress they reported (see also Frey, Havemann, & Rogner, 1983). Of course, it should not be inferred from these studies that excessive ruminations cause subsequent coping failure. Certain types of life crisis may be especially likely to induce state orientation, and also to present difficulties in coping. This is an issue that will be explored in more detail in a subsequent section of the chapter.

Positive Effects of State Orientation

A phase of state orientation can also have a positive impact on the coping process (Kuhl, 1981, 1984). In some cases, state orientation may be helpful in developing a full understanding of the crisis, and in formulating subsequent plans which lead to a resolution of the crisis. For example, full analysis of what caused a marital relationship to deteriorate may impede action in the short run, but may constitute a necessary first step in the slow process of rebuilding that relationship.

There is also the possibility that a period of state orientation can help a victim come to terms with a crisis for which no solution is possible. For example, a period of state orientation may help the bereaved reach a point where they no longer feel the need to avoid reminders of the loss, for fear of being flodded by grief and pain. Many psychoanalytic theorists, beginning with Freud (1914, 1920), have maintained that the repeated reviewing of an experience stems from a need to gain mastery over the event in question (see also Horowitz, 1976). Janis (1971) has proposed a similar notion – "worry work" – in which various thoughts about a potential crisis are mentally rehearsed to facilitate good coping.

By repeatedly focusing on the victimizing experience, it may be possible for the person to achieve a degree of understanding that would not have been possible if he or she had avoided or blocked thoughts about what happened. For example, Silver et al. (1983) report that as a result of reviewing their experiences again and again, some victims of incest were able to make sense of what happened. For example, many of the women they studied came to see the incest as stemming from particular consequences in the household, such as the death of their mother or discontinued sexual relations between their parents. Others made sense of the victimization by sympathizing with their father – for example, viewing his behavior as reflecting a desire for intimacy and love.

Similarly, in their research on bereavement, Parkes and Weiss (1983) have found that over time, most widows and widowers are able to reach a point where they no longer suffer "a continuous, oppressive awareness of loss and pain and no longer need to avoid thinking of the spouse's death in order to function . . ." (p. 157). According to Parkes and Weiss's research, a period of state orientation may be necessary to achieve this type of resolution: "For this state to be reached, there must be repeated confrontation with every element of the loss until the intensity of distress is diminished to the point where it becomes tolerable and the pleasure of recollection begins to outweigh the pain . . . it requires what appears to an observer to be a kind of obsessive review in which the widow or widower goes over and over the same thoughts and memories" (p. 157).

A feature of many life crises is that they can fundamentally alter a person's life goals. A bereaved spouse, for example, is faced with a situation where his or her plans, hopes, and dreams for the future may have little meaning. Another advantage of state orientation is that it might help victims to make the transition to the next life phase, in which a new identity and new goals must be formulated. Parkes and Weiss (1983) have suggested that the process of recovery can be aided by intervals in which the widow or widower draws strength and comfort from thoughts about the dead spouse. As they put it, "Movement toward a new identity is consistent with intervals in which the widow or widower feels and may even act as if the partner were still present. Widows . . . are strengthened by being able to 'talk with' their dead spouses ... They know the spouses are not there, but they find it comforting and sustaining to be able to talk through a problem, with the feeling that the spouse is there to listen" (p. 161).

Because of the disorientation that typically accompanies undesirable life events, the enactment of intentions is likely to be difficult, particularly if doing so requires complex reasoning. For this reason, those who attempt to enact their intentions at the height of a crisis may be very likely to fail. There is considerable evidence that those who make coping efforts that fail experience more distress than those who make no coping efforts at all (see Wortman & Brehm, 1975; Janoff-Bulman & Brickman, 1982). For example, Weiss (1971 a, b) has found that laboratory rats who sit and take electric shocks do not show as many stress symptoms as those who try to avoid the shocks, but obtain feedback that their coping responses have been ineffective. In other words, a high degree of action control may impede the coping process if the resulting action does not alleviate the problem it was meant to solve.

Moreover, because many of the coping tasks facing the victim are tied to important consequences, poor performance may create subsequent difficulties for the victim. A widow who makes a poor decision about her husband's financial holdings may jeopardize her well-being and security for years to come. Clearly, there are many cases where not carrying out one's intentions is preferable to carrying them out ineptly.

It should also be noted that in many cases, carrying out one's intentions may require a substantial amount of energy at a time when the victim is physically and emotionally exhausted. For example, it might take considerable energy for a widower to explore various child-care options for his children, or for a widow to resolve financial issues regarding her husband's estate. A period of state orientation may help to conserve energy until the victim feels capable of dealing effectively with such coping tasks, and until the chances of successful performance are enhanced.

These considerations about the positive and negative consequences of state orientation for the coping process result in a central question: is a high degree of action control adaptive for successful coping?

Adaptiveness of Action Control

Does a high degree of action control facilitate or impede successful coping? The previous illustration of the possible effects of action and state orientation made evident that this question cannot be answered in general terms. When people are preoccupied with the attempt to restructure their lives, there will be times when it is important to do something about a problem and other times when it is more important to think about it.

Especially during early stages of a life crisis, state orientation might be beneficial. As described above, there are some cases in which it may be useful to reflect on what has happened. Such reflection may help to clarify what might be done to improve the situation, or to provide some perspective on a situation that cannot be altered or undone. State orientation might also be desirable in those cases where a complex system of actions is necessary to reach one's intentions and where coping efforts are therefore likely to be unsuccessful.

On the other hand, some undesirable life events require the victim to perform certain actions which are crucial for improvement or even survival. For example, when confronted with an acute crisis situation such as a fire or flood (see, e. g., Tyhurst, 1951), a person's very survival may hinge on the ability to perform those actions that will lead him or her to safety. After the danger has passed, a period of state orientation might facilitate acceptance of what has been lost, as well as the formulation of new goals.

Earlier, we suggested that for many life crises, a period of state orientation might help the victim come to terms with what has happened and formulate new goals. It should be noted, however, that the potential adaptiveness of state orientation may not be confined to an initial period after the confrontation with the undesirable life event. Over time, the requirements of the situation may change and may make it necessary to reevaluate one's goals and set new ones if the old ones are no longer appropriate.

Until this point, we have been considering situations in which state or action orientation would directly facilitate the victim's performance of appropriate behaviors as well as their adjustment to a life crisis. However, it is important to recognize that neither action nor state orientation exists in a social vacuum. Others' reactions to the victim's state or action orientation may well have an important effect on the coping process. For example, there is evidence that others harbor a number of myths about the process of coping with undesirable life events (see Silver & Wortman, 1980; Wortman & Lehman, in press). Others often expect victims to be "recovered" from a crisis in a few weeks or months. Because recovery is expected, others are frequently intolerant of the victim's desire to discuss what has happened. Instead of allowing the victim to express his or her feelings about what has happened, others frequently encourage progression to the next life stage. For example, others may encourage a widow to begin thinking about remarriage "within a few days or weeks of the husband's death" (Maddison & Walker, 1967, p. 1063).

Because of their own myths about the coping process, then, others may react negatively to the victim who remains state orientated. They may communicate that he or she is coping poorly with the crisis, and may even suggest that the victim's continued distress is a sign of mental disturbance (see Wortman & Lehman, in press, for a more detailed discussion). Clearly, such messages from others can undermine the victim's confidence and reduce the likelihood that the crisis will be successfully resolved.

In short, the prevalence of state orientation that is likely to result from the experience of undesirable life events may not necessarily impede the successful resolution of the crisis. A high degree of action control can only be viewed as adaptive if it leads to the "right" action for reaching the "right" goal. In our society, much – maybe too much – emphasis is placed on being efficient, productive and goal oriented. This behavior pattern might be detrimental during certain phases of a life crisis, where victims either might be likely to pursue goals that are inappropriate and dysfunctional or may fail to reach their goals.

Thus, the central question is not so much whether action control is adaptive per se, but rather whether the goals that are reached by the intended action are adaptive goals for short- and long-term coping. Unfortunately, the model of action control does not take into consideration the context in which the intention in question has been developed and will be enacted. It is important to evaluate the successful performance of an intended action against this background, and ask whether it aids the individual in reaching a goal that facilitates subsequent adjustment.

Coping with Undesirable Life Events: Implications for the Theory of Action Control

In this paper, we have attempted to discuss several ways that the theory of action control can be applied to the process of coping with life crises. Until recently (c.f., Kuhl, 1983b), most of the empirical research on action control has been confined to laboratory settings or to behavior in nonstressful situations. For the most part, the situation facing most victims of undesirable life events is more rich and complex in terms of the emotions experienced and the coping tasks involved. We believe that by applying the theory of action control to the process of coping with life crises, it is possible to identify several areas where further elaboration of the theory would be useful.

Intention Formation

When an undesirable life event occurs, it may not only influence the relationship between intentions and behavior, but may also reduce the likelihood that the victim will develop intentions. Thus far, the explanatory power of the theory of action control is restricted to cases where victims have goals that they want to pursue and that can be pursued. These are stipulations that are not always fulfilled when people are attempting to cope with a life crisis. A crisis may alter a person's life situation so fundamentally that it is difficult for her or him to form intentions for future actions.

Unlike laboratory settings where the experimenter defines the problem and informs the subject exactly what should be done, many types of life crises require people to analyze a complex situation and determine the goals that should be pursued. However, there are several reasons why individuals who have experienced a life crisis may be unlikely to formulate intentions and plans.

First, the motivational basis for forming intentions might be low. Some theorists (e.g., Shontz, 1975) contend that during the early stages following exposure to an undesirable life event, the impact of reality seems overwhelming and the individual may be unable to plan, reason, and engage in active problem-solving to improve the situation. Furthermore, if the victim reacts with depression to the undesirable life event, as it is often the case (c.f., Silver & Wortman, 1980), intentions are unlikely to be formed. Beck (1977) describes "paralysis of the will" as one of the most prominent symptoms of depression.

Second, in some cases, the only goal that has meaning to the victim is an unrealistic one: getting back what has been lost by the event. Let us present an example: Joanne is a middle-aged woman who has recently been widowed. She finds herself in a situation where everything seems to be out of kilter. During the last 30 years there was hardly a day that she and her late husband had not spent time together. Now, many of the things she used to do no longer seem to be meaningful, and she derives no pleasure from activities that she enjoyed before. Friends encourage her to go out and meet people or develop some new interests, but she finds that she has no incentive for doing so.

In contrast to Sandra and Jim, whose hypothetical cases were described earlier, Joanne is unable to think of or plan activities that would help her adjust to her loss. Although she has a goal in mind (which is to have her husband back), there is no behavioral alternative available to reach it. Realizing that there is nothing she can do to get what she wants, she feels no incentive for doing other things. Kuhl (1984) would view Joanne's goal to have her husband back as a degenerated intention which hinders the enactment of other intentions. We think that focusing on this goal not only impedes her action control, but in addition keeps her from forming new, more appropriate intentions or goals.

In fact, we believe that a full understanding of how people cope with undesirable life events requires an explication of the process through which people are able to relinquish their former goals and develop new ones. Unfortunately, the theory of action control, as currently formulated, has devoted little attention to this problem. One possibility, suggested by researchers studying adaptation to life-threatening illness, is that individuals alter their goals gradually as they become more capable of dealing with reality. For example, a person who learns about a terminal illness may initially have the goal of getting well. Over time, this goal may shift to going into remission, living a few more years and, as the disease progresses, to having a day, or even an hour, without pain.

Some stressors confront victims with a life situation that is fundamentally altered, such as the loss of their spouse or permanent paralysis. In many other cases, however, the person is faced with a situation that is far more ambiguous. For example, consider the case of a couple who has been trying for many years to have children, but who have been unsuccessful. Should they abandon their hope and attempt to find other sources of satisfaction in their lives? Or should they intensify their efforts, perhaps seeking out new medical specialists or treatments? Should a man who is told by his wife that she loves somebody else plan to give her up and get a divorce, or should he work on improving their relationship in order to get her back? Should the woman whose husband is missing in action give up hope for his return and start a life on her own, which might include a new relationship? Or would it be more adaptive for her to continue to hope that her husband will return and thus refrain from dating other men? In these cases, conflict between two goals which demand completely different actions can impede the formation of intentions.

These examples illustrate that individuals who are facing a life crisis might have difficulties in developing precise goals for the future or in choosing between goals. Therefore we strongly feel that the theory of action control should devote more attention to the issue of how intentions are developed and how individuals shift from one goal to another. Recently, Heckhausen and Kuhl (1985) have done just that: They have elaborated the cognitive processes that mediate between wishes and the formation of intentions.

Determinants of Action Control

According to Kuhl (1981, 1983 a, 1984), the degree of a person's action control is determined by dispositional and situational factors. Unpredictable and uncontrollable outcomes are said to be situational characteristics which are especially likely to induce state orientation.

Many of the major classes of life events, such as loss of a loved one, criminal victimization, and serious illness, are frequently both unexpected and uncontrollable. On the basis of our experience, however, some of these sudden and uncontrollable events seem much more likely than others to induce the repeated ruminations and the inability to engage in goal-directed behavior, that are characteristic of state orientation. What factors might enhance the likelihood that a particular crisis will result in an extended period of state orientation?

We believe that when a victimizing experience shatters a person's prior assumptions about the world, state orientation may be especially likely to occur. The available literature leaves little doubt that life crises can have this type of effect. In their study of rape victims, for example, Scheppele and Bart (1983) report that their respondents "see their world as having changed in a major way" as a result of the attack (p. 21). Some women became less trusting of other people; others experienced almost constant feelings of vulnerability and fear; still others felt that they could no longer control what was happening in their lives. Some types of rape are more likely to shatter a woman's assumptions than others. For example, consider the case of a woman who is raped when she is trying to help a man by giving him directions, or a woman who is raped despite taking many precautions to avoid rape. In these cases, a woman's assumptions may be shattered to a greater extent than in cases in which a woman is raped while behaving carelessly. Scheppele and Bart (1983) found that a large majority of the women in their sample who experienced a "total fear reaction" – a profound disruption in their behavior and almost unbearable anxiety – were those women who believed they were safe – either because they had taken precautions to avoid being raped, or because they had coded the situation as safe (e.g., they were at home with the doors locked or were with a person they knew and trusted). We believe that because their assumptions have been shattered so fundamentally, such women may be especially likely to experience an extended period of state orientation in which they ruminate about what happened and are unable to engage in goal-directed behavior. Thus, the *degree* of incongruence between a victim's view of the world and the occurrence of an undesirable life event seems to determine the amount of state orientation. For a further elaboration on the determinants of action control, this aspect should receive more attention.

Anorther variable that would modify the degree of state orientation following a life crisis can be suggested by entertaining Kahneman and Tversky's (1982) availability principle. These investigators believe that an individual is especially likely to review the victimizing experience again and again if the possibility of not being victimized is *available in memory*. In order to illustrate their point, they ask the reader to consider the case of a man who becomes permanently paralyzed in an automobile accident on the way from work. They suggest that if the individual left work at the same time as usual, and took the same route home, he would be much less likely to ruminate about the accident than if he left work at an unusual time, or took a route home that was different from the one normally followed. According to Kahneman and Tversky (1982), the possibility of not having been paralyzed is more available in memory to the latter subjects than the former subjects and leads to a repeated review of the circumstances surrounding the accident. The latter individual may be very prone to experience repeated thoughts such as, "If only I hadn't taken that route home . . ." The former subjects, since they were not deviating from their normal routine, would be unlikely to experience such ruminations (see also Bulman & Wortman, 1977).

Thus far in this section we have been discussing characteristics of undesirable life events that may enhance the likelihood that state orientation is initially experienced. Another question of interest is the following: Once the victim has moved from a phase of state orientation to action orientation, are there conditions under which a period of state orientation is likely to be revoked? There is reason to believe that certain environmental cues can trigger a period of intense distress and frequent ruminations that may be indicative of state orientation. Among the bereaved, for example, symptoms of grief "tend to recur at various times, precipitated by anniversaries, geographical locale, etc." (Wiener, Gerber, Battin, & Arkin, 1975, p. 64). Parkes (1970) has suggested that during these times, "all feelings of acute pining and sadness return and the bereaved person goes through, in miniature, another bereavement" (p. 464). Silver et al. (1983) report that among the victims of incest that they studied, certain environmental stimuli trigger thoughts and feelings about the experience. For example, more than 80% of the victims reported that engaging in sex brought back memories of the incest experience.

If subsequent research substantiates the hypothesis that certain reminders of the past can trigger periods of state orientation, a number of intriguing questions could be raised. First, when possible, should individuals avoid coming into contact with cues likely to have this effect? Or is repeated contact with such cues desirable or necessary if resolution is to occur? In a recent television program about teenagers killed in drunk-driving accidents, parents were advised to choose a memorial home that was in a different city or town. It was suggested that by selecting a memorial home that one frequently passes, for example, that is on one's route to work, the sight of this home would repeatedly evoke painful and maladaptive feelings about the loss. At present, not enough is known about the impact of coming into contact with such cues to determine whether this constitutes good advice. Second, in those situations where it is difficult or impossible to avoid cues about the victimization (as in the case of sexual intercourse triggering cues of incest or rape), what can the victims do to enhance their ability to confront such cues with equanimity?

In this section, we have reviewed suggestive evidence that certain types of victimizing experiences may be much more likely to induce state orientation than others. If state orientation has important implications for the coping process, it would be desirable to have a more complete understanding of the specific characteristics of the situations under which state orientation is particularly likely to occur. For predicting when a person will become state oriented, and when action orientation will prevail, the theory of action control might have to include a variety of variables in order to account for individual differences in action control given the occurrence of unexpected and uncontrollable events.

Action-Facilitating Cognitions and Emotions

According to Kuhl's (1981, 1983a, 1984) model, a person is action oriented if his or her attention is focused on the present state, some aspect of the future state, a discrepancy between the two, and at least one alternative that could transform the present into the future state, thus eliminating the discrepancy. If any one of these elements is missing, the person is considered to be state oriented. Thus, any thoughts that do not involve all four of these elements are considered to be indicative of state orientation, and are considered to be equivalent in terms of the likelihood that they will lead to the realization of intentions. Victims of life crises experience a rich variety of thoughts and feelings as they attempt to come to terms with what has happened to them. In our judgement, some of these thoughts and feelings may be more likely to promote goal-oriented behavior than others.

After a life crisis occurs, victims may find themselves reviewing the past or contemplating the future. Their thoughts about what has happened might evoke anxiety, rage, or despair. Their recollections of the past may be accurate reflections of reality or self-serving distortions of the truth. In trying to settle on a cause for what has happened, some individuals may blame their character; others may focus on shortcomings of their behavior, and still others may attribute blame to other people in the environment.

In our judgement, these different cognitions and emotions may have different diagnostic value for predicting the likelihood that a person will engage in goal-directed behaviors. For example, a person whose primary emotional reaction is anxiety or depression may be less likely to take action than a person who is enraged about what has happened to him or her (c. f., Wortman & Brehm, 1975). A man who is aware of his shortcomings as a husband may be more likely to remain state oriented following the death of his wife than a man who is able to delude himself into thinking that he was a good husband (Parkes & Weiss, 1983). Finally, ascribing blame to one's behavior is likely to provide a feeling of mastery and control over the future that does not occur if blame is ascribed to one's character or to external factors (c.f., Janoff-Bulman, 1979; Wortman, 1976).

We agree that action orientation is likely to be present when the four elements identified above are also present. In the case of victims of life crises, however, the simultaneous or successive focus on these four elements may rarely, if ever be present. Yet some victims remain distressed and continue to ruminate about what has happened, while others are able to move on to the next phase in their lives (Silver & Wortman, 1980). It would be extremely valuable if further theoretical elaboration of the action control model made it possible to differentiate between those likely to remain in a phase of state orientation and those who are likely to initiate actions.

From State Orientation to Action Control

Earlier in this paper, we argued that in many cases, undesirable life events induce a phase of state orientation in the victim. Over time, the victim may gradually become more capable of performing intended actions. Thus far, however, the theory of action control has devoted little attention to the process by which this shift in orientation takes place. Information concerning how a person is transformed from state to action orientation would not only have important theoretical implications, but would provide useful information for the design of interventions to facilitate the process of coping with life crises.

In our judgment, a consideration of this topic raises a number of intriguing issues. There is considerable variability in the length of time required to resolve a crisis, and evidence suggests that this process often takes longer than the victim or others expect (Silver & Wortman, 1980). However, many victims do reach a point where the crisis loses its power over them. They are able to formulate goals and plans, initiate goal-directed behavior, and contemplate the prospect of an altered life without constant pain. They no longer need to avoid thinking about the victimizing experience in order to function effectively. Thoughts and feelings of distress may reemerge for brief periods when the person is confronted with reminders of the crisis – for example, when the widow passes the scene of an accident in which her spouse was killed, or when a mother finds a toy that belonged to the child that she has lost. But for the most part, the victim is able to keep feelings of distress within manageable limits, to experience some gratification in day-to-day living, and to contemplate the future with some degree of optimism and hope.

In some cases, the victim may reach this state by "working through" the crisis – by repeatedly focusing on what has happened until he or she is able to gain some perspective on the event, find some meaning in what has happened, or come to terms with what has been lost. For example, over time, victims of incest may come to regard themselves as strengthened by the experience. As one incest victim indicated, "I learned over the years that nothing as bad as what I had been through was going to happen again. Now I know there is virtually nothing I cannot overcome." (Silver et al., 1983, p. 90). In time, the bereaved may be able to appreciate the time that they shared with the person who is lost (Parkes & Weiss, 1983), and to value qualities in themselves, such as independence, that have emerged as a result of the crisis.

In other cases, however, no amount of state orientation or working through seems capable of providing a resolution to a crisis. As we mentioned earlier, Parkes and Weiss (1983) found that among the bereaved individuals who lost their spouse with little warning, over 60% expressed agreement with the item, "I still ask myself why it happened" at the final interview, which was conducted 2–4 years after the loss. Similarly, over 50% of the incest victims studied by Silver et al. (1983) reported that they were unable to make any sense out of what had happened. One victim emphasized, "There is no sense to be made. This should not have happened to me or any child" (p.89). As another victim expressed it, "I always ask myself why, over and over, but there is no answer" (p.89). Moreover, it was not the case that the longer these women searched, the more likely they were to find meaning. Time did not necessarily assist the victims in successfully resolving what had happened to them. These results suggest that in certain cases, repeated review of the experience may only serve to intensify the victim's distress. In such cases, it may be adaptive to accept that the crisis makes no sense – "that one's experience is, in effect, unexplainable" (Silver et al., 1983, p.97), and to block or avoid thoughts that are upsetting. As one incest victim suggested, "My strongest asset through all of my experiences was my ability to 'block out' whatever I didn't want to remember. If I didn't talk about them, or even think about them, I was able to survive" (p.97). In some cases, fostering the ability to interrupt or alter intrusive and distressing thoughts about the victimization might enable the victim to move beyond the past, and might facilitate action control.

In short, the available evidence suggests that in some cases, the repeated review of what has happened, which is characteristic of state orientation, will facilitate resolution of the crisis; in other cases, however, it is unlikely to lead to resolution. Two questions can be raised on the basis of this discussion. First, in the individual case, it may be very difficult to determine whether state orientation will be adaptive or maladaptive. How can we tell when continued ruminations about what has happened will provide new insight, and when such thoughts will simply enhance the victim's pain? Throughout this chapter, we have suggested that some types of victimizing experiences may be extremely likely to evoke state orientation, and may be especially difficult for the victim to resolve through repeated ruminations. But even in these cases, repeated review is sometimes effective. Some of the incest victims studied by Silver et al. (1983) were able to gain insight about what had happened by reviewing the past, although many were not. At what point should the victim attempt to stop reliving the past, and try to deal with state oriented thoughts by blocking them?

Second, are there interventions that will help victims of life crises move from state to action orientation? If the life crisis is one that is likely to precipitate repeated review of what hat happened (for instance, the sudden loss of one's spouse, a rape by a trusted friend, or repeated encounters with incest), is there any way to enhance the likelihood that this review will lead to a resolution of the crisis? Are there things that other people can say or do to facilitate movement from state to action orientation? In dealing with victims of life crisis, how do others typically respond to victims who express state oriented thoughts or feelings? Should others attempt to discourage state oriented comments and encourage action orientation? If so, will these support attempts be appreciated by the victim? Will they be effective?

There is substantial amount of evidence that others frequently become distressed when victims express state oriented thoughts or feelings (see, e. g. Coyne, 1976). As Parkes and Weiss (1983) have suggested, "the repeated review by which emotional acceptance is obtained can be painful to friends and relatives ... (they) may urge that the review be terminated long before the widow or widower has adequately come to terms with the past." (p. 159). There is evidence that others frequently try to block the victim's attempts to review what has happened (see, e. g., Kastenbaum & Aisenberg, 1972; Helmrath & Steinitz, 1978). These attempts to block the expression of state oriented feelings may take many specific forms (Wortman & Lehman, in press). Others may respond to state oriented thoughts with forced cheerfulness, encouraging the victim to focus on the positive elements of the situation and to remain as optimistic as possible about the future. Alternatively, others may make comments that minimize the extent of the crisis by suggesting that the situation is not as bad as it seems, or that things could be worse. For example, mothers who lose an infant are frequently told, "it was only a baby you didn't know – it's worse to lose a child you know" (Helmrath & Steinitz, 1978, p. 788).

Evidence suggests that others not only discourage state orientation, but also make efforts to encourage action orientation on the part of the victim. For example, friends of a widow may emphasize the importance of getting out more and of building a life of one's own. Among the bereaved, the topic of remarriage is often brought up by others quite early (Maddison & Walker, 1967; Glick, Weiss, & Parkes, 1974). Victims of life crises may also receive advice that is designed to encourage goal-directed behavior. For example, a widow may be encouraged to move out of her former home, and parents who have lost a baby may be advised to have another child in the near future (Wortman & Lehman, in press).

The available evidence is consistent in suggesting that both of these behaviors on the part of others – discouraging expression of state oriented cognitions and encouraging action orientation – are not regarded as helpful by the victim. Wortman and Lehman (in press) asked individuals who had lost a spouse or child to identify the most helpful and unhelpful responses of others. Giving advice about the future or minimizing the loss were almost always regarded as among the most unhelpful things that others had done. Similarly, Glick et al. (1974) have reported that "widows invariably found early suggestions that they consider remarriage unpleasant and even jarring" (p. 222). According to Maddison and Walker (1967), "conversations which aroused interest in new activities, development of new friendships, resumption of old hobbies or occupations" during the first 3 months after the death of a spouse were not appreciated by the bereaved.

On the contrary, evidence from many studies suggests that victims of life crises desire opportunities to express state oriented thoughts and feelings to others, that such discussions are perceived as helpful by the victim, and that open discussion of feelings is associated with positive health outcomes (see Silver & Wortman, 1980, for a review). For example, Wortman and Lehman (in press) found that providing an opportunity to express one's feelings about the loss was commonly identified as the most helpful thing anyone had done by the bereaved people they studied, and was never identified as unhelpful. Similarly, Silver et al. (1983) asked victims of incest whether they had "at least one person that you can confide in about the incest experience" (p. 89). Those women who did have a confidant were significantly more likely to report having made sense of their incest experience than those who did not.

In summary, there is suggestive evidence that others' attempts to block state oriented cognitions, or to move the victim to a phase of action orientation, are very likely to fail. What seems to be more helpful to the victim is permitting the expression of state oriented thoughts. There is little doubt that it is extremely difficult to listen to such thoughts and feelings. As Parkes and Weiss (1983) have indicated, "it takes great tolerance for another's pain" to be able to do so (p. 159). There is no evidence that conversations in which the victim expresses feelings must be focused on the desired future state, and how to achieve it, in order to be useful. In fact, in their study of bereavement, Maddison and Walker (1967) have concluded that "only a small handful of widows found any sort of detailed discussion about their future to be of value to them."

If others' direct exhortations to stop dwelling on the past and to begin building a future life are doomed to failure, are there any ways to help the victim relinquish the past and move toward a successful resolution of the crisis? One possibility suggested by the literature is to induce the victim to become involved in activities that may provide a brief respite from the distressing ruminations and emotional distress associated with the crisis. Parkes and Weiss (1983) refer to this phenomenon as "distancing"; as an example, they discuss the case of the widow who throws herself into cleaning the house "because she must do something to take her mind away from her bereavement" (p. 158). Parkes and Weiss suggest that by temporarily directing their attention away from the loss, widows and widowers are able to achieve some relief from the intense pain produced by their spouse's death. They suggest that while this defensive manuever is difficult to maintain and easy to interrupt, "many find it necessary if they are not to be exhausted by pain" (p. 158). Similarly, in discussing the intrusive thoughts experienced by Vietnam veterans, Williams (1983) has suggested that denial or emotional numbing can "provide distance from the memories when the working through process becomes too intense or threatening for the individual" (p. 12). Horowitz and Wilner (1980) have reported that those who respond most adaptively to distressing events tend to experience intrusive thoughts about the stressor that are alternated with periods of denial. These investigators suggest that it is the vacillation between these responses that enables an individual to resolve the crisis successfully. A helping strategy frequently identified as "the most helpful thing anyone has done" among the bereaved studied by Wortman and Lehman (in press) was to involve the bereaved in social activities, perhaps because such involvement provided a temporary respite from their grief. As one respondent indicated, "they asked me to go fishing . . . this helped me to get my mind off it for a while." Individuals who are able to achieve such brief periods of respite may be able to return to their thoughts with increased energy, a better frame of mind, or a slightly different perspective. They may be more likely to make slow but steady progress in coming to terms with their loss, rather than getting "stuck" in a state of chronic, unresolved grief.

At present, the theory of action control focuses on two states: state and action orientation. During some periods of state orientation, the victim may be focused on the life crisis or his or her feelings about it. During other periods, attention may be directed elsewhere – for example, the person may immerse him or herself in work or may engage in other activities that enable him or her to block painful thoughts. These latter periods would not be considered to be action oriented, since they are not really designed to reduce a discrepancy between a present state and a future state. In subsequent theoretical and empirical work, it may be useful to draw a distinction between these two types of state orientation, and to elaborate on their implications for the coping process.

Another way to help the victim gain some perspective on what has happened, and thus make a transition from state to action orientation, may be to encourage interaction with others who have also been victimized and who are coping successfully with what has happened (see Dunkel-Schetter & Wortman, 1982, for a more detailed discussion). In their study of bereavement, Wortman and Lehman (in press) found that contact with a similar other was frequently mentioned as "the most helpful thing" in coming to terms with the loss of their spouse or child, and was never mentioned as unhelpful. Parkes (1972b) has argued that contact with similar others facilitates effective coping among individuals who have lost a limb: "It is only when he meets someone more seriously mutilated than himself who appears to be coping well and cheerfully with his disability that it becomes possible for him to look to the future in an optimistic and realistic manner. It is for this reason that people who have successfully come through major transitions in their lives are often best able to help other individuals" (p. 112). Although carefully controlled studies are rare, there is some evidence in the literature that contact with similar others indeed facilitates successful adjustment (see, e. g., Vachon, Lyall, Rogers, Freedman-Letofsky, & Freeman, 1980).

Conclusions and Implications

By attending to the processes that mediate between the generation of an intention and the implementation of an action, the model of action control makes a unique and significant contribution to the understanding of the coping process. The analysis of coping responses in terms of the metacognitive processes of action and state orientation raises the question of whether the performance of intended actions furthers successful coping. Above, we have described how undesirable life events frequently evoke state oriented coping responses. According to the theory of action control, state orientation impedes the enactment of intentions by reducing a person's degree of action control. Thus, it can provide a possible explanation for individual differences in speed and success of recovery from an undesirable life event: The crisis will be resolved more slowly, or not at all, if the victim is state oriented during those phases of the coping process where a high degree of action control is required for the performance of actions that would further successful coping.

The application of the theoretical perspective of action control to the analysis of people's attempts to cope with an undesirable life event suggests a number of areas for subsequent research. As discussed above, we believe it would be worthwhile to investigate the conditions under which state orientation hinders and under which it facilitates the successful resolution of the crisis. Before this question can be addressed, methods will have to be developed for the assessment of the discrepancy between a person's degree of action control and the situational requirements for it. We also feel that it would be important to elaborate those characteristics of undesirable life events most likely to induce state orientation, and those factors most likely to aid the victim in becoming action oriented.

The theory of action control also provides a new perspective from which to consider several conceptual variables which have been discussed in the literature as possible determinants of effective coping. These variables may have a rather indirect impact and influence the coping process by facilitating or impeding action control. For instance, under certain circumstances, social support might be helpful most of all because it facilitates an individual's action control by focusing his or her attention on the tasks that have to be performed. In a similar way, concomitant stressors might impede adjustment to a particular life crisis primarily because they create new phases of state orientation. The search for a meaning in a life crisis may only be adaptive during phases of the coping process where a high degree of action control is not important, because there are no goals available to pursue.

We hope that the issues raised in this paper will encourage subsequent intellectual exchanges among investigators in the action control and coping fields. In our judgment, such exchanges would be highly beneficial in guiding further theoretical elaboration of the theory of action control, and in setting new research agendas in the field of coping with life crises.

References

Abramson, L. Y., Seligman, M. E. P. & Teasdale, D. (1978). Learned helplessness in humans: Critique and reformulation. *Journal of Abnormal Psychology, 87,* 49–74.

Atkinson, J. W. & Birch, D. A. (1978). *Introduction to motivation.* New York: Van Nostrand.

Bandura, A. (1977). Self efficacy: Toward a unifying theory of behavioral change. *Psychological Review, 84,* 191–215.

Beck, A. T. (1977). *Depression: Causes and treatment.* Philadelphia: University of Philadelphia Press.

Bulman, R. & Wortman, C. B. (1977). Attributions of blame and coping in the "real world": Severe accident victims react to their lot. *Journal of Personality and Social Psychology, 35,* 351–363.

Cornwell, J., Nurcombe, B. & Stevens, L. (1977). Family response to loss of a child by Sudden Infant Death Syndrome. *The Medical Journal of Australia, 1,* 656–658.

Coyne, J. C. (1976). Depression and the response of others. *Journal of Abnormal Psychology, 85,* 186–193.

Cranach, M. M. von, Kalbermatten, U., Indermühle, K. & Gugler, B. (1980). *Zielgerichtetes Handeln*. Bern: Huber.

Dunkel-Schetter, C. & Wortman, C. B. (1982). The interpersonal dynamics of cancer: Problem in social relationships and their impact on the patient. In: H. S. Friedman & M. R. Di Matteo, (Eds.), *Interpersonal issues in the health care*. New York Academy.

Ferster, C. B. (1973). A functional analysis of depression. *American Psychologist, 28,* 857–870.

Ferster, C. B. (1974). Behavioral approaches to depression, In: R. J. Friedman & M. M. Katz (Eds.), *The psychology of depression: Contemporary theory and research*. Washington, D. C.: Winston-Wiley

Freud, S. (1914). Remembering, repeating, and working-through. In: J. Strachey (Ed.), *Standard Edition*. London: Hogarth Press, 1958, *12,* 145–150.

Freud, S. (1920). Beyond the pleasure principle. In: J. Strachey (Ed.), *Standard Edition*. London: Hogarth Press, 1953, *18,* 7–64.

Frey, D., Havemann, D. & Rogner, O. (1983). Kognitive and psychosoziale Determinanten des Genesungsprozesses von Unfallpatienten. Final report for the Deutsche Forschungsgemeinschaft. Universität Kiel.

Glick, I. O., Weiss, R. S. & Parkes, C. M. (1974). *The first year of bereavement*. New York: Wiley.

Hacker, W. (1973). *Allgemeine Arbeits- und Ingenieurpsychologie*. Berlin: Deutscher Verlag der Wissenschaften.

Heckhausen, H. (1980). *Motivation und Handeln: Lehrbuch der Motivations-Psychologie*. New York: Springer.

Heckhausen, H. & Kuhl, J. (1985). From wishes to action: The dead ends and shortcuts on the long way to action. In: M. Frese & J. Sabini (Eds.), *Goal-directed behavior: Psychological theory and research on action*. Hillsdale, N. J.: Lawrence Erlbaum.

Helmrath, T. A. & Steinitz, E. M. (1978). Death of an infant: Parental grieving and the failure of social support. *The Journal of Family Practice, 6,* 785–790.

Herrmann, C. (1981). Attribuierungsmuster Depressiver. In: M. Hautzinger & S. Greif (Eds.), *Kognitionspsychologie der Depression*. Stuttgart: Kohlhammer

Herrmann, C. (1984). Was sind „depressive Attributionsstile"? Probleme und Möglichkeiten der Berücksichtigung von Attributionen in der Depressionsforschung. In: M. Hautzinger & R. Straub (Eds.) *Psychologische Aspekte depressiver Störungen*. Regensburg: S. Roderer Verlag

Horowitz, M. H. (1976). *Stress response syndrome*. New York: Aronson.

Horowitz, M. H. (1979). Psychological response to serious life events. In: V. Hamilton & D. Warburton (Eds.), *Human stress and cognition: An information-processing approach*. London: Wiley.

Horowitz, M. J. & Wilner, N. (1980). Life events, stress and coping. In: L. Poon (Ed.), *Aging in the 1980's: Psychological issues*. Washington, D. C.: American Psychological Association.

Janis, I. L. (1971). *Stress and frustration*. New York: Harcourt Brace Jovanovich.

Janoff-Bulman, R. (1979). Characterological versus behavioral selfblame: Inquiries into depression and rape. *Journal of Personality and Social Psychology, 37,* 1798–1809.

Janoff-Bulman, R. & Brickman, P. (1982). Expectations and what people learn from failure. In: N. T. Feather (Ed.), *Expectations and actions*. Hillsdale, N. J.: Lawrence Erlbaum.

Kahneman, D. & Tversky, A. (1982). The simulation heuristic. In: D. Kahneman, P. Slow & A. Tversky (Eds.), *Judgment under uncertainty: Heuristics and biases*. New York: Cambridge University Press.

Kanfer, F. H. (1970). Self-regulation: Research issues and speculations. In: C. Neuringer & J. L. Michael (Eds.), *Behavior modification in clinical psychology*. New York: Appleton-Century-Crofts.

Kastenbaum, R. & Aisenberg, R. (1972). *The psychology of death*. New York: Springer.

Klinger, E. (1975). Consequences of commitment to and disengagement from incentives. *Psychological Review, 82,* 1–25.

Klinger, E. (1977). *Meaning and void: Inner experience and the incentives in people's lives*. Minneapolis: University of Minnesota Press.

Kuhl, J. (1981). Motivational and functional helplessness: The moderating effect of state versus action orientation. *Journal of Personality and Social Psychology, 40,* 155–170.

Kuhl, J. (1983 a). *Motivation, Konflikt und Handlungskontrolle*. Heidelberg: Springer.

Kuhl, J. (1983 b). Motivationstheoretische Aspekte der Depressionsgenese: Der Einfluß von Lageorientierung auf Schmerzempfinden, Medikamentenkonsum und Handlungskontrolle. In: M. Wolfersdorf, R. Straub & G. Hole (Eds.), *Der depressiv Kranke in der psychiatrischen Klinik – Theorie und Praxis der Diagnostik und Therapie.* Frankfurt: Ciba-Geigy.

Kuhl, J. (1984). Volitional aspects of achievement motivation and learned helplessness: Toward a comprehensive theory of action control. In: B. A. Maher (Ed.), *Progress in experimental personality research,* (Volume 13). New York: Academic Press.

Kuhl, J. (In press). Integrating cognitive and dynamic approaches: A prospectus for the unified motivational psychology. In: J. W. Atkinson & J. Kuhl (Eds.) *Motivation, thought, and action: Personal and situational determinants.* Hillsdale, N. J.: Lawrence Erlbaum.

Lacoursiere, R., Godfrey, K. & Ruby, L. (1980). Traumatic neurosis in the etiology of alcoholism: Vietnam combat and other trauma. *American Journal of Psychiatry, 137,* 966–968.

Lewinsohn, P. M. (1974). A behavioral approach to depression. In: R. J. Friedman & M. M. Katz (Eds.), *The psychology of depression: Contemporary theory and research.* Washington, D. C.: Winston-Wiley.

Maddison, D. & Walker, W. (1967). Factors affecting the outcome of conjugal bereavement. *British Journal of Psychiatry, 113,* 1057–1067.

Miller, G. A., Galanter, E. & Pribram, K. H. (1960). *Plans and the structure of behavior.* New York: Holt, Rinehart & Winston.

Miller, D. T. & Porter, C. A. (1983). Self-blame in victims of violence. *Journal of Social Issues, 39,* (2), 139–152.

Parkes, C. M. (1970). The first year of bereavement: A longitudinal study of the reaction of London widows to the death of their husbands. *Psychiatry, 33,* 444–467.

Parkes, C. M. (1972 a). Bereavement: Studies of grief in adult life. New York: International Universities Press.

Parkes, C. M. (1972 b). Components of the reaction to loss of a limb, spouse or home. *Journal of Psychosomatic Research, 16,* 343–349.

Parkes, C. M. & Weiss, R. S. (1983). *Recovery from bereavement.* New York: Basic Books.

Rees, W. D. (1971). The hallucinations of widowhood. *British Medical Journal, 4,* 37–41.

Scheppele, K. L. & Bart, P. B. (1983). Through women's eyes: Defining danger in the wake of sexual assault. *Journal of Social Issues, 39* (2), 63–80.

Shontz, F. C. (1975). *The psychological aspects of physical illness and disability.* New York: Macmillan.

Silver, R. L., Boon, C. & Stones, M. H. (1983). Searching for meaning in misfortune: Making sense of incest. *Journal of Social Issues, 39* (2), 81–102.

Silver, R. L. & Wortman, C. B. (1980). Coping with undesirable life events. In: J. Garber and M. E. P. Seligman (Eds.), *Human helplessness.* New York: Academic Press.

Silver, R. L., Wortman, C. B. & Klos, D. S. (1982). Cognitions, affect and behavior following uncontrollable outcomes: A response to current human helplessness research. *Journal of Personality, 50* (4), 480–514.

Singer, J. L. (1978). Experimental studies of daydreaming and stream of thought. In: K. S. Pope & J. L. Singer (Eds.) *The stream of consciousness.* New York: Plenum.

Taylor, S. E., Wood, J. V. & Lichtman, R. R. (1983). It could be worse: Selective evaluation as a response to victimization. *Journal of Social Issues, 39* (2), 19–40.

Tyhurst, J. S. (1951). Individual reactions to community disaster: The natural history of psychiatric phenomena. *American Journal of Psychiatry, 107,* 764–769.

Vachon, M. L. S., Lyall, W. A. L., Rogers, I., Freedman-Letofsky, K., Freeman, S. I. I. (1980). A controlled study of self-help intervention for widows. *American Journal of Psychiatry, 137,* 1380–1384.

Volpert, W. (1974). *Handlungsstrukturanalyse als Beitrag zur Qualifikationsforschung.* Köln: Pahl-Rugenstein.

Volpert, W. (1982). Das Modell der hierarchisch-sequentiellen Handlungsorganisation. In: W. Hacker, W. Volpert & M. v. Cranach (Eds.) *Kognitive und motivationale Aspekte der Handlung.* Bern: Huber.

Weiss, J. M. (1971 a). Effects of coping behavior in different warning signal conditions on stress pathology in rats. *Journal of Comparative and Physiological Psychology, 77,* 1–13.

Weiss, J. M. (1971 b). Effects of punishing the coping response (conflict) on stress pathology in rats. *Journal of Comparative and Physiological Psychology, 77,* 14–21.

Wiener, A., Gerber, I., Battin, D. & Arkin, A. M. (1975). The process and phenomenology of bereavement. In: B. Schoenberg, I. Gerber, A. Wiener, A. H. Kutscher, D. Peretz & A. C. Carr (Eds.), *Bereavement: Its psychosocial aspects*. New York: Columbia University Press.

Williams, C. C. (1983). The mental foxhole: The Vietnam veteran's search for meaning. *American Journal of Orthopsychiatry, 53* (1), 4–17.

Wortman, C. B. (1976). Causal attributions and personal control. In: J. H. Harvey, W. J. Ickes & R. F. Kidd (Eds.), *New directions in attribution research*. Vol 1. Hillsdale, N. J.: Lawrence Erlbaum Associates.

Wortman, C. B. (1983). Coping with victimization: Conclusions and implications for future research. *Journal of Social Issues, 39,* (2), 195–222.

Wortman, C. B. & Brehm, J. W. (1975). Responses to uncontrollable outcomes: An integration of reactance theory and the learned helplessness model. In: L. Berkowitz (Ed.), *Advances in experimental social psychology* (Volume 8). New York: Academic Press.

Wortman, C. B. & Dunkel-Schetter, C. (1979). Interpersonal relationships and cancer: A theoretical analysis. *Journal of Social Issues, 35,* 120–155.

Wortman, C. B. & Lehman, D. (In press). Reactions to victims of life crises: Support attempts that fail. In: Sarason, I. G., B. R. Sarason (Eds.), *Social Support: Theory, Research, and Applications*. The Hague: Martinus Nijhof.

Part III

Problem-Solving and Performance Control

Chapter 9

Mechanisms of Control and Regulation in Problem Solving*

Rainer H. Kluwe and Gunnar Friedrichsen

According to most cognitive models of human action, behavioral life is determined by a series of decisions between several goals and action alternatives. These goals and action alternatives are considered well-defined and stable throughout the decision-making process. Within this theoretical framework, the degree of cognition-behavior consistency depends on whether or not an actor sticks to the goal chosen and enacts the action alternative most appropriate to attaining that goal (see Parts I and II of this volume). This chapter is based on a different theoretical perspective. We believe that many episodes in an individual's behavioral life entail a *problem* in finding a set of action alternatives to choose from. In many situations, people do not even have a clear-cut conception of the goal they are striving for. Someone may decide, for example, to remodel his home, but he does not have a clear idea of the changes he wants to make. Equally important is the fact that many situations require the consideration of alternative goals as well as the handling of competing goals. When goals, or action alternatives or both are ill-defined, a theory claiming to close the cognition-behavior gap has to account for processes involved in defining and solving the relevant problems.

This chapter deals with some cognitive activities that are conducive to the organization, and to the course of human thinking, for example, when solving a difficult problem. Cognitive psychology offers general models for human problem solving, for example the model of means-end-analysis (Newell & Simon, 1972; Lüer, 1973; Dörner, 1974). Such models or rather some of their components have been prepared by German psychologists like Selz (1913, 1922), and especially by Duncker (1935). However, on a less abstract level we find considerable evidence for individual differences in problem solving (see Simon, 1975). Individuals organize and direct their cognitive endeavours in different ways and to different degrees. Therefore, we have to ask about those mechanisms that may contribute to such differences.

* *Acknowledgment.* Parts of the work reported in this chapter have been financially supported by a grant of the Foundation Volkswagenwerk, Hannover, awarded to the first author.

In addition, individuals do not always cope with a problem in a straightforward manner. There are changes in strategies, modifications of plans; goals are reevaluated, or are even abandoned, goal-directed actions are terminated and so forth. Problem situations are states of uncertainty. Not only does a person lack the necessary knowledge in order to act in such situations, he also does not really know how to find a solution path (i. e., actions). The same problem situation may allow for different approaches to find a plan for action. Consequently the solutions may also look very different. Quite often a person will only be aware that something special has to be done, that a certain cognitive effort is required. The right way to tackle the problem, that is, the way a person copes with the situation mentally must first be found. This uncertainty requires monitoring of one's own cognitive efforts, of the effects of these efforts, and it requires decisions, according to certain criteria, with regard to the course and goals of one's own cognitive initiative. Furthermore, the execution of plans must be monitored and regulated, in order to insure that one's actions are directed towards the desired goal state.

Requirements of a Theory of Problem-Solving

It will be one goal of this chapter to show that in searching for the solutions to problems, decisions are made that refer to the course, the organization, and the direction of one's own cognitive endeavors. A second goal is to show that thinking as it is expressed in behavior may undergo many shifts, switches, and modifications; hence, a person's behavior entails considerable intraindividual flexibility. The precondition is the monitoring of one's own approach in the search of a solution to a problem, of its organization, its effects, and so forth. Such cognitive activity, that is monitoring and subsequent regulation of one's own cognitive efforts, will be referred to as *executive control*. The notion of an executive component in simulation programs is closely related to this conception. It refers to a main program that controls and regulates the work of subroutines that are used in order to search for the solution to a given problem.

Another related concept from developmental psychology is metacognition. It refers to a person's knowledge about thinking, and to cognitive activities dealing with one's own cognitive state and with one's own cognitive efforts. An important cognitive skill with respect to this domain is *planning*. The intention to make a plan for the search for a solution, be it a mathematical or social problem, may lead to a plan which is a frame that guides action (i. e., a program for action). A plan reduces uncertainty with regard to future action, and it requires a considerable amount of decision making about the organization and the direction of one's own approach (i. e., monitoring and regulating). It demands selecting and evaluating strategies for the solution of a problem, the prediction of outcomes of one's own activities in the solution search, the setting up and the evaluation of goals for action, the monitoring of the execution of a plan, and so forth. This chapter will focus on executive control and planning.

The Question of Consistency

As emphasized above, problem situations are states of uncertainty. Aside from problems like the "Tower of Hanoi", arithmetic tasks etc. in everyday problems the desired goal state may be uncertain and illdefined (for example how to manage a small group of people in a business; how to arrange a lesson at school; how to arrange the communication with a neighbor). Thinking about how to reach such goals, and thinking about whether to pursue such goals may be accompanied by actions that do not always seem terribly consistent, even for the observer. Following a statement by Norman (1978) one has to assume that "a lot goes on in the mind" of a problem solver. Thinking about one's own efforts in coping with a problem, about the effects of one's own actions, the usefulness of a plan, and the rationality and the importance of goals may give rise to many changes. These changes are not always apparent to the observer. A professor who does not meet deadlines for a publication in one case, but does in another or who rejected an invitation to give a talk, but later accepts it may be considered as inconsistent with regard to his or her behavior. However, we do not know much about the professor's perception of certain problem situations, the evaluation of goals and of the strategies used to cope with certain demands. For an observer such changes of action are hard to understand, and may be easily judged as inconsistent. Yet there may be nothing wrong in such changes and they do not necessarily suggest a lack of consistency between cognition and behavior. It may well be the case, as we will see with planning models, that a person generates a plan that provides for alternatives of action. The shift from one type of action to another does not necessarily mean inconsistency.

To search for invariants in a person's behavior (e. g., prosocial behavior) in such cases is usually misleading. It is tempting to take such invariants or abstractions as cognitive entities that bring about such behavior. The cognition-behavior-consistency question is basically wrong when it is directed at the lack of overlap between an assummed, for example "prosocial demon", and the corresponding behavior.

In order to cope with a certain problem a person has available knowledge about facts in this domain and procedural knowledge about how to generate a plan and how to organize the solution search in order to reach the desired goal state. There are no content-specific agents that activate only prosocial behavior or conscientious behavior. There are no procedures that are a priori, socially or ethically, more or less valuable. However, the amount and the type of data that are subject to processing may vary within and between individuals as may the amount and manner of processing itself. What we can ask is the following:

1. How well does a person consider and balance competing goals of different importance in a problem situation? (Many people do not even realize that there are competing goals.)
2. How well does a person's approach correspond with certain given demands? (For example, given an important problem with high risks, does a person make a plan before starting solution search?)

The first is a question of consistency within a person: one may become aware that now a goal X has become more important than the competing, usually higher ranking goal Y. This requires actions that probably do not satisfy goal Y conditions.
⚹

Related to this point is another point: thinking does not end with action. A person that generates a plan for how to convince another person to buy a house may for several reasons abandon this plan while executing it. The situation may have changed since the plan was generated, or the other person may react in an unexpected way. There is no reason to assume any inconsistency with regard to the person's behavior. However, action and its effects are used as data in evaluating the initial plan.

At the core of these changes of mind is the notion of *executive control*. The amount and type of control and of regulatory decisions while thinking about action as well as when acting contributes to the intraindividual variability and flexibility of thinking as it is expressed in behavior. The goal of this chapter is to describe such steps in problem-solving processes, that is, when decisions are made about the course of the solution search. It will be evident that the controlling and the regulatory activity is an important source for important changes, modifications of initial approaches, of intentions and goals, of perspectives taken, and of evaluations.

The second point refers to the question of how appropriate a solution approach is with regard to a desired goal state. One may find a person who clearly follows incorrect strategies, for example for solving a conflict with a neighbor. The point here is: how well do people tune their own solution efforts to task demands? Strongly related to this question is the problem of assumed cognitive competence and corresponding performance. Do people use and apply the available knowledge? It is the purpose of this chapter to show possible conditions that may prevent people from selecting actions that match the available knowledge.

The chapter is organized as follows: in Part 2, general assumptions and models referring to control and regulation of cognitive activities will be discussed. Part 3 deals with the essential attributes and components of recent computer planning models. Finally, in Part 4 we will focus on available empirical findings with regard to executive activity during problem solving and cognitive performance.

Models of Cognitive Control in Problem Solving

The Concept of an Executive

Pylyshyn (1980) states that the "appropriate way to functionally characterize the mental activity that determines a person's behavior is to provide an initial representational state; then a sequence of operations is assumed to transform the initial state into intermediate states. All intermediate states constitute claims about the cognitive process." (p. 120). What we are interested in are those activities that control and regulate the flow of processing as hypothesized above. With regard to this problem Pylyshyn has discussed desiderata, that is executive requirements that were not met by psychological models ten years ago (Pylyshyn, 1974). The question has been reformulated in terms of an analogy. Pylyshyn compares control processes in human organizations with control processes in information processing. In models of information processing the main work with regard to this point (i. e., control and regulation) has to be done by an *executive component*. Earlier, Neisser (1967) selected as a title for one chapter in his book "The problem of the executive." The reason for this headline has been the philosophical discussion of a homunculus invoked by the invention of an executive component in information processing models. Neisser emphasized that the problem of infinite regress can be solved. The simulation models available at that time, for example Newell and Simon (1963) and Reitman (1965), supported Neisser's position. The concept of an executive component, in terms of a main program, achieving control and regulation of subprograms working on the solution of the problem made the assumption of an executive part plausible.

Today, it seems to be widely accepted that a hypothetical executive component is postulated in models for human information processing in order to take into account those steps in problem solving that achieve control and regulation of one's own cognitive endeavors. Executive routines in simulation programs usually are expected to fulfill the following tasks: (a) call routines at lower level; (b) keep track of the programs at work; (c) evaluate progress (from feedback of results); (d) interrupt the solution effort; (e) shift to another subgoal (if a more important goal is given or if no progress is made); (f) register when a subgoal is reached. Compared to the subprograms the executive component works with other information. Reitman (1973) put it as follows: the executive component does not actually solve the problem, "it does not dirty its hands itself ..." (p. 58). Concepts like *executive control, executive function* (Butterfield & Belmont, 1975), and *central processor with current plan and executive monitor* (Bower, 1975) all refer to some controlling and regulating instance in simulation models. Hilgard (1976) assumes that there is an *executive ego* necessary in order to decide about competing subprograms: "At the top is an executive ego or central control structure that has the planning, monitoring, and managing functions that are required for appropriate thought and action" (p. 146). According to Bower (1972) the notion of an executive has become one of the fundamental constructs of cognitive theory "that controls the internal show of information processing" (p. 107).

The question of whether human beings indeed need such a mechanism in order to solve problems successfully does not lead very far, as Bower points out. This part of the program does not necessarily correspond to anything inside an individual. The executive routine organizes and regulates a program's search for a solution in a way which is assumed to be similar to a human problem solver's procedure: "The humming of the well-tempered computer is proof positive that the executive is not an occult entity" (Bower, 1972, p. 108).

There are, in other fields of psychological research, hypothetical central mechanisms for the control of cognitive activity that are very much related to the concept of an executive. A hypothetical mechanism is described by Luria (1973) with regard to the human brain and is based on physiological assumptions and observations made with brain lesions. Luria assumes that there are three functional units in the brain: "A unit for regulating tone or waking, a unit for obtaining, processing and storing informations . . ., and a unit for programming, regulating and verifying mental activity" (1973, p.43). The last one of these units would correspond to the concept of a central, executive component as described above with respect to computer programs. Luria localizes this unit in the frontal lobe.

In developmental psychology there is also a distinction between schemata which is comparable to the main idea discussed here. Pascual-Leone (1976), referring to the central concept of a schema in the developmental theory of Piaget assumes three types of schemata: figurative schemes, that represent stored data about the domains of the reality; operative schemes that correspond to procedural knowledge for the transformation of figurative schemes; and finally, Pascual-Leone describes executive or operational schemes ("operatoire" in Piaget's work): ". . . executive schemes are functional entities which monitor (i. e., the control, planning functions) the combination and temporal order or activation of schemes so as to produce a given complex goal directed performance" (Pascual-Leone), 1976, p.114).

With regard to the domain of intelligence, recently Sternberg (1979, 1980) has proposed that executive mechanisms should be taken into account with respect to intelligent performance. Sternberg, explicitly referring to the concept of metacognition in the work of Brown and Flavell extended his componential theory of intelligence by *metacomponents*. These are higher order control processes that are assumed to control and regulate the work of lower level components. Sternberg (1980) assumes that there are metacomponents responsible for the selection of lower order components, for the selection of strategy, for the decision about speed-accuracy trade-off, and so forth. This list is to some degree similar to the set of executive decisions offered by Kluwe (1981, 1982). Furthermore, Sternberg and Brown as well as Kluwe assume that important individual differences in cognitive performance may be traced back to the functioning of such executive control mechanisms.

All these approaches have in common the conception that considerable parts of individuals' thinking when solving a problem consist of directing, organizing, and evaluating their own cognitive endeavors. Decisions have to be made such as whether or not to generate a plan, how to generate and execute that plan, whether to pursue a goal or not, and so forth. To a considerable degree thinking is not automatized and, therefore, it requires deliberate organization and composition. According to our view, the developmental research on metacognition during the last ten years has had much influence on promoting this position.

The variety of situational demands and the numerous alternatives in responding to a problem situation require a flexible and selective use of one's own thinking ability. In addition, the human organism also needs executive procedural knowledge in order to maintain and to guarantee cognitive activity of a system that is subject to emotional and motivational oscillations.

Metacognition

Assumptions About Metacognition

Metacognition was introduced to psychology by John Flavell (1979). The concept refers (a) to subjects' declarative knowledge about cognition, among others about their own cognitive activities and capacities, and (b) to procedural knowledge, that is, processes that may be activated in order to control and regulate one's own thinking (cognition about cognition; see also Kluwe, 1981, 1982). The origins of this domain have been systematic observations reported by Flavell and his coworkers with regard to children's cognitive performance. Obviously younger children lack knowledge about cognition; they engage less in monitoring their own cognitive states and activities.

Flavell and Wellman (1977) offered a taxonomy of metacognitive knowledge mainly referring to memory. This taxonomy broadened our view with regard to the question "what is memory development the development of?". In 1981, Flavell proposed a model of "cognitive monitoring." This model represents the first approach to make more precise the relation between cognition and metacognition. Of special interest is Flavell's notion of metacognitive strategies in this article. According to this approach, metacognitive strategies are cognitive actions directed at the monitoring of cognitive progress (compared to cognitive strategies applied for making cognitive progress). The knowledge about both, cognitive and metacognitive strategies, is referred to as metacognitive knowledge. In addition, Flavell introduces here the notion of metacognitive experiences: they refer to feelings about one's own cognitive states and cognitive activities (e. g., feeling puzzled by a result). The model assumes that metacognitive experiences may trigger metacognitive activity. Though Flavell himself feels that a more precise concept of metacognition is necessary, his theoretical and experimental work has initiated an immense body of research in this domain, and it has raised many questions for psychological research.

Another line of research was developed by Brown (see Brown, 1978; Brown & De Loache, 1978). She refers mainly to what Flavell calls metacognitive strategies. It was an important advance when Brown conceived of *metacognitive skills* as executive functions, referring to the architecture of simulation programs. Brown (1978) gives the following list of executive functions: ". . . (a) predict the system's capacity limitations, (b) be aware of its repertoire of heuristic routines and their appropriate domain of utility, (c) identify and characterize the problem at hand, (d) plan and schedule appropriate problem-solving strategies, (e) monitor and supervise the effectiveness of those routines it calls into service, and (f) dynamically evaluate these operations in the face of success or failure so that termination of strategic activities can be strategically timed. These forms of executive decision making are perhaps the crux of efficient problem solving because the use of an appropriate piece of knowledge or routine to obtain that knowledge at the right time and in the right place is the essence of intelligence" (1978, p. 82). Individual differences with respect to intelligent behavior may thus be traced back to the variability of executive activities in information processing.

These cognitive activities may also be considered as central tasks of an executive program in simulation models (see p. 187).

Bringing together assumptions about metacognition and executive control with some principles of the architecture of simulation programs, the following *general framework for executive processes* directed at the *how* of thinking is possible.

A Framework for Executive Processes

If one accepts the distinction between declarative and procedural knowledge (see Ryle, 1949; Anderson, 1976) then there are different types of knowledge that might have an impact on problem solving.

Considering the declarative part, we distinguish between domain knowledge and cognitive knowledge. By domain knowledge we mean stored data about the domain of reality where the problem has to be solved (e. g., economy, arithmetic, social relations, and so on). Corresponding to that we distinguish economical knowledge, mathematical knowledge, social knowledge, and so forth. "*Cognitive knowledge* refers to an individual's stored assumptions, hypotheses, and beliefs about thinking" (Kluwe, 1982, p. 203). A taxonomy of different categories of cognitive knowledge had been proposed by Kluwe (1981; 1982); Flavell & Wellman (1976) refer to this knowledge type as metacognitive knowledge.

General cognitive knowledge means, for example, an individual's assumption that difficult problems require many resources and a long solution time; also, an individual's belief in having, in general, difficulties in finding solutions for social problems, belongs to this category. *Domain-specific cognitive knowledge* refers to a person's assumptions and beliefs held for a certain domain, for instance social relations, with regard to cognitive activities. For example, someone may be sure that he will be able to solve every electrical problem, however, given problems in the domain of social communication he will feel helpless. Or a person may know that social problems require careful analysis before action, because of the web-structure of such problem situations which may have undesired side effects for example, for persons not involved. Both types of data are necessary in problem solving; the amount as well as the precision of such data may contribute to the manner in which a person approaches a problem.

With regard to the procedural knowledge, one may distinguish between *solution processes* and *executive processes* (see also Schoenfeld's distinction between tactical and strategic decisions, 1981). Solution processes are applied for the problem at hand in order to transform problem states until the goal is reached. Executive processes control and regulate the selection and execution of solution processes. *Executive control* may include the following basic activities: (1) Classification: Determination of the status, the type of certain ongoing cognitive activity and of one's own cognitive states; (What am I doing? What is required? What is my problem?). (2) Checking: Determination of the progress and the success of solution activity; determination of the plausibility of results; keeping track of the solution activity; determination of conflicts between different goals. (3) Evaluation: Determination of the quality of the solution process, and of the attained state with regard to certain criteria (efficiency, risk, etc.), determination of the costs of alternative strategies, the importance of goals; (4) Prediction: Determination of the possible course of solution processes, possible alternatives, determine possible outcomes, side-effects. These control-actions are invoked when certain conditions, that is, states are given; they correspond to specific subroutines that achieve classification, checking, evaluation, and prediction.

Regulatory executive decisions may concern the following aspects of problem solving: (1) The amount of cognitive capacity a person is willing to invest in the solution of a problem; (2) The subject of information processing, that is, which segment of a problem is to be worked on; (3) the speed and (4) the intensity of cognitive activity, that is, the frequency of operations, the amount of information which is processed, the duration of processing in cases of failure. Regulatory decisions with respect to point four determine to a considerable degree whether information processing is exhaustive or not. Sternberg (1979) reports that this, among others, is an important source of developmental differences.

Executive processes, considered as procedural knowledge, allow for: *(a)* The acquisition of information about one's own ongoing cognitive activity; *(b)* The maintenance, continuation or transformation of cognitive activity. Executive procedural knowledge may be conceived of as stored condition-action-connections (see also Chi, 1984). Conditions correspond to symbolized representations of cognitive states during problem solving like the amount of solution effort, the duration of solution search, the type of problem given, the distance to goal, and so forth. Actions would correspond to subprograms directed at *(a)* the control of the solution process, and *(b)* the modification, that is, regulation of the solution process. Executive decisions, then, are rules for the control and the regulation of cognitive activity given in certain states. An example would be the following rule: "*if* very long solution time" (condition) – "*then* check distance to goal" (action). An action like this may call a subprogram in order to check the reached state and previous activity; or it may call a subprogram to estimate the effort that has still to be invested. It is a control-decision. Regulatory decisions are represented in the same fashion. An example: "*if* no success" (condition) – "*then* increase resources for information processing" (action). Again, this action will invoke a certain subprogram that would achieve this goal, for example, by selection of further strategies, and by focussing on relevant parts of the problem situation.

Regulatory executive decisions may be activated directly as a consequence of certain perceived states that result from the application of solution operations, and that match a certain condition. For example, "*if* effects of operators are not reversible" (condition) – "*then* increase intensity of solution search" (e.g., by slowing down, making a plan, action).

Other states, however, may be the result of specific executive control: "*if* long solution time – *then* check solution approach." The outcome of this checking procedure may be the identification of an approach as inadequate. This state may match a condition which is connected with appropriate regulatory activity, like search for other strategies, activation of an alternative strategy, or even termination of the solution approach.

As a consequence of this framework, executive decisions thus involve two types of knowledge: (a) the identification of a certain cognitive state out of many as relevant for the solution (identification of a condition). There has to be declarative knowledge available, for example about the approximate duration of learning and solution processes, that allows such an identification of a given cognitive state, for example, searching for a solution too long, or searching without a plan; (b) the subsequent action connected with this state, directed at control or regulation. This requires procedural knowledge about how to control, how to evaluate, how to increase speed, how to generate a plan, and so forth. Both types of knowledge are acquired during cognitive development and during experience with problem solving. Also, the connections between identified conditions and appropriate actions for control and regulation have to be acquired.

Selz (1913, 1922) in his earlier work on problem solving already described processes that are quite similar to executive processes discussed here. For example, he describes processes like the evaluation of strategies or the checking of solutions. Interesting enough, he postulates a readiness for executive control that may be reduced, for example in case of easy problems. If a problem solver erroneously believes in the easiness of a solution for a problem this may lead not only to wrong approaches, but also to persistent search in a wrong direction (Kluwe, 1983).

Executive Decisions and Action

Changes of Mind

The degree to which a person monitors and regulates his own search for a solution in order to reach a desired state may be a source for shifts and changes in the overt behavior that erroneously may be judged as inconsistent.

Strategy shifts means that the initial goal is still valid, however essential attributes of the approach towards the goal have changed. To the observer the resulting behavior is not easy to match with the assumed goals of the acting individual. *Switching goals* means that initial goals may no longer be valid or desired. In addition, previous goals may receive different weights during the process of attaining them. More common are situations where competing goals have to be delt with. In this case the evaluation of a previous goal comes into conflict with the evaluation of new goal, and this may be a source for changing behavior. As a consequence, checking and evaluation of one's goals, strategies and effort in solving problems may lead to considerable behavioral changes: goals may be abandoned, effort may suddenly be increased, organization of strategies may be altered, and so forth. Such changes of mind expressed, then, in behavioral shifts are not always easy to understand for the observer.

Incomplete Procedural Knowledge

Another discrepancy occurs when the observable approach in solving a problem is obviously not appropriate for the situation at hand and does not correspond to a person's assumed knowledge. The question is: why is a certain knowledge not expressed in behavior. According to our view there may be three possible reasons for that: (a) *Incomplete executive decision rules:* an executive decision (as described on p. 191) has a condition and an action component. Imagine the communication between two persons *(A* and *B);* person *A* might become aware that the dialogue and future communication has to be very carefully arranged in order to avoid misunderstandings and unnecessary risks with regard to this desired communication. However, person *A* may lack the knowledge for how to regulate his own social communicative behavior given this desired goal: He may therefore simply terminate the communication or continue it in in unusually distanced manner. This example indicates that the condition-side of an executive decision may be given (classification of a certain state, e. g., as riskful), but the person lacks the action part for regulation. He has the knowledge that a certain classified state requires special efforts, has to be handled in a different way than usual, but at the same time this person lacks the appropriate knowledge for how to regulate and/or how to modify his own behavior in order to make it more attuned to the perceived state. Thus, the behavior of this person may be judged inappropriate or inconsistent with regard to a knowledge, which he or she in fact has. However, the knowledge that a certain situation has to be classified as risky does not mean that the corresponding knowledge is available on how to organize behavior in order to guarantee reaching the desired goal (i. e., continuing communication in this case). To know that a situation requires careful planning of one's own activity, and the intention to make a plan, this does not guarantee that a good plan will indeed be made. As a consequence we will meet individuals, who are aware of a given state during problem solving, that is who have knowledge about features of the situation and about requirements of the situation, however they lack regulatory knowledge.

(b) *Reduced readiness for control:* Executive knowledge might be available, however not used. The reason for this is that the encoding of the problem situation does not match the conditions of executive decision rules directed at control. Consequently, for example, there may be less monitoring and regulating activity. From cognitive psychology we know that individuals may persist with erroneous solution approaches for some time because their perception and understanding of the problem is incomplete (see Selz, 1913, 1922; Kluwe, 1983). The problem solver may search in the wrong direction being convinced that he or she is right. Selz emphasized that the perception of a problem may be incomplete and therefore the subsequent solution process follows a wrong direction. In a similar way, the assumed easiness of a problem, that is the assumed availability of the solution path may lead to neglect of control and regulation. Selz (1913, 1922) emphasizes that in situations of assumed easiness the readiness to monitor one's own cognitive efforts and the effects of one's own behavior is reduced. The encoding of the situation makes those executive decisions that trigger control actions less likely. That is why an apparently wrong direction for reaching a desired state may be pursued so persistently.

(c) *Lack of cognitive knowledge in a domain:* A third interesting case has been described by Miller, Galanter, and Pribram (1960) as being "trapped into loops." That means that a problem solver may search for the solution for a problem in a wrong direction, however, he does not become aware of this. The reason is that the problem solver is concerned with the solution of subtasks which are obviously correct but do not lead to the goal. Students have this experience with mathematics when they are performing computations over many pages, which may be correct, but which are not related to the main goal. It is important to note that the computational knowledge is there, however the performance does not correspond to that. Take as an example a person who wants to improve the work and achievement of a human organization. He will initiate certain changes in order to increase the attendance of the members, to improve the flow of information between the members, etc. All these subgoals may be reached to a satisfying degree; however they may not lead to a higher level of working output and achievement. The problem is the relation between subgoals and the main goal. This person, however, reinforced by reaching the subgoals, may still pursue similar goals in order to reach the desired state. Miller, et al. at this point refer to reference values a person may consider and which are tolerated before the execution of a plan is terminated. "The 'halting problem' is unsolvable. The only thing we can do is to adopt an arbitrary criterion that reflects the value to us of the outcome. We can say, "I will try to do it in this way, and if, after N years, I have not succeeded, then I will stop." The problem is then to determine N as a function of the importance of the problem." (1960, p. 171). Indeed a problem solver may lack such reference values which again may be a function of the amount of experience in a domain. As a consequence, his ongoing unsuccessful performance is not in accord with his cognitive competences since he lacks the knowledge of how to untie the loop. He may for a long time not become aware of that. The knowledge referred to here is called cognitive knowledge (see p. 190). It means knowledge a person holds with respect to solving problems in certain domains (e. g., it will take me an hour to convince this person; this problem will require long time and much effort). In addition, Miller et al. point out with regard to this problem and referring to Lewin, that aspects like persistance, satiation, or level of aspiration may play an important role here.

One may ask to which degree assumptions about the control and regulation of cognitive efforts are implemented in simulation programs. These programs are taken here as models of cognitive activity in the sense of Herrmann (1983), that is, as *useful fictions.* One may consider such models as guidelines for psychological research. In the following we will focus on planning processes in problem solving. Planning requires cognitive activities like prediction of outcomes and states, selection and evaluation of strategies, monitoring the execution of the plan, and so forth. Planning incorporates much cognition about cognition, and it results in a frame for behavior which again may be used to monitor the progress of one's own solution effort. There are simulation models that explicitly focus on planning problem-solving activity. We will start with a discussion of the general problem solver (GPS) with regard to executive functions and planning; following that will be a description of ABSTRIPS and NOAH developed by Sacerdoti (1974, 1977), and of the OPM-model by Hayes-Roth and Hayes-Roth (1979). Finally a discussion of extensions follows, including Wilensky's PANDORA-model (1981) and the paranoia-model proposed by Colby (Faught, Colby & Parkinson, 1977). The goal will be to describe the central components of the different models and to examine the models with respect to the psychological applications. Of central interest are those decisions involved in planning that indicate control and regulation of the problem solver's own approach.

Models of Planning: A Metacognitive Activity

The General Problem Solver (GPS)

The General Problem Solver (Newell & Simon, 1972) consists of an executive program and of heuristic methods. The executive program calls the subroutines corresponding to heuristic methods in order to set up, evaluate, and to attempt goals. The basic heuristic procedure for working on a problem is described as means-end-analysis. Control and regulation of processing is achieved by the executive routine. The heuristic methods are the tools of the executive, that is, the strategies for action. The activation of these methods is the case when the given state is evaluated to match the condition for a method. All important decisions are made by the executive program on the basis of the results fed back by the subroutines. The GPS is a serially working centralized system, that is, there is only one process active at a given time. When a method is activated, then the executive loses control for a while until the segments of a method have been terminated.

The main criticism of GPS with respect to the issue we are addressing (i. e., metacognitive skills) has been forwarded as early as 1973 by Reitman. Reitman questions the psychological plausibility of GPS with regard to the beginning of the problem solving process. Contrary to the program, a human problem solver has to figure out "what is my problem" (see also Duncker, 1935, Konflikt-Analyse), that is, to identify and to evaluate his or her own cognitive state [see also the executive control process of "classification" on page 190; and the first meta-component in the list offered by Sternberg (1980, p.575): "Decision as to just what the problem is that needs to be solved"]. According to this step a problem solver would have to tune himself with respect to the perceived cognitive demands. The GPS, however, "knows automatically what to do from the way it is written" (Reitman, 1973, p.60). Essential metacognitive activities like estimating the time for working on a solution, or estimating the difficulty of the demands, evaluating the risk and the need for intensive thinking are not required. An important point are criteria for search and choice built into the program (e.g., to evaluate subgoals). The evaluation of goals in GPS is based only on a list of differences within the task domains. The differences are ordered according to their difficulty. Most difficult differences are at the top and will be worked on first. Reitman points out, that this requires (a) a large amount of knowledge in a task domain, and (b) the possibility to order the differences. This again is the basis for determining progress. The question then is whether a subject indeed always starts with the most important difference towards the goal, that is in the strategically most reasonable way. This may be confounded with motivational factors that lead to preferences with regard to lower level differences to begin with. An example would be if a person has to meet three other persons in order to talk with them about a raise of the salary. The most reasonable step from a tactical point of view would be to talk to person A, since A has influence on B and C. However, A is known as aggressive and the person may decide to begin by talking with B or C, even if the plan originally provided for meeting person A first. The behavior of the person thus does not match from a rational point of view the intentions of that person. As far as we know there is no research on this problem.

One of the most frequently raised criticisms related to GPS is that the program does not elaborate a plan. Reitman (1973) argues that this is due to the "perfect rationality" of the system: compared to the human problem solver this model assumes "that you have all possibilities in front of you and that you simply walk through the space of possibilities and select the best one. So there is no need for planning here, because you have it all there, and because planning is only required when there is a limited system" (Reitman, 1973, p.60). Actually, GPS has two heuristics, means-end-analysis and a planning method (Newell & Simon, 1972, p.428). Means-end-analysis is not a planning device; instead it is a step-by-step strategy that does not generate an overall plan before a solution is tried. Since means-end-analysis is "seeing only one step ahead" (Newell & Simon, 1972) this method is weak with respect to the criterion of a flexible problem solver that is able to evaluate and revise its approach. What would be necessary is the elaboration of an overall plan before executing detailed solution attempts. The planning method discussed by Newell and Simon (1972) represents such an approach. The basic steps of the method are: (1) abstraction of the problem (distinguish essential/inessential differences and ignore inessential differences); (2) solution for the abstract formulation of the problem; (3) application of the solution as a plan for the original problem; and (4) execution of plan to solve the original problem. This method works in a top-down manner which must not necessarily be the case in human planning problem solving.

Human thinking-aloud protocols were the basis for this planning model. Newell and Simon (1972, p.557) report that especially successful subjects show planning steps similar to those of this method. However, it is admitted that subjects are more flexible: "There is a stiffness and deliberateness in GPS' movement back and forth between the problem space and planning space that contrasts with the subjects' flexibility" (1972, p.557). There is some evidence in the subjects' protocols that they are planning in a top-down fashion; however, one is struck by the fact that subjects start solution attempts without having a complete plan for a solution, and without even knowing if some steps of the plan will be practicable (see, for example, 1972, p.559). Contrary to this observation the GPS-planning method is very formal. However, it has proliferated the basic principles for planning models like ABSTRIPS and NOAH as elaborated by Sacerdoti (1974; 1977).

Hierarchical Planning Models

The fact that the GPS-model does not start the search for a solution with an over-all strategic plan (see also STRIPS; Fikes & Nilsson, 1971) means that the details for a specific solution have to be worked out step-by-step. This has several disadvantages: First, a problem solver working without a plan presumably needs more time for his solution attempt (see Sacerdoti, 1974, for a comparison of STRIPS and ABSTRIPS with respect to this point); second, a problem solver without the ability to plan has difficulties in determining in advance if a problem may be solved or not. The elaboration of an overall plan allows for the appraisal of the question of whether there will be a solution to a problem, at least within limits.

An overall strategic plan for reaching a goal may have the following components: a general direction for the solution, and a set of general subgoals that connect initial state and desired state. An example would be to sell the old car, to get additional money, to buy a new car. This includes the application of general operators like advertisement of the old car, negotiation with the bank, and the decision about a new car. The details of operators (e. g., which newspaper for the advertisement, text of the advertisement, etc.), and the more detailed subgoals are not included. In addition an overall strategic plan might consider guidelines like the avoidance of risk, and avoiding time consuming actions which determine the search for appropriate solutions.

Sacerdoti (1974) proposed a planning program (ABSTRIPS) which represents a further elaboration of the basic ideas given with the planning-heuristic discussed by Newell and Simon (1972). The main principle underlying ABSTRIPS is to start the search for a solution in a simplifying abstraction space. The program determines the goal and tries to achieve a state, a new "world model", the conditions of which match the goal statement. This is attempted at the highest abstraction level until a skeleton plan is available. Then the next lowest space is determined by the executive in which planning is needed: again, the subproblems in the general plan are solved. The executive program invokes itself until a complete plan is available, that is, until the subproblems on all levels of the original, abstract plan are solved.

The program obviously works in a top-down, length-first (breadth-first) manner. The abstraction level is determined by criticality values that are assigned a priori to the preconditions of operators. High values correspond to high criticality, that is, high importance, low values correspond to less important details of a given state. Each action has a list of *if*-conditions, only parts of these are considered in the highest space. Only the most critical details of a domain will be attended to in the highest abstraction space. That is, ABSTRIPS' hierarchical planning activity is based on the distinction between important and less important details in a problem space. In ABSTRIPS, criticality values are partially predetermined; a problem solver, however, has to decide what is considered to be relevant or irrelevant in a given situation.

The fact that ABSTRIPS works in a length-first manner implies that the planning process is continued on each abstraction level until the goal state is true. This means that an unsuccessful plan may be detected in an early state during the search process, before working out any details on lower levels. In case of failure, control is returned to the executive and the search is continued in the abstraction space prior to that subproblem.

One of the problems with ABSTRIPS is that independence of subgoals is assumed, that is, independent achievement of subgoals must be possible. However, quite often subgoals interact; they may have an impact on the preconditions of other subgoals or may even prevent the achievement of other subgoals. An example would be a person asking for a raise; he is talking to person A, gets support, and then talks to B. B, however, may compete with A, and does not usually accept A's decisions. In this case the person should have first talked to B and then to A. The neglect of dependencies of subgoals thus may lead to a failure of the initial plan, the detection of dependencies may lead to the modification of initial plans, too.

NOAH, a 'Network of Organized Action Hierarchies' (Sacerdoti, 1977), therefore uses ordered subgoals in order to overcome this weakness. The model is provided with specific domain knowledge, and with general problem solving and planning knowledge. The program uses this knowledge to work out a plan for a specific task. The basic organization underlying the planning activity is the development of a procedural net. Each node of the net corresponds to an action connected with the prerequisite features of a situation that are necessary to perform this action and with consequences that will occur when the action is executed. Knowledge stored in this manner is used to develop a plan where the consequences of action have to match the preconditions of action performed later. Procedural knowledge here refers to domain knowledge, that is knowledge about the execution of a sequence of elementary operations in order to perform an action. Plan knowledge is implemented declaratively in the contents of the nodes, that is, the input and output states of an action. The relations between nodes are given among others, by parent-child relations. Each node in the net may refer to more detailed 'child-nodes' that represent more detailed subactions. The execution of 'child-nodes' in a certain stored order leads to the output-state of the 'parent-node'. The nodes, or actions at each level of the hierarchy are connected by input-output links. The resulting time sequences represent plans for action with a certain level of detail. A plan then is elaborated as a sequence of actions, starting with a simple, abstract plan that is expanded in a top-down fashion.

What the planner does according to NOAH, is connect actions by examining inputs and outputs of these actions. This is done on different levels of abstraction, starting on a simplifying highly abstract level. What the planner needs according to NOAH is considerable knowledge about actions and about the sequence of actions. NOAH indicates that planning may be poor if a problem solver has insufficient knowledge in a domain.

In contrast to ABSTRIPS, there is a check of the plan before it is executed. "The planning system must take on overall look at it to ensure that the local expansions make global sense together" (Sacerdoti, 1977, p. 28). This is reached by a set of critics that, among others help to avoid the problem given with interacting subgoals. A final examination is necessary since the expansion of plan nodes on more specific levels is local; the final global overview should ensure that all actions of the plan are compatible and may be executed according to a certain order (for example, when somebody is planning to build a house). The plan may include buying land in a certain area of the town and then building a house according to an architect's ideas. On a more detailed level this plan may not be successful since there is not enough money left for building the house after buying the land. This requires a debugging of the plan (for example, changing the area, changing the plan for the house, etc.). Sacerdoti's invention of critics would detect such interacting subgoals.

Also, in contrast to ABSTRIPS, the planning system NOAH may not always necessarily lead to successful solutions at lower levels. In ABSTRIPS, planning continuously narrows the solution space in a manner that makes a solution at lower levels possible. NOAH may develop higher-level solutions that do not lead to successful solutions at a more detailed level. This is an important modification along with the introduction of critics. NOAH recognizes such failures of a plan at detailed levels by using a checking mechanism that ensures that the plan at the higher level is still carried out, and it may "debug the failing approach" (p. 53).

A major feature of NOAH is its distinction between planning and execution of the plan; the system may shift from one activity to the other. The important advantage of the model is that for planning and execution monitoring there are common representations. However, the planning part of the system processes the procedural net as a hierarchy of partially linearized plans (breadth first), while the execution monitor processes the net as a sequence of action hierarchies (depth first, see Figure 16 in Sacerdoti, 1977). The execution monitor asks for the accomplishment of actions. In case of error NOAH tries to create a new plan in order to achieve the goal of the erroneous action.

Sacerdoti considers Noah to be an answer to Miller et al. (1960) who ask for mechanisms like plan generation, execution monitoring, and planning abstract spaces. It is a first attempt to model a control mechanism (i. e., planning).

Planning mechanisms as modelled in NOAH may help in understanding some assumed discrepancies in the behavior of individuals: (1) The system requires a considerable amount of domain knowledge; that is, a person may have the knowledge of how to generate plans, may have definite intentions, however, may not be able to devise a plan for action in a certain domain because there is not enough knowledge about facts, relations, and so forth in this domain. Actions towards reaching the desired goal state may be erroneous or unsystematic. This fits in with the earlier notion that the intention to make a plan does not yet guarantee the generation of a really good plan. Lack of knowledge in a domain may be one reason for that. (2) Sacerdoti's concept of critics is an important point with regard to the checking decisions described earlier. The completion of the goal after planning satisfies the condition for an executive decision (i. e., to check the plan). A person who does not have this knowledge may detect during plan execution that there are subgoal dependencies that may even lead to a termination of the execution and to abandonment of the previous goal. (3) There is a further checking mechanism directed at detecting incompatibilities between local solutions on lower levels and the overall plan. This mechanism may help to prevent falling into loops. People who lack such executive knowledge may proceed in the wrong direction for a long time, and finally will have to realize that their actions do not lead to the initial goal, which again may result in the abandoning of a goal.

The Opportunistic Planning Model (OPM)

Hayes-Roth and Hayes-Roth (1979) proposed a model of planning activity, the basic assumptions of which are similar to those of Sacerdoti. Planning activity predetermines ordered sequences of actions on different levels of abstraction with respect to a given problem situation. Contrary to Sacerdoti's NOAH, however, planning does not follow a mere top-down, breadth-first search alone, starting with a complete plan in a simplifying problem space. Instead, it is assumed that a problem solver may not always proceed in a clear top-down manner. The planning activity in human problem solving may vary with respect to the abstraction levels on which planning occurs, with respect to the focus on planning phases (the planner may, for example, refocus earlier states of his planning, may not necessarily plan forward in time), and with respect to the degree of specification reached at different points in the course of planning. In sum, "people's planning activity is largely opportunistic. That is, at each point in the process, the planner's current decisions and observations suggest various opportunities for plan development." (Hayes-Roth & Hayes-Roth, 1979, p. 276). This does not mean, however, that the course of planning in terms of successive refinement of high level goals and action sequences is rejected by Hayes-Roth and Hayes-Roth. Planning activity of this type is just one possible form of planning that may occur. The authors assume, that the "relative orderliness of particular planning processes presumably reflects individual differences among planners as well as different task demands" (p. 276). The OPM-model thus opens the possibility for the analysis of individual differences by allowing both top-down as well as bottom-up planning. This is well in accord with assumptions of Brown (1978).

The basic components of the OPM-model are, in brief, the following:

(1) Cognitive 'specialists' make 'decisions' directed at the elaboration of a plan. These 'specialists' correspond to sets of condition-action rules in the program. (2) All decisions are recorded in a common 'data structure' which allows interaction and communication between the specialists. (3) It is assumed that different decisions are recorded separately; the data structure mentioned in (2) is subdivided into categories of decisions; Hayes-Roth and Hayes-Roth thus provide a taxonomy of planning decisions (see p. 286): (a) plan decisions about explicit actions that a problem solver intends to carry out; (b) plan abstraction decisions about desired attributes of plan-decisions (like beginning with the most difficult subgoal); plan-abstraction decisions do not specify an explicit sequence of actions; they are decisions about desirable kinds of actions; (c) knowledge-based decisions about specific environmental aspects that bear on the planning process; for example, the registration that a certain part of a problem is the most

difficult part may lead to an instantiation of an initial plan-abstraction decision, namely, to begin with the most difficult subproblem.

Decisions in these three categories, plan, plan-abstraction, and knowledge-base "determine features of the developing plan" (p. 287). Decisions of two further categories are of more interest in the context of executive control since they determine the course, and the organization of the planning process itself, and they influence the general approach to the problem: (d) Executive decisions determine which parts of the plan will be developed, and which type of decisions (specialists) will be made; (e) Meta-plan decisions refer to the general approach to the problem, that is the understanding of the problem, the methods to be applied, and the criteria for the evaluation of a possible solution. There is considerable overlap between types of metaplan decisions and the notion of executive decisions as outlined on (p. 190) (see Kluwe, 1981; 1982; see also the concept of meta-components by Sternberg, 1980).

Hayes-Roth and Hayes-Roth are able to identify decisions of this type in the thinking aloud protocol of one subject. Compared to pure top-down models of planning, the OPM-model is multidirectional. A plan is successively accumulated: "... planning is incremental and, therefore, will rarely produce complete plans in the systematic fashion described above. We assume that people make tentative decisions without the requirement that each one fit into a current, completely integrated plan. As the planner relates each new decision to some subject of his previous decisions, the plan grows by incremental accretion" (1979, p. 304). Furthermore, Hayes-Roth and Hayes-Roth abandon the idea of a simple hierarchical structure for planning. Instead, the OPM-model is best described as a heterarchical plan structure. The different decisions made contribute to the plan, however, they do not form a single hierarchical structure.

As far as we can see, control of plan execution has not been elaborated yet by Hayes-Roth and Hayes-Roth. This would be necessary since all models imply the sequence of planning and then execution of the plan. The shift between planning and execution of a plan as well as the shift between execution and reconsidering the initial plan is of considerable importance in human problem solving. It is by no means clear that planning always precedes execution of the plan so that only a complete, compromising plan is executed.

The OPM-model incorporates some of the components of the framework that has been proposed earlier in this chapter with respect to executive control and planning: For example, the criticism with respect to the GPS-model is no longer valid. Reitman (1973) argued that the GPS-program does not formulate the problem by itself in contrast to the OPM-model (see meta-plan decisions). Furthermore, the decision to start working on the most serious difference between present state and desired state is just one possibility. It depends on the problem solver's plan-abstraction decisions, and on his policy. He may, therefore, postpone unpleasant parts of the planning process. The invention of executive decisions, and of meta-plan decisions comes close to the notion of a "meta-plan" offered by Miller et al. (1960), that is, a plan for forming other plans. However, it is not taken into account that a problem solver may decide to proceed without making a plan first. The model in its details shows how much procedural knowledge a problem solver might need in terms of rules for executive decisions in order to create a successful plan. Lack of such knowledge, especially with respect to categories (d) and (e) will seriously call into question the success of planning, and of the subsequent action. Executive and metaplan decisions are monitoring and regulating mechanisms. Lack of such rules or not applying those rules may lead to wrong classification of the problem and to execution of erroneous plans without checking before. As a consequence, the behavior of the problem solver might be in conflict with other important goals and might lead to costs that are not in accord with the initial goal.

Brown, Bransford, Ferrara, and Campione (1983) conclude that the development of models like GPS, NOAH, and OPM show that "with increasing sophistication, information-processing and artificial intelligence models have gained more power by paying increasing attention to the metacognitive aspects of thinking" (p. 105).

Extensions

In the following we will discuss several problems given with the models described above that have been taken into account in other psychological models. An interesting elaboration of what Hayes-Roth and Hayes-Roth (1979) call meta-planning is provided by Wilensky (1981). He aims at the development of a computer model for understanding plans. This requires that the model be provided with some knowledge about planning. Such knowledge is referred to as meta-planning knowledge by Wilensky. It is knowledge about how to plan; there are two main components, namely, knowledge about meta-goals and knowledge about meta-plans. Meta-plan knowledge is used to achieve meta-goals. This type of knowledge enables the system to understand the plan of another system, for example, a human problem solver, that is communicating with the system.

Wilensky distinguishes meta-themes like "Do not waste resources", "Achieve as many goals as possible" in planning (see also meta-plan decisions in the OPM-model). Given the dominance of certain themes, the system will detect states in the problem space like overlapping goals, conflicting goals, and so forth. If such states are detected in the problem space general meta-goals are set up like "combine plans", "resolve goal conflict". Connected with these goals is procedural knowledge, that is, meta-plans that may be activated in order to satisfy the meta-goal at hand.

One of the major advantages of this meta-plan model is that it allows the explicit handling of conflicting goals, for example, in a complex problem situation. This is an important point: In the course of problem solving a human subject often has to handle conflicting goals; decisions are necessary regarding which goal to pursue. This point has been discussed as one possible source of discrepancies in an individual's actions. Procedural knowledge for detecting and handling conflicting goals prevents a problem solver (a) from actions that are inconsistent with goals and (b) from being helpless, that is having no action available to satisfy a desired state when there are several conflicting goals. The critics in the model of Sacerdoti (1977) were necessary to resolve interactions between subgoals only, not between conflicting goals in the course of solving a problem (like common dilemma). McDermott's notion of policy comes close to this concept of meta-planning knowledge. However, meta-goals in the sense of Wilensky are more general, less domain specific than policies in the sense of McDermott (1978).

Wilensky introduces another highly interesting component that, as far as we know, has been neglected. The general architecture of a planner provided with meta-planning knowledge according to Wilensky is as follows (see also 1981, p.215):

(1) A goal detector: this system component determines the goals of the planner; it has access to the system's "likes and dislikes", to the state of the environment and to the plan structures; (2) A plan generator, which proposes plans for detected goals; (3) An executor, which carries out the steps of a plan and detects errors.

Wilensky claims that the goal-detection component is really new here. It is necessary for a number of reasons and it overcomes the criticism stemming from the fact that "most planners do not worry about where their goals come from" (Wilensky, 1981, p.216). This is true for models like GPS, NOAH or OPM. We will later see that the paranoia-model offered by Colby and his group considers this problem, too. Goal detection is implemented by a mechanism called the *noticer*. This mechanism is designed to register changes of the system's state as well as of the system's environment. This component has some similarity with the knowledge-base decisions in the OPM-model of Hayes-Roth and Hayes-Roth (1979). Those changes in the external environment are registered which are connected with themes and meta-themes of the system. According to our view this enables the system to act in a bottom-up, data-driven manner. More important, the achievement of a goal may be abandoned in order to satisfy a more prominent goal. We will later see that Colby's simulation model for paranoia explicitly includes this possibility (see also Simon, 1967, and the notion of a noticing program with respect to goals and goal hierarchies).

The goal-detector mechanism incorporated in the model of Wilensky meets an important requirement for model construction that has been formulated earlier (e.g., Reitman, 1965).

What is necessary for a system satisfying this requirement is the monitoring of the environment with respect to information that may be important for the system. This has been described as another source for shifts in actions, that is, new goals that have to be evaluated and compared to the goal at hand. The lack of monitoring one's goals in problem solving, that is, evaluating the goal against other goals and the lack of continuously checking one's own perception of the problem situation may lead to behavior that is highly inconsistent. This is especially true for social situations. Colby (1978) in his critique of mental models also emphasizes the detection of goals. Many AI-programs are, in Colby's opinion, of "slave-mentality"; they do not have purposes of their own. Colby argues that any system with its own set of goals and motives, needs a way to judge the efficiency of its own work. This makes a control mechanism for evaluation necessary. The global structure of such a program would be that an object program is monitored by a meta-program. This meta-program achieves self-evaluation and self-regulation. Colby points out that these programs are not physically separate; they use the same hardware, the same interpreter, however, the rules of these programs work on different levels (meta-rules, object-rules; see also Chi, 1984 for this distinction).

In his own work (see Faught, Colby & Parkinson, 1977) Colby tries to take into account such requirements. The most important part in the paranoia-model elaborated by Colby and his colleagues is the affect mechanism. This component of the system makes the system a purposive one. It represents the needs and desires of the system; its processes correspond to rules for interpreting the information held in the belief system of the model and contribute to the modification of affects. This guarantees that the performance of the system is always related to the needs and desires of a system. By inference, beliefs held by the system may change affect values when evidence is found that these beliefs are true. The affect component then computes which affects have to be changed and to which degree. New values for affects may lead to a state where a certain affect, for example fear, requires attention. The system then evaluates if the situation is appropriate to satisfy goals connected with this affect or if the situation has to be changed. Then an intention may be activated which again leads to appropriate actions in order to meet the requirements as given with the present state of affects. Thus, in Colby's paranoia model, a system component is incorporated that makes the system a purposive one, and that evaluates and monitors the system's performance.

Faught et al. (1977) in their model explicitly provide that affects may dominate the cognitive components of the system. The "affect process" may "override" the execution of the other processes that is, of the inference processes and of the intention processes. If one of the affects in the system (e. g., fear) acquires high value then it requires immediate attention. In this case there is no further selection of goals in order to satisfy the needs, using the available knowledge about the world and the situation at hand. Instead, a certain intention or action pattern is directly triggered by the high-affect value. This is true for the paranoid mode of the system. In the paranoid mode, actions are executed like attacks as a consequence of strong affect. In addition, high affect values of the system may lead to erroneous inferences in the model. Inferences are made about the current situation with less evidence than usual in order to add support to beliefs held by the system. Consequently, the system may select action patterns that are not in accordance with the situation, but due to lack of correct inference match parts of the belief system.

The system thus may cause behavior that is not appropriate with regard to the situational features. It may even contradict the one's own belief system; however, strong affect values prevent evaluation, selection and decision making with respect to actions, and they prevent the drawing of correct inferences about the situational demands.

A difficult problem is the relationship between planning and execution of plans. NOAH requires a successful and complete plan before execution. "The problem with doing things this way is that the proper plan for a task usually depends upon the state of the world when the task is to be executed, and this state is difficult to compute in advance" (McDermott, 1978, p. 77). McDermott's important solution in his own model is that parts of the plan or subtasks are executed as soon as scheduling rules allow for this. There are no distinct phases such as first planning and then execution; both parts are entwined. There is no sequence equal to reduce-criticize as in the NOAH-model by Sacerdoti (1977). We consider this to be an important advantage compared with the models of Sacerdoti and Hayes-Roth and Hayes-Roth. Some of the inadequacies of human action given certain problem-situations may indeed arise from the time passed between planning and execution. Especially in social environments, one has to take into account the dynamics of situations. This makes the connection between planning and execution necessary.

Several models have been discussed that deal with the control and regulation of thinking during problem solving. The goal has been to show that a problem solver makes decisions that are directed at acquiring information about the effects, the progress, and the efficiency of one's own cognitive enterprise. The results may give rise to regulations, that is changes and modifications of one's own cognitive endeavor. The resulting picture is a potentially highly flexible course of thinking that is expressed in behavior. To the observer changes of action are not always easy to understand. On the other hand, lack of such decisions, that is, failure to monitor and to regulate one's own cognitive endeavor may result in behavior that is not appropriate with respect to goals, available knowledge, and the situation at hand.

Empirical Results: Evidence for Executive Control

Results from Analyses of Thinking-Aloud Protocols

The instruction to think aloud as a method for collecting data about cognitive processes is questionable. Despite the extensive discussion of Ericsson and Simon (1980), there is considerable evidence for methodological problems with regard to the reactivity of the method, and to the completeness of protocols.

If one is willing to apply this method taking into account its constraints, then it is reasonable to look for verbal statements that may indicate regulatory and monitoring activity, that is, executive decisions directed at the course of the subject's own thinking process. The verbal statements may indicate that the subject is attending to information with regard to his own problem solving process that matches the conditions for certain decision rules, directed at checking, evaluating, planning and regulation.

It is interesting that in earlier work on problem solving relying on thinking-aloud protocols, such verbalizations have been widely ignored. Among others, we analyzed the thinking aloud protocol for a cryptarithmetic task published in Newell and Simon (1972). Sixty percent of the statements judged by five subjects as "metastatements" have not been coded by Newell and Simon, or have been provided with a questionmark according to their coding system. In a recent study by Simon and Simon (1978) such statements with regard to the subject's own approach are explicitly referred to as metastatements, and they gain considerable importance in the discussion of differences between experts and novices.

Metastatements do refer to the course, the organization, the efficiency and the quality of one's own thinking. There are some studies that, relying on thinking-aloud protocols, reveal important individual differences with respect to cognitive performance. Such differences may lead to variations in the regulatory activity as indicated by verbalizations.

Good and Poor Problem Solvers

In Germany Dörner and his research group studied the differences between successful and weak subjects handling a simulated complex system (Dörner, Kreuzig, Reither & Stäudel, 1983). The subjects had to role-play the mayor of a town. The results of the individual's interventions were taken as the basis to separate poor and good mayors. The authors selected central variables, constituting the final states of the town and decided about the success of their subjects.

A comparison of these two groups has been made on the basis of thinking-aloud protocols. As a general result one can conclude that successful problem solvers are significantly more inclined to control their own cognitive endeavors. The following central differences with respect to executive control can be found: (1) Successful problem solvers try to remember to reconstruct earlier states and initial data. This increased checking does not only enhance a more precise storage of one's own trace of problem solving but also ensures that one is still on a promising track. (2) Successful problem solvers do significantly more often evaluate, that is, analyze and judge their own solution activity. This is in accord with point (1). (Compare for (1) and (2) the checking procedure in NOAH p. 198); see also the problem of "falling into the loops", p. 194). (3) Successful problem solvers do more often anticipate and predict possible directions and states, as well as evaluate possible approaches and consequences. This supports the view that successful problem solvers plan their own activity more carefully and, with regard to the execution of plans, more often simulate the subsequent plan execution.

Poor and Good Learners

Results similar to those of Dörner et al. (1983) are reported by Thorndyke and Stasz (1980) who studied poor and good learners. The subjects' task was to memorize and recall a map. As a dependent variable for the distinction between good and poor learners the completeness of the map-reproduction was selected. Again, thinking-aloud protocols were taken as a basis to analyze differences with regard to the learning process. Thorndyke and Stasz report differences between good and poor learners that again correspond to our notion of executive decisions. (1) Good learners regulate in a systematic manner the selection of the contents to be memorized. Good learners explicitly decide about the segments of the map that will be subject to memorizations, and they systematically determine sequences of segments. Poor learners however follow a less planful procedure. (2) Good learners monitor and check their own cognitive state during learning more precisely than poor learners. In the work of Dörner et al. (1983) there was evidence for a higher frequency of checking activity with good problem solvers; here we have data that indicate more precise checking activity. In the study of Thorndyke and Stasz, this is shown by the fact that good learners have more correct information about segments of the map already stored and segments not yet stored perfectly (1980, p. 155). Furthermore, good learners focus their memorization activity on those segments of the map that are not yet available for recall. Poor learners on the other hand only check what they already know. (3) Compared to poor learners good learners show a regulation of information processing in terms of higher intensity. This is close to the notion of exhaustive information processing (see for example Sternberg, 1979, 1979; Kluwe & Schiebler, 1984). Successful learners attempt to memorize one segment of the map completely before they proceed with the selection of the next segment. Poor learners, however, obviously terminate the memorization activity too early (see Sternberg, 1979 for this distinction with respect to age differences referring to the solution of analogies). They proceed with the next segment of the content to be memorized before the preceding segments have been stored safely. The processing of information is not exhaustive in the following sense: The frequency of necessary memorization operations may be too low, or the amount of information which is subject to the memorization operations is restricted.

It is interesting that Reither (1980) reports the same difference with respect to the problem solving procedures of experts and novices. "When analyzing the protocols we found significantly more thematic 'jumpings' with our novices. That means that experts commonly stick to a topic until they find a, sometimes tentative, decision, whereas novices jump from one topic to another, treating all superficially" (p. 11).

In sum, good problem solvers and good learners presumably show more executive control of their own cognitive initiative as indicated by metastatements in thinking-aloud protocols. Good problem solvers and good learners regulate their own thinking in a manner which is marked by more intensive, that is, more exhaustive information processing. These differences with respect to executive activity go along with performance differences.

Experts and Novices

The analysis of differences between experts and novices with regard to the processes during problem solving is important since we may acquire information about learning processes and finally, about the improvement of thinking. The intellectual performance of the expert may be considered to be the goal state of a learning process; accordingly, the intellectual functioning of the novice may be conceived of as the initial state that has to be developed. Simon and Simon (1978) compared two problem solvers, one novice and one expert, with regard to their solution procedures. The basis for the comparison is given by the thinking aloud protocols of both subjects. One of the major differences reported by Simon and Simon is the frequency of metastatements in the protocols. The novice's protocol contains remarkably more metastatements, that is, verbalizations of planning and of other activities, referring to the course of one's own solution endeavor (for example evaluation). This does not mean, however, that the expert is working in a less goal-directed and less planful way. Instead, the expert subject knows how to solve the problem and may apply some of the solution steps in an automatic manner. However, given harder problems, the expert's solution protocol shows more metastatements as well (see 1978, p.331). Obviously the novice in the study reported by Simon and Simon was a good problem solver. The reported solution process as indicated by the meta-statements shows an increased readiness for planning, checking, and evaluation. This is, given the novice state, a reasonable procedure. The expert simply does not need this amount of control.

The findings reported by Dörner et al. (1983) with regard to good problem solvers are in accordance with the results mentioned above. Good problem solvers more often show control of their own solution course. Similar differences are reported in the study by Reither (1980). He compared experts and novices with regard to the problem of improving the life quality of people in certain areas of Africa. The system has been simulated with a computer. Reither, like Simon and Simon, reports that novices show more verbalizations referring to their own solution process. An interesting observation is reported by Simon and Simon (1978). They assume that "physical intuition" may be a major distinguishing feature between novices and experts. The notion *physical intuition* refers to the hypothetical process of constructing a *perspicuous representation* of a problem. "By a perspicuous representation, we mean one that represents explicitly the main direct connections, especially causal connections, of the components of the situation" (Simon & Simon, 1978, p.337). This process corresponds to the construction of a schema that represents the central, critical components and relations of a problem situation. The question is whether such a representation is used as a basis for action, or as a basis for developing a plan. This relates to a question raised earlier that individuals may execute actions towards a goal without a complete plan (see p.196).

The perception of the problem, that is, the feature of the initial representation of a problem developed by a subject may guide the planning and solution process. This is shown in a study by Jeffries, Turner, Polson, and Atwood (1981). Their central assumption is that experts have available a higher order knowledge structure or schema. Given a specific problem, it is mapped into this higher-order structure. Jeffries et al. offer a general model for software design processes. They assume a design schema that is used for the elaboration of solutions for design problems. This schema is a higher-order knowledge structure that governs the search for an appropriate design. It includes abstract knowledge about the overall structure of designs, the structure of good designs and about the design process. Furthermore it includes knowledge about processes to generate software designs. The authors assume that the design schema allows an efficient management of a designer's resources. Given a specific design problem the principal and critical components of the situation are specified; the complex domain is reduced and organized. This is presumably what Simon and Simon mean by perspicuous representation. The basic procedure for the schema driven generation of a design that finally leads to a plan for the solution is to decompose the initial problem into subproblems. This assumption is very much in accordance with the principles of NOAH, the model offered by Sacerdoti (1977). The decomposition process breaks up the problem into a set of subtasks that will solve the problem. The decomposition method then is applied again to the subproblems until a solution is available that can be executed. Jeffries et al. compared the processes of the designing of software from subjects with different amounts of experience. The bases for the comparison were the thinking-aloud protocols of these subjects. Despite considerable variation of design solutions within and between expertise levels, this analysis revealed striking similarities. Obviously, almost all subjects applied the same top-down, breadth-first strategy: decomposing the initial complex problem, that is, breaking it down into subproblems. "They began with an initial sketchy parse of the problem which we have called the solution model" (p. 270). This abstract and higherlevel solution plan is elaborated on a more detailed level. The authors report that none of the subjects continued with the execution of a solution. Instead, a decomposition into subproblems and possible solution paths on several levels followed.

Novices and experts did not differ remarkably with respect to this global procedure. However, the novices did not come up with correct or complete solutions as did the experts. Although similar to the experts with regard to the global decomposition strategy for planning, the novices were lacking the appropriate strategies for handling subproblems. Experts, however, were more effective in using the decomposition method; the protocols reveal a "polished design schema" (p. 271). The reason for novice's inferiority obviously rests in the lack of knowledge constituting a design schema (e.g., knowledge about the overall structure about a good design). Jeffries et al. report that only one of the five novices showed a planning hierarchy of more than two levels. Furthermore, novices did not consider criteria for their design such as efficiency, aesthetics etc. during the planning procedure; that is, they lack the subtle details of a design schema. Another difference is that experts evaluate alternative plans for solutions.

In sum, this analysis supports Sacerdoti's position of a breadth-first, top-down planning procedure. In addition, it also shows that the amount of knowledge which is available for planning obviously determines the amount of planning, in terms of hierarchically ordered levels and in terms of considering alternative directions.

Theory-Guided Studies of Executive Control

In a more systematic manner than in the studies mentioned above executive control during problem solving is studied in the work of Hayes-Roth and her coworkers. There are three studies available where the post hoc manner of studying executive control is abandoned. Instead, the OPM-model is used to derive as-

sumptions about the planning process and about individual differences with respect to planning and control processes. The data basis, however, is again mainly
verbal data from thinking-aloud protocols.

Goldin and Hayes-Roth (1980) required subjects to schedule a route for a
day's errands. The subjects were provided with a map of a town and had to elaborate a plan for a day that should include errands to be accomplished, the order
of errands and the corresponding route. The five subjects were instructed to
think aloud while working on the problem.

The goal of this study was to detect features of the planning process that distinguish between good and poor planners. The OPM-model provided the conceptual framework for the analysis of the thinking aloud protocols. Obviously,
the general assumption has been that good and poor planners differ with regard
to the set of decision categories provided by the OPM-model.

The main results reported by Goldin and Hayes-Roth are as follows:
Good planners make more executive and metaplan decisions than poor planners. It is important that Goldin and Hayes-Roth explicitly connect metaplan
and executive decisions with metacognitive activity: "These two planes describe
meta-cognitive activity: awareness and control of planning process" (1980,
p. 19). The result is in accordance with the findings about executive control and
learning reported by Thorndyke and Stasz (1980) discussed above.

In terms of the OPM-model good planners in addition make more high-level
decisions, that is, they are concerned with the more global aspects of the plan design. They make decisions about the overall design of the plan and its intended
outcomes. Poor planners, however, operate on a more detailed level. Furthermore, good planners shift more often from one plan to another, that is, from one
type of decision to another type. The authors consider this result as evidence for
greater attentional flexibility of good planners. Of particular interest are the individual differences with respect to executive decisions, since these decisions determine the sequences of decisions belonging to all other categories of the data-
structure in the OPM-model. The higher amount of executive decisions reported
for good planners indicates that they proceed more purposefully and that they
ponder and justify the planning process. With regard to metaplan decisions, i.e.,
decisions that determine something like the policy of the plan (for example, use
time efficiently), good planners again showed a higher number of such decisions. This indicates a planning process that is strongly guided, and evaluated
with regard to several criteria. In sum, good planners show more decisions, and
their decision patterns indicate a more purposeful procedure, they review and
evaluate previous decisions more often. This corresponds with findings discussed earlier referring to good and poor learners, and good and poor problem
solvers.

In a second study (Hayes-Roth, 1980), the subjects' control of their own planning activity has been studied in more detail. Central to this question is the notion of executive decisions in the OPM-model. An executive strategy is conceived of as a set of executive decisions that determines the overall organization
of a planning and problem solving process, and that determines the specific
cognitive operations at a certain point in the solution process. Executive strate-

gies determine what data a planner considers, and which planning decisions are made. [The distinction is similar to Schoenfeld's distinction between strategical and tactical decisions (1981); see also the distinction made by Kluwe (1981, 1982) between executive processes and solution processes]. Individuals may adopt different executive strategies and they may switch between different executive strategies.

With respect to the errands-task mentioned above Hayes-Roth distinguishes two alternative strategies that go along with different sets of executive decisions. The "traveling salesmen" strategy is basically a bottom-up strategy, the "scheduling" strategy a top-down strategy. The findings in Part 1 of this study do not confirm the hypothesis of the author. Contrary to the expectation that under time pressure the subjects should adopt the scheduling strategy, most subjects according to their self-reports, adopted the traveling salesmen strategy under both limited as well as sufficient time conditions. This result is interpreted as indicating a certain preference for the bottom-up strategy.

In Part 2 of the study Hayes-Roth instructed subjects to use one of both strategies, under time pressure or not. The instruction for the time pressure condition was to use the scheduling strategy; the instruction for the sufficient time condition was to use the traveling salesmen strategy. Half of the subjects in each group did not receive any hint for a strategy. The results show that subjects can adopt the scheduling strategy, too. In fact, the top-down strategy leads to a reduction of planning time when under time pressure compared to subjects with no instructions at all.

In Part 3 of the study the transfer of successful executive strategies has been studied. Subjects working successfully on a traveling salesmen problem as a primer applied this strategy to subsequent problems of both categories (i.e., scheduling problems, too). The same finding is true for scheduling problems as a primer and the transfer of the top down strategy. It is interesting that the subjects do not distinguish between the different transfer problems. This is accompanied by minor disadvantages due to applying the less appropriate strategy for the transfer problem (e.g., higher planning time).

In Part 4, subjects were instructed to use two executive strategies. Then they worked on problems which required either the scheduling strategy or the traveling salesmen strategy. In fact, the type of problem determined the subject's choice of executive strategies as inferred from the self-reports of the subjects. Most subjects adopted the appropriate strategy when given a certain type of problem. Again the choice of an appropriate strategy goes along with less planning time.

Finally, in Part 5, Hayes-Roth wanted to know whether subjects are able to learn to adopt the appropriate strategy. Subjects worked on primer problems of each type; then half of the subjects worked on transfer problems requiring the scheduling strategy, half of the subjects on problems requiring the traveling salesmen strategy. Indeed, the problem type influenced the choice of the strategy; after experience with both strategies, subjects selected the appropriate strategy given a certain problem type. However, the data also show the clear preference for the bottom-up strategy.

These studies indicate that subjects can adopt different executive strategies for planning solutions; however, they also demonstrate a clear preference towards executive decisions leading to a bottom-up procedure during planning. Successful application and experience with executive strategies may lead to appropriate subsequent use of executive strategies. Probably more important, also with respect to the OPM-model, is that the choice of an inappropriate strategy entails disadvantages with regard to the planning process (time) and the result. This has been predicted on the basis of the OPM-model.

Hayes-Roth and Thorndyke (1980) report findings from three experiments dealing with decision-making during planning. Two aspects were of central interest: Do planners make decisions on different levels of abstraction? Do a planner's previous decisions influence subsequent decisions opportunistically, that is, regardless of the abstraction level? The model of an only top-down planner is doubted.

With respect to the first question, the authors found that subjects in a sorting task in fact distinguish between decisions on different levels of abstraction. Statements taken from thinking aloud protocols were hierarchically sorted as predicted on the basis of the OPM-model.

The second question was tested by providing subjects with the errand problem and with additional constraints. Under one condition a high-level decision was made and a choice had to be made between two subsequent alternative low-level decisions. Under the second condition, one low-level decision was made and the subject had to select one of the two subsequent high-level decisions (high-level: "You have decided to make a circle around the twon, doing each errand as you come to it"; low-level: "You will go to the card and gift shop.").

According to the assumptions of the OPM-model, the subjects should select those decisions that represent reasonable completions of previous decisions: Given a high-level decision first, the subsequent low-level decision should fill the gaps of this previous decision; given low-level decisions first, the subsequent high-level decision should incorporate the constraints of the former. The authors predicted the choice of certain decisions. Indeed, they can show that the previously made decisions have an impact on the selection of subsequent decisions. They consider this finding a confirmation of the assumption that planning includes both, top-down and bottom-up decisions. It is unclear however, how subjects would have decided, if the alternative decisions would also have included decisions of the same type. A similar question has been studied in Experiment 3. Contrary to Experiment 2 one of the alternative questions to be selected was above the level of the prior decision, while the second one was below that level. The prior decisions made one of the two alternatives more reasonable. The subject's decision then represented a top-down or a bottom-up sequence. Clearly, both directions could be shown, that is planning includes top-down decisions as well as bottom-up decisions.

The findings reported by Hayes-Roth and her coworkers support some assumptions of the opportunistic planning model and – a substantive point – suggest that features of the executive decisions in planning procedures contribute to important individual differences with regard to planning performance. This leads to the question of whether one could improve problem solving performance by the training of planning strategies, whereby the focus might be on executive control and regulation. A necessary step in research would be to allow for execution of plans. This was not the case with the studies by Hayes-Roth. However, with regard to our discussion of planning, plan execution, and of monitoring the execution, this would be an interesting point to study.

Training of Executive Control

The empirical findings discussed above support the assumption that training of executive control might contribute to the improvement of the search for solutions in problem situations.

Usually, training programs directed at the enhancement of problem solving are organized around the instruction of problem-solving strategies. Self-regulative aspects are widely neglected. However, it is plausible to assume that the mere instruction of a strategy does not guarantee the flexible application of this procedure given varying cognitive demands. Problem solvers still have to decide when to select a strategy, how to apply it, which segments of the problem space to work on, they have to evaluate the efficiency of the strategy, predict possible outcomes, check results. These executive control activities are not acquired as a by-product. Instead, they have to be learned. Of course, the training of executive decision making alone will not ensure improved cognitive performance. The cognitive operations and strategies that are subject to executive control have to be trained together with monitoring and regulating activities.

There is little research with respect to adult subjects. However, there are two training approaches that explicity aim at the development of executive control with adults. Sternberg (1980) offers a componential "subtheory" of human intelligence. Components are basic units of information processing, cognitive operations. Sternberg distinguishes components with regard to the level of their functioning. Of central interest for our discussion is the notion of *metacomponents*. These are "higher-order control processes used for executive planning and decision making in problem-solving" (1980, p. 575). Sternberg identifies six metacomponents: "1. Decision as to just what the problem is that needs to be solved . . . 2. Selection of lower-order components . . . 3. Selection of one or more representations or organizations for information . . . 4. Selection of a strategy for combining lower-order components . . . 5. Decision regarding speed-accuracy trade off . . . 6. Solution monitoring" (1980, p. 575 ff.). If one compares this list with the one of metacognitive skills offered by Brown (1978) or the framework provided on p. 190 (see also Kluwe, 1981, 1982) one will easily find that there is considerable convergence. We will not discuss the details of Sternberg's componential theory here. Instead, we will refer to his conception of *intellectual skills training* (1983). As a central prerequisite for a training program Sternberg (1983) requires a theory. This theory, here provided by his componential theory of intelligence, allows for justifying the training program, it allows for elaborating a training program on a preliminary basis, and it finally provides criteria for the evaluation of a training program. Sternberg shares our view that both, solution processes and executive control require training. According to his theory Sternberg (1983) proposes the training of executive processes which are obviously equivalent to the metacomponents formulated earlier. He distinguishes nine such executive processes (compared to six metacomponents offered in 1980) that should be trained: 1. Problem identification, 2. Process selection, 3. Strategy selection, 4. Representation selection, 5. Allocation of resources, 6. Solution monitoring, 7. Sensitivity to feedback, 8. Translation of feedback into an action plan, 9. Implementation of the action plan (p. 9). However, Sternberg does not give an outline of how to train subjects with regard to these skills. Open questions are, for example, whether one should train components separately, or if one should provide a subject with a strategy including components that may be identified during training.

A different view with regard to this point is taken by Schoenfeld (1979). Schoenfeld aims at the development of an executive control procedure instead of providing separate executive components. The starting point is that a student provided with a heuristic alone is not necessarily able to apply this heuristic efficiently; he may not even select this heuristic though it might be appropriate. Schoenfeld demonstrates that the instruction of heuristics alone requires training in many contexts, and extensive application. However, a "student may learn to employ a series of individual heuristics, but this does not guarantee that he can solve a heterogenous collection of problems. He may demonstrate a subject-matter competence and still lack an efficient means of sorting through the heuristics at his disposal and of selecting the appropriate one. He needs a means of assessing and allocating his resources; he needs a *managerial strategy*. Without this, the student may lose the possible benefits of his heuristic resources" (1979, p. 320). Schoenfeld refers to a study that clearly shows the advantages of an experimental group instructed to use a managerial strategy. At the core he aims at providing the student with an *executive program* (p. 321). The corresponding procedure consists of five main steps: Analysis of the problem (understand the problem), design (plan), exploration (search for a solution), implementation (execution), and verification (test of solution). For each step Schoenfeld provides a list of heuristics (for example, analysis: examine special cases, etc.). The problem solving training included a thorough introduction and discussion of all components of this managerial strategy.

The *central component* of it is the design-process embedded in the executive procedure. It represents the *intention of making a plan* and has to be connected with *planning methods* (as corresponding heuristics for this step). That is, this part of the general executive strategy incorporates planning as a metacognitive skill as discussed earlier in this chapter. "Design" means developing an overall plan, being able to justify one's approach, to predict outcomes, to evaluate one's progress, to monitor the own solution process, keeping track of the level of one's endeavor.

In sum, Schoenfeld proposes to provide students not only with heuristics or separate executive skills, instead, he asks for the training of both, that is, of heuristics as well as of executive control. The major advantage we see is that he aims at elaborating an overall executive strategy. This might enable the student, provided with heuristics, to tune himself with regard to perceived cognitive demands.

Both authors, Sternberg as well as Schoenfeld agree that the present state of teaching problem solving to be a primitive one requiring still more effort, and allowing only modest expectations. However, as the model of Faught et al. (1977) shows problem solving is not solely a matter of strategical and tactical decisions. Affects and motivations intervene to a great degree. This requires further skills like meta-volitional operations in order to persist.

It "should never be forgotten that the source of the ability to solve problems may lie outside of the cognitive world entirely, being related instead to the ability to manage intellectual resources – to devote attention to the task, to actually carry through mental calculation. All this I have simply summarized in the term discipline" (Newell, 1980, p. 187).

The Methodological Status of the Concept of Executive Control

Executive control in problem solving is conceived of as a theoretical phenomenon which is not directly observable. We assign this ability rather theoretically to the mental processes of a problem solver. The concept of executive control is an *interpretational construct*. It constitutes a terminologically and analytically true link between the prototypical, necessary and sufficient application conditions and an interpretational component (i.e., the ability of executive control imputed to a problem solver). When the empirical component (representing the necessary and sufficient application conditions) is corroborated one is allowed to conclude that the nonempiricial, interpretational component is valid. Because we attribute the ability of executive control to a problem solver with regard to the validity of the terminological link, the interpretational construct of executive control *cannot* be empirically disproved. This validity of the construct is the reason for the impossibility of (empirically) refuting the construct of executive control or any other theoretical construct. However, it may be evaluated with regard to its *usefulness* in describing the performance of subjects engaged in cognitive activity (see also Lenk, 1978; Herrmann, 1980, 1982; Friedrichsen, 1983; Friedrichsen & Birkhan, in press; Birkhan & Friedrichsen, 1983).

Thus, the notion of executive control is a classification construct. It does not represent a nomological if-then statement. However, the notion of executive control can be included as if- or then-components of nomological assumptions. But, nomological assumptions must incorporate further elements, namely, a description of the features of a *situation,* which, for example, causes a person to perform his ability of executive control. Only when a situational description is combined with, for example, the notion of executive control do the combined components constitute a nomological statement (where the concept of executive control has to be defined independently of the situational description). For example, the following statement is a nomological assumption with regard to the notion of executive control: If a subject, engaged in learning from a test, uses a self-testing strategy during learning, then executive control in terms of "metacomprehension" takes place. The notion of executive control as a component of nomological assumptions are demonstrated as useful and appropriate when empirical facts are correctly explained and predicted by means of such nomological assumptions. In contrast to terminologically true statements, nomological assumptions can be empirically refuted.

Consequently, the theory of executive control is composed of analytically true statements (i.e., the construct of executive control), as well as of nomological if-then statements where a priori true statements are used heuristically for generating nomological statements (see also Herrmann, 1976, 1980, 1982). Research procedures characterized in this way reflect the actual methods of psychologists and researchers more closely than the idealized conception of science formulating only empirical and empirically confirmed statements.

References

Anderson, J. (1976). *Language, memory, and thought*. Hillsdale, N.J.: Erlbaum

Birkhan, G. & Friedrichsen, G. (1983). Handlungstheorien im Lichte der Strukturalistischen Theorienauffassung. In: G. Lüer (Ed.), *Bericht über den 33. Kongreß der Deutschen Gesellschaft für Psychologie in Mainz 1982. Vol.1*. pp.453–456. Göttingen: Hogrefe.

Bower, G.A. (1972) A selective review of organizational factors in memory. In: E. Tulving & W. Donaldson (Eds.), *Organization of memory*. pp.93–137). New York: Academic Press.

Bower, G. (1975) Cognitive Psychology: An introduction. In: W.K. Estes (Ed.), *Handbook of Learning and Cognitive Processes. Vol.1*. (pp.25–79) Hillsdale, N.J.: Erlbaum.

Brown, A. (1978) Knowing when, where, and how to remember: a problem of metacognition. In: R. Glaser (Ed.), *Advances in instructional psychology*. (pp.77–165) Hillsdale, N.J.: Erlbaum.

Brown, A., Bransford, J., Ferrara, R. & Campione, J. (1983) Learning, remembering and understanding. In: J.H. Flavell & E. Markman (Eds.), *Handbook of child psychology. Vol.III*. (pp.77–166). New York: Wiley.

Brown, A. & DeLoache, J.S. (1978) Skills, plans, and self-regulation. In: R.S. Siegler (Ed.), *Children's thinking: What develops?* (pp.3–35). Hillsdale, N.J.: Erlbaum.

Brown, A.L. & Smiley, S.S. (1977) Rating the importance of structural units of prose passages: A problem of metacognitive development. *Child Development, 48,* 1–8.

Butterfield, E.C., & Belmont, J.M. (1975) Assessing and improving the cognitive function of mentally retarded people. In: J. Bailer & M. Sternlicht (Eds.), *Psychological issues in mental retardation*. Chicago: Aldine.

Chi, M. (1984) Bereichsspezifisches Wissen und Metakognition. In: F.E. Weinert & R.H. Kluwe (Eds.), *Metakognition, Motivation und Lernen,* (pp.211–232). Stuttgart: Kohlhammer.

Colby, K.M. (1978). Mind models: an overview of current work. *Mathematical Biosciences, 39,* 159–185.

Dörner, D. (1974) *Die kognitive Organisation beim Problemlösen*. Bern: Huber.

Dörner, D., Kreuzig, H.W., Reither, F., & Stäudel, Th. (1983). *Lohausen*. Bern: Huber.

Duncker, K. (1935). *Zur Psychologie des produktiven Denkens*. (1st ed. 1935) Berlin: Springer.

Ericsson, K.A., & Simon, H.A. (1980) Verbal reports as data. *Psychological Review, 87,* 215–251.

Faught, W.S., Colby, K.M., & Parkinson, R.C. (1977). Inferences, affects and intentions in a model of paranoia. *Cognitive Psychology, 9,* 153–187.

Fikes, R.E., & Nilsson, N.Y. (1971). Strips: A new approach to the application of theorem proving to problem solving. *Artificial Intelligence, 2,* 189–208.

Flavell, J.H. (1979) Metacognition and cognitive monitoring. *American Psychologist, 34,* 906–911.

Flavell, J. H. (1981) Cognitive monitoring. In: W. P. Dickson (Ed.), *Children's oral communication skills.* (pp. 35–60). New York: Academic Press.

Flavell, J. H., & Wellman, H. (1977) Metamemory. In: R. Kail & J. Hagen (Eds.), *Perspectives on the development of memory and cognition.* (pp. 3–33). Hillsdale, N. J.: Erlbaum.

Friedrichsen, G. (1983) Strukturalistischer Rekonstruktions-Versuch einer Sprechhandlungs-Theorie. *Bericht aus dem Fachbereich Pädagogik (Abt. Psychologie), Hochschule der Bundeswehr Hamburg.* Vortrag gehalten auf der 25. Tagung experimentell arbeitender Psychologen in Hamburg, März 1983.

Friedrichsen, G., & Birkhan, G. (in press). Versuch einer strukturalistischen Rekonstruktion von psychologischen Handlungstheorien. *Psych. Institut II der Universität Hamburg. Institutsbericht.*

Goldin, S. E., & Hayes-Roth, B. (1980) Individual differences in planning processes. Santa Monica, Rand Corporation, *N-1488-ONR,* June 1980.

Hayes-Roth, B. (1980). Flexibility in executive strategies. *Rand Note N-1170-ONR,* Rand Corporation, St. Monica, Ca., Sept. 1980.

Hayes-Roth, B., & Hayes-Roth, F. (1979) A cognitive model of planning. *Cognitive Science, 3,* 275–310.

Hayes-Roth, B. & Thorndyke, P. W. (1980). Decisionmaking during the planning process. Rand Corp. St. Monica: *N-1213-ONR,* Oct. 1980.

Herrmann, Th. (1976). *Die Psychologie und ihre Forschungsprogramme.* Göttingen: Hogrefe.

Herrmann, Th. (1980) Sprechhandlungspläne als handlungstheoretische Konstrukte. In: H. Lenk, *Handlungstheorien – interdisziplinär. Band 1.* (pp. 361–380). München: Fink.

Herrmann, Th. (1982) *Sprechen und Situation.* Berlin: Springer. (1982).

Herrmann, Th. (1983) Nützliche Fiktion: Anmerkungen zur Funktion kognitionspsychologischer Theoriebildung. *Sprache & Kognition, 2,* 88–99.

Hilgard, E. (1976) Neodissociation theory of multiple control systems. In: G. E. Schwartz & D. Shapiro (Eds.), *Consciousness and self-regulation. Vol. 2.* (pp. 137–171). New York: Plenum.

Jeffries, R., Turner, A. A., Polson, P. E. & Atwood, M. (1981) The processes involved in designing software. In: J. Anderson (Ed.), *Cognitive skills and their acquisition.* (pp. 255–283). Hillsdale, N. J.: Erlbaum.

Kluwe, R. H. (1981). Metakognition (Positionsreferat). In: W. Michaelis (Ed.), *Bericht über den 30. Kongreß der Deutschen Gesellschaft für Psychologie in Zürich 1980.* (pp. 246–258). Göttingen: Hogrefe.

Kluwe, R. H. (1982) Cognitive knowledge and executive control: metacognition. In: D. Griffin (Ed.), *Animal mind – human mind.* (pp. 201–224). New York: Springer.

Kluwe, R. H. (1983) Beweglichkeit des Denkens. In: L. Montada (Ed.), *Denken und Handeln.* (pp. 127–145). Stuttgart: Klett.

Kluwe, R. H., & Schiebler, K. (1984). Entwicklung exekutiver Prozesse und kognitive Leistungen. In: F. E. Weinert & R. H. Kluwe (Ed.), *Metakognition, Motivation und Lernen.* (pp. 31–59). Stuttgart: Kohlhammer.

Lenk, H. (1978). Handlung als Interpretationskonstrukt. In: H. Lenk (Ed.), *Handlungstheorien – interdisziplinär. Vol. 2.* (pp. 279–350). München: Fink.

Lüer, G. (1973). *Gesetzmäßige Denkabläufe beim Problemlösen.* Weinheim: Beltz.

Luria, A. R. (1973) The working brain. London: Penguin.

McDermott, D. (1978) Planning and acting. *Cognitive Science, 2,* 71–109.

Miller, S. A., Galanter, E., & Pribram, K. (1960). *Plans and the structure of behavior.* New York: Holt, Winston & Rinehart.

Neisser, U. (1967). *Cognitive Psychology.* New York: Meredith.

Newell, A. (1980). One final word. In: D. T. Tuma & F. Reif (Eds.), *Problem solving and education.* (pp. 175–189) Hillsdale, N. J.: Erlbaum.

Newell, A., & Simon, H. A. (1963). Computers in Psychology. In: R. D. Luce, R. R. Busch & E. Galanter (Eds.), *Handbook of Mathematical Psychology, Vol. 1,* (pp. 361–428). New York: Wiley.

Newell, A. & Simon, H. A. (1972). *Human Problem Solving.* Englewood-Cliffs, N. J.: Prentice Hall.

Norman, D. A. (1978) What goes on in the mind of the learner? In: McKeachie, W. J. (Ed.), *Cognition, college teaching and student learning.* San Francisco: Jossey-Bass.

Pascual-Leone, J. (1976). Metasubjective problems of constructive cognition: forms of knowing and their psychological mechanism. *Canadian Psychological Review, 1,* 2, 110–125.

Pylyshyn, Z. W. (1974). Complexity and the study of artificial and human intelligence. *Res. Bulletin No. 300, Dept. Psychology, University of Western Ontario.* London, Canada.

Pylyshyn, Z. W. (1980). Computation and cognition: issues in the foundation of cognitive science. *The Behavioral and Brain Sciences, 3,* 111–169.

Reither, F. (1980). *About thinking and acting of experts in complex situations.* University of Bamberg (FRG).

Reitman, W. (1969). *Cognition and thought: An information processing approach.* New York: Wiley.

Reitman, W. (1973). Problem solving, comprehension, and memory. In: G. J. Dalenoort (Ed.), *Process models for psychology.* (pp. 51–77). Rotterdam: University Press.

Ryle, G. (1949). *The concept of mind.* Harmondsworth: Penguin.

Sacerdoti, E. D. (1974) Planning in a hierarchy of abstraction spaces. *Artificial Intelligence, 5,* 115–135.

Sacerdoti, E. D. (1977) *A structure for plans and behavior.* Amsterdam: Elsevier.

Schoenfeld, A. (1979). Can heuristics be taught? In: J. Lochhead & J. Clement (Eds.), *Cognitive process instruction.* (pp. 315–338). Philadelphia, Pa.: Franklin Inst. Press.

Schoenfeld, A. (1981) Episodes and executive decisions in mathematical problem solving. *Paper presented at the 1981 AERA Annual Meeting,* Los Angeles, Ca.

Selz, O. (1913). *Über die Gesetze des geordneten Denkverlaufs. 1. Teil.* Stuttgart: Spemann.

Selz, O. (1922). *Über die Gesetze des geordneten Denkverlaufs. 2. Teil: Zur Psychologie des produktiven Denkens und des Irrtums.* Bonn: Friedrich Cohen.

Simon, H. A. (1967) Motivational and emotional controls of cognition. *Psychological Review, 74,* 29–39.

Simon, H. A. (1975). The functional equivalence of problem solving skills. *Cognitive Psychology, 7,* 268–288.

Simon, D. P. & Simon, H. A. (1978) Individual differences in solving physic problems. In: R. S. Siegler (Ed.), *Children's thinking: What develops?* (pp. 325–348) Hillsdale, N. J.: Erlbaum.

Sternberg, R. (1979). The nature of mental abilities. *American Psychologist, 34,* 214–230.

Sternberg, R. (1980). Sketch of a componential subtheory of human intelligence. *The Behavioral and Brain Sciences, 3,* 573–614.

Sternberg, R. (1983). Criteria for intellectual skills training. *Educational Researcher, 12,* 6–12.

Thorndyke, P. W., & Stasz, C. (1980). Individual differences in procedures for knowledge acquisition from maps. *Cognitive Psychology, 12,* 137–175.

Wilensky, R. (1981). Meta-planning: Representing and using knowledge about planning in problem solving and natural language understanding. *Cognitive Science, 5,* 197–233.

Chapter 10

Thinking and the Organization of Action

Dietrich Dörner

The Organization of Behavior: A General Picture

Thinking is one of the possible ways to organize and to control action but not every form of action incorporates thinking processes. Under which circumstances does thinking play an essential role for the control of action?

We assume that thinking plays an important role for the control of action if appropriate memory schemes are not at hand or have proven unsuccessful. Thinking plays the role of a "trouble-shooter" for the organization of action. We shall be concerned with this role of thinking as a trouble-shooting system on the following pages.

Before talking about the role of thinking in the organization of action one must first elucidate the ways human behavior and actions are generally organized. It may be supposed that human behavior and action is more or less organized as shown in Fig. 10.1. A motivation exists which is directing the behavior. This motivation usually is not the only possible motivation, but is the one that is at present the strongest. It may be the strongest because the need that constitutes the motivation (for instance hunger, thirst) is the strongest one; or because a given time limit has expired, before which the goal (contained in the motivation) should have been reached (e. g., a store closing, the departure of a train); or because the chance of success in reaching the goal is high. Expectancy-value models of motivation specify this point (Heckhausen, 1980, p. 216).

The existence of a motivation means that there exists (a more or less exact) idea of a goal. Moreover, a motivation means that a certain course of behavior is activated to reach that goal (Heckhausen, 1980, p. 27). How does this happen in detail?

The regulation of behavior is supposed to comprise two major levels, that is, the level of *automatic* behavior (units 1-2-6-5, in Fig. 10.1) and the level of heuristic processes (3-4, in Fig. 10.1). These levels are subdivided in many ways. On the level of automatism, the regulation of behavior entails searching memory for a way that may lead from the momentarily given situation towards one of the pos-

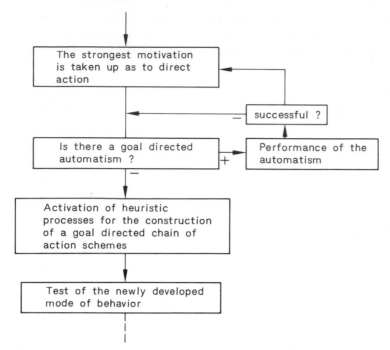

The strongest motivation is taken up as to direct action

successful ?

Is there a goal directed automatism ?

Performance of the automatism

Activation of heuristic processes for the construction of a goal directed chain of action schemes

Test of the newly developed mode of behavior

Fig. 10.1. A two-step organization of the structure of human behavior

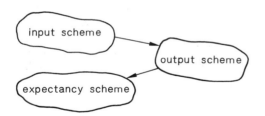

input scheme

output scheme

expectancy scheme

Fig. 10.2. Structure of an action-scheme. See explanation in text

sible goal situations. (The structure depicted in Fig. 10.1 must be completed by an "interrupt system" which is sensitive to unexpected or new events in order to give a rough, rather complete schema of human behavior regulation. See Dörner, 1982.)

One may think of the memory as a network of *action schemes*. Fig. 10.2 shows such an action scheme. In the simplest form, it consists of three components: an *input scheme*, an *output scheme*, and an *expectancy scheme*.

The input scheme can be imagined as a kind of matrix which fits in a certain group of outer or inner situations. In the simplest case such a scheme might concern a certain group or class of objects, for instance, *apples* or *chocolate*. More complicated forms of input schemes may refer to a more general class of situations such as threatening or obscure situations.

The output scheme could be imagined in the simplest case as a program for the performance of certain coordinated movements of the muscles, such as *to grasp* or *to walk* or *to look at*. Such motor programs involve complicated coordinations. An obvious example is the action of grasping just mentioned.

The expectancy scheme, on the other hand, is like the input scheme: a matrix for a class of objects and situations. It fits to those objects and situations which are expected as an effect of the activation of the output scheme. In grasping the pencil in front of you, you expect a specific sensation of touch. The concept of an action scheme is closely related to the "script"concept of Schank and Abelson (1977). An action scheme is a kind of atomic script.

Action schemes form networks. The expectancy scheme of one actionscheme might be the inputscheme for a second or third one. In this way action schemes may establish large networks.

The regulation of behavior on the level of automatism involves checking whether the momentarily given situation (start situation) and one of the goal-directed situations striven for are located in the beginning or the end of a chain in the network of action schemes. If this is the case, the corresponding chain will be activated. If this is not the case, the control of behavior goes over into the level of heuristic processes. Heuristic processes serve to produce a chain of action schemes which lead to the goal. In the following we shall have a closer look at these heuristic methods.

There exist very simple as well as extremely complicated heuristic methods. Durkin (1937) already distinguished "trial and error," "gradual analysis," and "sudden reorganization." Blind trial and error behavior is the simplest form of heuristic process. This consists in applying the action schemes known to a person in the given situation in an optional arrangement and free sequence.

In its more elaborated forms, trial and error behavior turns into systematic trials, in which, for instance, the repeated use of the same operation for the same situation is avoided. Typically, there is some preselection in the choice of an operation. Not everything will be tried out in a situation, but first that which promises success according to certain (sometimes rather vague) criteria.

Such elaborated forms of trial-and-error behavior already presume certain *operations of control*. A sort of recording memory is needed in order to file the actionschemes already proven successful, thus avoiding unnecessary repetitions. Before the operation is performed, the recording memory has to be "looked through" to find out whether the operation in question has been used before.

The internalization of controlled trial- and error behavior might have been the first form of thinking. Later, it will be described how, by introducing an increasing number of processes that delimit the search area investigated, the process of trial- and error behavior gradually becomes an analytic problem-solving process.

Another primitive form of heuristic behavior is *searching behavior*. The surroundings are being searched for initial points of goal-leading chains of action schemes. The cakefork is missing at the coffeehouse. Where is the waitress to get you one?

Heuristic Processes: Their Elements and Determinants

The Organization of Thinking

In the preceding chapter we looked at the role the heuristic process plays within the general organization of the regulation of behavior. We shall now turn our attention to some special components of thinking.

Thinking while solving problems involves several intertwining phases. Everyday problems are usually defined inexactly. First of all a process of *goal-analysis* is necessary. The problem-solver has to develop a clear conception of what s/he wants to achieve.

If the sphere of reality with which the person is surrounded is not familiar, a phase of information gathering about the possible alternatives of action and their effects has to follow next. Once a clearly defined picture of the sphere of reality does exist (or at least is thought to), the next step will be the trying of possible actions with the purpose of constructing a way to reach the goal.

Depending on the characteristics of the problem and the problem-solver, some phases can be very short or may even be omitted from the process. Stepbacks are also possible. Unsuccessful efforts in finding a way to the goal may make another goal analysis necessary and a new goal definition may demand more gathering of information.

Finally, the solving of problems often renders processes of *self-regulation* necessary. One has to think about whether the applied strategies are the right ones or whether others might be more suitable. Thinking activities have to be planned in advance and time has to be scheduled.

All these phases consist of elementary steps of thinking. Such elements of a thinking process may be: associations and classifications, memory-finding processes, logical and inductive inferences, the divisions of entities into their single parts (analysis of components), internal trials, that is, probative acting in a fictitious model, analysis of the determinants and consequences of certain events, etc. In this paper we shall not be dealing with the finer details of these basic operations of thinking which are partly conscious and also unconscious. The larger organization of action concerns us here: the macro-structures of thinking; information about the micro-structure may be taken from other sources (see for instance Lompscher, 1972; Selz, 1913; Dörner, 1979, p. 111).

Thinking, which consists of the conscious arrangement of elementary phases as mentioned above, is a serial process, one step of thinking following the other; the process can usually be verbalized fairly well.

Here's an example taken from the game of chess: "This is a variant of a Royal-Indian opening!" (Classification of a situation). "In such a situation, Capablanca reacted in 1925 with the move X!" (Remembrance of a possible reaction to the given situation; completion of complex, according to Selz, 1913). "If I do x now, my opponent will probably react with the move y!" (Internal problem action and analysis of the possible consequences; maybe because of the classification of the opponent as an aggressive or a defensive player.)

The conscious action planning of a chessplayer could partly look like this. But even in such a sequence of conscious steps there are elements which the person will not be aware of. Exactly how the given situation on the chessboard is classified by the brain as "Royal Indian" and various processes of trial and comparison will not be fully conscious acts of thought in its fine detail. For example, even the identification of a "castle" will be an unconscious process.

Beside the sequence of conscious processes (which are paralleled by unconscious processes at the molecular level of elementary processes of identification and classification), there supposedly exist unconscious processes which may accompany the conscious processes of thinking. Freud distinguished primary and secondary processes of thinking. Secondary processes of thinking are the conscious ones that may also be verbalized, whereas dreams, daydreamings and the unconscious processes which prepare sudden ideas belong to the sphere of primary processes.

The structure and organization of the conscious processes and of the primary processes is complicated. With regard to the primary processes, further research is needed. Microstructures of the conscious processes of thinking are examined in Newell and Simon (1972) and Dörner (1979). The latter also contains several speculations about structures of unconscious processes in connection with the phenomena of sudden inspiration (Dörner 1979, p. 91).

With regard to the theme "thinking and the organization of action" we shall now consider with closer analysis the above mentioned phases of goal analysis, accumulation of information, planning of action, and self-regulation, giving special regard of their usual procedure and their specific susceptibility of disturbances.

Within the total context of psychic processes one relation demands special attention, namely the relation of thinking and emotion. Of course, thinking is also connected with other mental processes. Perception supplies the material, without which thinking would take another direction; motivation provides a goal for thinking; learning takes place while thinking – but the relation of thinking and emotion is of special importance because emotional processes also guide the course of thinking, quite independent from the specific contents and goals of the respective act of thinking.

Considering for instance the impact that anger has on thinking, this will be easily understood. Generally, anger lowers the level of thinking. For example, a chessplayer may try to annoy an opponent – by holding a purring cat in his arms, well knowing that the opponent hates cats. A chessplayer might behave extravagantly, arrive at the last minute, argue for ridiculous conditions, hoping such acts will excite his opponent.

Emotions are dependent on the absence or presence and the loss or regaining of control. Loss of control over a sphere of reality may relate to sorrow, resignation, annoyance or rage. Regaining control will produce joy or triumph.

Thinking has a lot to do with control. The necessity of thinking shows up in situations in which a direct control over the situation is no longer given – otherwise one would not need thinking. Thinking is guided and motivated by the hope for regaining control. Ineffective thinking produces annoyance, rage, anger, and sorrow (respectively). Successful thinking produces triumph, joy, comfort. These emotional phenomena, which thinking produces itself, again establish the basis for the course of thinking, just as rubble and sand, by which a river is moved, are on the other hand, the new conditions for this very movement.

Stress, time pressure, and an overload of information produce a more or less strong situation of activation, the feeling that control is slipping away accompanied by corresponding feelings. Thus, the course of thinking is influenced by them. In the following investigation, therefore, a respective analysis of the influence of emotional conditions on the course of thinking shall be taken into account.

Kuhl (1983) presents an interesting theory of the influence of fear and anger and congenial negative emotions on the primary and secondary processes of thinking as they interrelate with each other. He assumes that the emotions mentioned above produce a switch from an intuitive-holistic mode of thinking to a more sequential and analytic kind of thinking – in other words – smash the balance between sequentially-analytic and primary-processual, intuitive-holistic processes of thinking.

Without adopting Kuhl's argumentation totally, concerning the disturbance of balance, one can agree with him as far as the effects are concerned – at least up to a certain point: in view of a great extent of fear and anger the sequential-analytic thinking will disappear completely, giving over to a primitive holistic reaction.

Stages of organization of action will now be considered.

Goal Analysis

Most of the everyday problems occurring need a goal analysis. A goal analysis consists of the division of a global goal into separate, better defined parts. The global goal: "My room should become more comfortable" develops into: "I need brighter curtains," "instead of a ceiling light I need a floor-lamp in one corner and a lamp on that wall."

The elementary cognitive operations of a goal analysis are the *components analysis* and *dependency analysis*. A components analysis consists of getting an impression of the elements of a factual situation. What are the components of a comfortable room? Lighting, wallpaper, furniture, and their composition as a whole. What is a good position for an industry? One that offers good traffic conditions and cheap transporting facilities for goods and materials in addition to a large pool of qualified labour, favourable tax regulations, reasonable rates of rent, expenses, etc.

A dependency analysis consists in determining the interdependence of certain variables. Why are some firms subject to good tax conditions and some to bad? What effects an item's decline in sales? What causes a high rate of illness at a factory? Applying components analysis and dependency analysis we go from a global goal, for example: "I am seeking a suitable site for establishing my plant" to a more differentiated catalogue of goals: "Certain tax conditions must be given, a special work force should be available, certain transportations facilities, and a number of subcontracting firms should exist in the area, etc."

Goal analysis is a laborious task; coming down to a point is always difficult. Time pressure and stress may lead to perfunctory goal analysis which does not actually belong to the indispensable problem-solving activity of thinking. However, partial or even total neglect of goal analysis results in a number of consequences for the process of thinking:

(1) Thinking occurs according to a "repair-service" principle. The first occasion is picked up and made into an object of thinking: "I still have to get prepared for the examination in General Psychology; I just found a book here about perception psychology, I'll read this now!"

(2) Thinking changes quickly from one object of thinking to another. The relations between partial goals and total goal remain obscure because of a missing goal analysis; their significance remains obscure as well. That means that the most salient events will gain greatest importance for the subject, pushing others aside: "A friend told me, he had just read an interesting article about schizophrenia and thinking. Good Lord! The subject *thinking* is also on the program for my examination, so let's do away with the book about perception psychology and turn to the article about thinking!" The tendency towards a rapid change of intention has, elsewhere, been described as "thematic vagabonding" (Dörner, 1980). One can show that this appears far more often with people with little self-esteem than with those who have greater confidence in their own abilities of coping with complicated problems. Of course, people with little self-esteem tend to get more easily into a situation of helplessness and stress.

Figure 10.3 shows the mean number of thematic changes within a two hour session in a simulation game for a group of 10 subjects with low self-esteem and a group of 10 subjects with high self-esteem.

Within this simulation game, subjects had to gather information about a community and had to propose measures to solve the most urgent problems of that community (unemployment, lack of apartments, etc.). The differences between the groups were statistically significant.

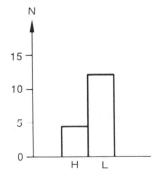

Fig. 10.3. Mean number of thematic changes within a two hour experimental session for a group of subjects with high self-esteem (H) and a group with low self-esteem

Accumulation and Examination of Information

Once one knows what one wants, that is, after goal analysis, one has to look for a way to the goal. For that purpose different processes of thinking are necessary. In an unknown field of reality it may first of all be necessary to inquire and to accumulate information in order to develop a clear picture of the respective field of reality.

The way an idea of an unknown sphere of reality is developed depends on the circumstances. Sometimes it is enough just to read about it or to ask. Sometimes proper research and experimental programs might be necessary.

An important means for generating missing information is analogical reasoning. One deals with the unknown sphere of reality *as if* it behaved like a familiar one. Thinking about electrical circuits, we may take the system of water supply as an analogy; a resistance thus being a narrow point in a water supply, etc.

The importance of analogical thinking for the development of science is undoubted and frequently emphasized (for instance, Dreistadt, 1968). One may think here of the analogy drawn between the atomic structure and the planetary system and the role it played for the development of the atomic model of Bohr and Rutherford. One should also be reminded of the analogy between the two-dimensional order of the elements in the periodic-law, which as is known played quite a role in the discovery of the periodic law by Mendelejew. Mendelejew was a passionate patienceplayer (Sergejew, 1970).

Formally, analogical thinking means to extract the relational structure of a known situation and to transfer it to a new situation. By considering a pack of cards arranged according to colour (club, spade, hearts, diamonds) and to value (ace, king, queen, . . .) the idea of a system of arrangement in returning periods is won. This idea is transferred on to chemical elements which might be imagined arranged according to validity (in accordance to the value of the card) and to atomic weight (thus not totally corresponding with the colour of the card; as this remains the same within a "period" while the atomic weight increases within a period).

There are some preconditions for the ability of a person to produce analogy transfers. One precondition is the availability of differentiated and concrete knowledge of various spheres of reality. For Mendelejew it was important not only to have a vast and very exact knowledge of the qualities of chemical elements but also of the characteristics of card games.

Knowledge of many different spheres of reality is an important condition for the ability of analogical thinking. This condition indeed is necessary, but not sufficient. A further condition has to be added: the concrete realms of knowledge available have to be of the kind to be connected with each other over *abstracta*. In the case of the Bohr and Rutherford analogy drawn between the atomic structure and the solar system, it was not enough to recognize the latter as a concrete fact (Dreistadt, 1968). It had to be understood as an abstract system of masses circulating around a mutual centre.

Where does comprehensive and differentiated knowledge of many and very different spheres of reality come from? It could be imagined that persons with a high readiness for frequent exposure to new stimuli and new situations, that is, persons with great curiosity, bring along one precondition for gaining comprehensive knowledge of divergent circumstances. However, a further precondition is necessary. The individuals must possess a high sensibility for disharmony and cognitive dissonances. This results in the readiness to modify and enlarge memory structures up to the point where a differentiated structure free of contradiction is attained. Therefore, it is not sufficient just to expose oneself to new situations and circumstances; an effort to integrate new knowledge is indispensable.

Certain other preconditions are helpful for the free and open accumulation of information and for analogical thinking. High self-esteem in an individual is of great advantage. The accumulation of new information and its integration into the individual's "picture of the world" postulates that the individual is willing to adjust his prior conception of the world, if necessary. This is against a central human motive, namely, the desire to feel control over one's surroundings. Accepting new information may produce uncertainty. Therefore, integration of new information is connected with self-confidence. Little self-confidence decreases the tendency to accumulate information. Greater self-confidence makes it easier for the individual to expose her or himself to new and uncertain situations. Being "success-motivated", means that one tends to rely on one's capacity to cope with the uncertain.

In this connection some quite informative results were obtained from an experiment, in which testpersons were to run a community as mayor (LOH-HAUSEN-experiment; Dörner, Kreuzig, Reither, Stäudel, 1983). In this experiment it was very important for the subject to get, as fast as possible, exact information about the unknown sphere of reality. Stäudel (1983) found that people with little self-confidence put significantly less "why-questions" in the test than persons with high confidence. Rch (1983) showed that the depth of analysis of subjects in search of solutions was significantly lower in cases of individuals with lower self-confidence than with others. Depth of analysis means the extent to which subjects follow the possibilities of influencing a variable (dependency-analysis, for instance: "How can I increase sales?"); and to what extent they consider the consequences of certain events (effectance analysis, for instance: "What else results from a price-cut – apart from an increase of sales?").

Beside the accumulation of information for obtaining an image as adequate as possible of the respective sphere of reality, it is important to expose one's own suppositions concerning this sphere of reality to critical examination. This requires a distanced relationship to one's own hypotheses; which should not be cherished too much, and eventually even dropped. Such a distance postulates a high degree of self-confidence and heuristic competence, as a critical test of one's own suppositions may endanger the basis of one's own actions.

According to that, a critical, cool distance towards one's own suppositions is found only rarely in the case of incompetence. Hence incompetent subjects very seldom test their own hypotheses, and their suppositions tend to remain unquestioned. Stäudel (1983) found that subjects whose actual competence was rather low detected fewer discrepancies between new information and their memory (e. g. "That is quite different from what I imagined") than test persons with greater competence. It could be that lower competence parallels a reluctance to state discrepancies between one's own suppositions and reality. It could also be the case that subjects with smaller actual competence simply have fewer memory images; the data available, however, do not allow any conclusion as to whether the effect was produced by one or the other, or both factors together. Of course, avoiding forming a hypothesis is an effective means to prevent being forced to acknowledge discrepancies. Low differentiation of the image of the sphere of reality may therefore be a kind of protection against unwanted refutations which might endanger one's competence. The statement: "You can look at it this way or another!" is always correct. Accordingly Stäudel found that the presumably less competent subjects tended to renounce a greater differentiation of memory images.

Planning of Action

If enough information about a certain sphere of reality is available, the next task will be the construction of a goal-directed chain of operations. There exists a startingpoint, namely a given situation and a goal. Between these exists a gap, otherwise one would not have a problem. Often the gap will be closed simply by collecting information. If the accumulation of information is not sufficient, the attempt is made to construct a chain of operations which may lead from the startingpoint to the goal.

The construction of a chain of operations leading to the goal is the central subject of the problem-solving research (Newell & Simon, 1972; Dörner, 1979). It must be emphasized that the construction of a chain of operations is only one part of thinking.

The construction of a chain of operations leading to a goal can be done in different ways, which depend to a high degree on the respective sphere of reality. The most primitive form is trial and error. The composition of a puzzle will sometimes be done this way; different possibilities of operation are just played or tried.

It can be argued, however, that a completely free trial and error behavior in the sense of chance behavior does not occur with human beings. With humans there always exist certain criteria of selection which guide the choice of actions.

On the other hand, free trial and error behavior would be a completely inadequate strategy in many spheres of reality because the number of possible operational combinations is so large that the probability of reaching the goal within any given time is near zero. An obvious example is chess. A systematic trial of *all* possibilities could, with the help of computers, last millions of years (Osterloh, 1983, p. 250).

Therefore, it is necessary to make a preselection within the field of the possibilities. Certain criteria guide such a selection process. A simple and obvious strategy, intuitively adopted, consists of selecting operators according to their aptitude to remove the differences between startingpoint and goal. One begins with an analysis of the differences between startingpoint and goal and considers only those operators which are suitable for the elimination of differences. This strategy is so obvious that it is usually not recognized as such. It is needless to say that one proceeds like this. In its single points however, this strategy is not without difficulties and pitfalls. For instance, it is important to consider the fact that the removal of certain differences causes side-effects while other obstacles can be eliminated without complications (Ernst & Newell, 1969).

The neglect of side-effects is a cardinal mistake in the planning of actions and is commonplace. Nobody anticipated finding DDT in mother's milk when it was first used for eliminating insects. Townplanners who improve and extend an arterial road may think of the removal of traffic problems. But they might not think of the fact that this very improvement may encourage greater use of cars. So they not only remove traffic problems, but – maybe at other places in town – create new ones and that increases the financial deficit of public transportation (see Wiegand, 1981), for a vivid description of such effects of urban planning).

Another way to confine the area of investigation is aiming at interim goals. This is especially convenient if a total planning up to the final goal is not possible. Think for instance of such a limited sphere of reality like a chess game. In every sphere of reality there are special points which are quite suitable as interim goals. These are situations which leave a high degree of freedom for further activities offering many different options. In a chess game such an interim goal is the domination of the centerfield. Another point is the well developed pawn-position.

Oesterreich (1981) calls such situations, with maximal freedom of choice regarding diverse prospective courses of action, points of high efficiency-divergence. Points of high efficiency-divergence are especially appropriate as interim goals.

A third way of confining the area of investigation is the combination of working forward and working backward. Intuitively, most subjects proceed according to a working forward strategy. They consider how they might go from a given starting point to the goal. But principally one could analyze how to get from the final goal back to the starting point. If someone wants to go from Berlin-Grunewald to Berlin-Lichterfelde he may reflect which busline, subway or railway leads from Berlin-Grunewald eastward. He might as well consider which lines lead from the west to Lichterfelde.

Working backward provides more clues for the direction of planning than may be obtained by simply aspiring toward a goal point. The reflected and flexible use of methods of confining the area of investigation is the essential point of a good planning method. In the next chapter some factors will be regarded that sometimes disturb the development of such methods.

Planning of Action and Emotional Stress

Problems are to a larger or smaller extent always stress-situations. Having a problem means not to know at the moment how to reach the goal, thus lacking control of the situation. Depending on the importance of the problem and the failures in trying to solve it the stress of a problem situation might increase and lead to feelings of lowered competence in mastering the situation. One effect of deficient competence is sometimes a *short-cut reaction:* The first available, approximately suitable way of action is chosen; one muddles through ("muddling-through behavior", Lindblom, 1964). The possibilities for action are not elaborated as far as possible. It can be proven that incompetent testpersons take less measures to realize a certain intention than competent persons (Dörner et al., 1983, p. 223), who, on the contrary, tend to use complex catalogues of measures for reaching their goal. Certainly the incompetent persons do not dispose of less time at all, but obviously they feel continuous pressure to settle the matter.

Short-cut behavior in the elaboration of measures involves a situation in which *secondary* and *long-range effects* are not being adequately calculated. This means that the different attempts at solving the problem are uncoordinated and, under certain circumstances lead to mutual suspension: the solution of one problem may produce another (possibly a larger) one.

A further characteristic of deficient planning is the exact opposite to short-cutbehavior, namely *overplanning*. Here, one tries to take into account each detail and each possible incident in order to be prepared for anything that might happen. This is, however, impossible to achieve even in a less complex sphere of reality. Usually time is only wasted in such attempts of total planning, but this is not the most dangerous point with total planning. Discovering an unexpected obstacle, a person who planned everything in advance may be frustrated more easily and feel more incompetent than those who expect the unexpected. Someone, however, who assumes that unforeseen events will happen may not feel so destabilized when they do.

It may be supposed that the renunciation of planning ahead, that is, extreme "muddling-through behavior" as well as overplanning share similar roots. In both cases feelings of incompetence are indicated, that is, a feeling of not being able to cope with things. Overplanning is a kind of countermeasure against such feelings. Insecurity is removed partly by elaborate planning. In the case of short-circuit ad-hoc-behavior, however, the person in effect gives up and only reacts instead of acting.

It is difficult to say something general and normative about the amount of planning ahead and of muddling-throughbehavior which is appropriate for a given sphere of reality. Usually a mixture may be necessary. On the one hand, one may proceed according to the principle of planning ahead and on the other hand according to the napoleonic: "On s'engage et puis on voit!"

In addition to short-cut behavior and excessive planning ahead, a further effect due to lowered competence is to be seen in the inclination towards *primitive terminal reactions*. Three major models of such emotional-cognitive modes of reaction may be discerned: *aggression, regression* (in the sense of escape), and *resignation*. Lazarus (1966, 1968) also distinguished three possible reactions to a threatening situation: *attack,* associated with a feeling of annoyance; *avoidance,* associated with fear; and finally *resignation,* associated with hopelessness and depression.

In the course of problemsolving, aggressive tendencies appear in the form of an extreme dosage of measurements on the one hand (e. g., "all of the money left is to be spent for the P. R. campaign!"), or in the preference for those actions that result in very obvious effects (i. e. showing that something is done). Think of the "robbing of figures" in a chess game; this action momentarily produces the illusion of competence – sometimes the whole chess board may even be thrown to the wall, thus producing a powerful effect in addition to thoroughly "solving" the problem.

Regressive tendencies come up in the attempt to push the whole problem aside or at least to postpone its solution ("Today I am not fit at all! Tomorrow . . ."). Apparently "delegating" belongs to this category ("Here a specialist is needed!") and quite particular encapsulations in a well controlled section of the problem. One "oversolves" a well-controlled partial problem, for example one develops extensive computerprograms for the analysis of a research project which will never be started; one takes ballpoint pens apart or sharpens pencils, but the planned article will never be written, etc. *Resignation* finally means that nothing is done at all; submission to the inscrutable flow of current events takes place.

Which controls, aggression, regression or resignation predominate in a given situation? Presumably this depends on the degree of competence experienced. In a case where considerable competence is still felt, the acting person will tend towards aggression; otherwise, towards regression or resignation. Learned tendencies of behavior may play a role as well as controlling factors.

In the preceding we have seen what planning behavior in general looks like and how it might be influenced by feelings of incompetence. Now we will consider the negative influence of emotions based on high estimation of one's own competence. Too high an estimation of one's own competence, however, may lead to an underevaluation of difficulties. The actual behavior can verify the assumption whether certain actions would lead to certain results. In the case of "internal action on trial" things are different: the latter is performed in the memory. But memory contents are poorer in characteristics and are more abstract than the things they copy. Therefore, the frictions of the real action (Clausewitz, 1880, p. 78) do not become apparent. Hence, this might lead to an optimism in planning. Real action is always bound to various conditions and something does always come across: "Today I am going to develop the design, tomorrow I shall test it and the day after tomorrow I shall start with the experiments!" Such a planning does not take into consideration that a power breakdown (does something like this ever happen?) might spoil the half-fed program, or that the finally fed program would come to a broken data-logger and could not be called off anymore, and so forth.

Already with Rumpelstilzchen, a person in an ancient German fairy-tale, the reckoning did not work out well in the end: "Today I shall bake, tomorrow I shall brew . . ."), the frictions were against it. Too high an estimation of one's own competence may cause difficulties; "Pride will have a fall," as the proverb says, thus contributing a hypothesis about the relationship between emotion and cognition.

Nevertheless, a high estimation of one's own heuristic competence makes individuals engage themselves in problems where others regard any solution as hopeless from the very beginning. Often important and new experiences may enlarge one's competence yet and justify a high esteem of one's competence after all. The influences of negative and positive emotions apparently have the property of *self-stabilization* through positive feedback which in the case of negative emotions can be fatal. Lowered competence leads to emotional stress which in turn leads to failure in problem-solving, for which competence is diminished and so on. High competence, causing an engagement in difficult spheres might lead to success, thereby justifying the assumed competence after all.

Self-Regulation

Even with relatively restricted processes of thinking, self-regulation plays an important part. The term self-regulation in this case comprises all thinking activities aimed at the organization of one's own thinking process, thus being a meta-process of thinking. The thinking person constantly has to come to decisions such as: "Should I first deal with partial-problem A or problem B?" – "By which criterion can I find out whether my decision is right?"

An essential element of self-regulation is self-reflection, that is, the reflection about one's own thinking. The procedure is in a limited sense the following (Dörner, 1978): The thinking person makes the *memory records* of his own previous thinking the object of analysis. This analysis may lead the person to discover "loops" (i. e., circular movements in his different starts). He may as well find that the various unsuccessful attempts to solve a problem all had the same attributes in common, which should therefore be given up.

In general one can say that self-reflection is an important method to keep thinking flexible and to adapt it to adequate forms and for new requirements. Self-reflection – which means reflecting upon one's own forms and procedures of thinking – is rare. It seldom occurs spontaneously in experimental-psychological investigation (Dörner, 1974, p. 156 ff). One can suppose that those subjects will refer to it who, due to many ineffective beginnings have a feeling of futility. But on the whole it looks as if this happens more seldom than would be necessary. Several experiments show that just instructing the experimental subjects to critically analyze their own way of thinking from time to time results in a distinct and statistically significant improvement of thinking (Reither, 1979; Hesse, 1979).

Figure 10.4 shows the mean number of trials in two experimental groups of subjects which had to solve 10 problems with a switch-throwing apparatus. The subjects had to produce light combinations by throwing certain switches. This was rather complicated since the switches were dependent on each other in ways the subjects did not know beforehand.

The subjects of the experimental group were instructed to reflect on the course of their thinking from time to time. These self-reflecting subjects were superior in achievement to the subjects of the control group. This result suggests that the means of self-reflection is not always employed spontaneously when it can be useful.

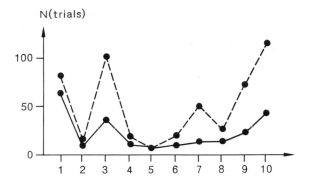

Fig. 10.4. Mean number of trials for 10 switch-throwing problems for subjects urged to reflect the course of their thinking (——) and for a control group (-----). See text

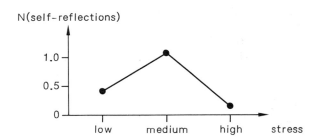

Fig. 10.5. Average number of spontaneous self-reflections in dependence of self-rated stress in a problem solving task. See text

Striking is the strong internal barrier installed by many problem solvers against self-reflection. This becomes visible in another investigation concerning complex problem solving in ecological spheres of reality (Reither, 1981). The subjects were urged to self-reflection; many of them admitted that they had felt rather uneasy about this method. From the very beginning emotional reactions seem to have caused a certain defensive behavior which impeded greatly the spontaneous performance of such reflection. Figure 10.5 represents the average frequency of spontaneous self-reflection in an experiment of dependence on self-estimated emotional stress.

Besides the generally small scale of values, it is striking that the greatest readiness to reflect upon one's own proceedings is found under medium emotional stress. In the case of high emotional stress, one very often encounters only self-evaluation in the sense of feelings of responsibility or guilt. Reither (1979) found under high emotional stress an average of 4.3 self-evaluations of that kind, which remained without visible significance for the reorganization of the behavior. It may be assumed that high emotional stress is found especially with those subjects whose competence is low. Similarly, Stäudel (in Dörner et al., 1983) found that incompetent subjects carried out self-reflection aimed at a change of behavior to a much lower degree than subjects with greater competence.

What sense does it make from the psychological point of view that subjects block themselves against a mode of self-regulation that seems to have many adaptive functions (Scheele, 1981)? The following explanation seems to be plausible: Especially in stressful situations people seem to have an increased fear of making negative findings in respect to the self-estimation of their own competence in the course of analyzing their own behavior. This supposition is supported by the experiments of Scheier and Carver (1982). These authors were able to show that the process of self-attention which provides the material for analysis through self-reflection leads to losses of capacity with subjects who are uncertain of their own competence. Consequently, the risk of an even higher loss of competence is indeed considerable and is feared in regard to the increasing danger of self-stressing. A behavior which from the very beginning avoids such a risky maneuver is meaningful under these circumstances. We do not believe that such a defense against self-reflection is done consciously; it is a behavior of unconscious avoidance.

Leaving out critical inspection of one's own thinking may result in a loss of repair possibilities. Another negative aspect of such diminished self-regulation is that coordination of the respective thinking processes suffers. Reflection about one's thinking regulates how the parts of thinking are connected, how they follow and supplement one another. If self-reflection is missing, coordination is missing too; thinking is reduced to an isolated attempt.

Conclusion

In the present paper we have given an idea of the processes of thinking in the planning of action. We started with depicting a global picture of the organization of human action and came to discern the major phases of thinking in the organization of action. These phases were goal analysis, accumulation of information, action planning and self-regulation. With regard to the total process of thinking self-regulation plays the part of a subprocess. It operates for the purpose of repair, that is, in situations when a process of thinking turns out to be ineffective.

Processes of thinking occur in many variations; their development is not very preplanned. Through self-reflection, for instance, the course of thinking and its patterns, might be changed. Emotions modify the characteristics and the course of one's thinking. Anger and fear, happiness and pride change the course of thinking significantly. Emotions, on the other hand, depend in their processual characteristics on the success or failure of action. But emotions also depend to a high degree on the image a person possesses of himself and especially on his confidence in his ability to overcome difficulties.

References

Clausewitz, C. von (1880). *Vom Kriege*. (4th edition, Part 1). Berlin: Dümmlers Verlagsbuchhandlung.

Dörner, D. (1974). Die kognitive Organisation beim Problemlösen. Bern: Huber.

Dörner, D. (1978). Self reflection and problem solving. In: Klix, F. (Ed), *Human and Artificial Intelligence*. Berlin: Deutscher Verlag der Wissenschaften.

Dörner, D. (1979). *Problemlösen als Informationsverarbeitung*. (2nd edition) Stuttgart: Kohlhammer.

Dörner, D. (1980). On the difficulties people have in dealing with complexity. *Simulation & Games, 11,* 87–106.

Dörner, D. (1982). The ecological conditions of thinking. In: D.R.Griffing, (Ed.), *Animal Mind – Human Mind. Dahlem Konferenzen 1982.* Berlin: Springer.

Dörner, D., Kreuzig, H.W., Reither, F., & Stäudel, T. (Eds.) (1983). *Lohhausen: Vom Umgang mit Unbestimmtheit und Komplexität.* Bern: Huber.

Dreistadt, R. (1968) An analysis of the use of analogics and metaphors in science. *Journal of Psychology, 68,* 97–116.

Durkin, H.E. (1937). Trial and error, gradual analysis and sudden reorganization. *Archiv für Psychologie.*

Ernst, G.W., & Newell, A. (1969) *A case study in generality and problem solving.* New York: Academic Press.

Heckhausen, H. (1980). *Motivation und Handeln.* Berlin: Springer.

Hesse, F.W. (1979). *Zur Verbesserung menschlichen Problemlöseverhaltens durch den Einfluß unterschiedlicher Trainingsprogramme auf die heuristische Struktur.* Dissertation (unpublished), TH Aachen, Aachen, West Germany.

Kuhl, J. (1983). Emotion, Kognition und Motivation: II. Die funktionale Bedeutung der Emotionen für das problemlösende Denken und für das konkrete Handeln. *Sprache & Kognition, 28,* 215–227.

Lazarus, R.S. (1966). *Psychological stress and the coping process.* New York: McGraw-Hill.

Lazarus, R.S. (1968). Emotions and adaptations: Conceptual and empirical relations. In W.Arnold, (Ed.), *Nebraska Symposium on Motivation.* Lincoln, NE: University of Nebraska Press.

Lindblom, C.E. (1964). The Science of "Muddling Through". In H.J.Leavitt & L.R.Pondy (Eds.), *Readings in managerial psychology.* Chicago, IL: University Press.

Lompscher, H.J. (Ed.) (1972). *Theoretische und experimentelle Untersuchungen zur Entwicklung geistiger Fähigkeiten.* Berlin: Volk und Wissen.

Newell, A., & Simon, H.A. (1972). *Human problem solving.* Englewood Cliffs, N.J.: Prentice Hall.

Oesterreich, R. (1981). *Handlungsregulation und Kontrolle.* München: Urban & Schwarzenberg.

Osterloh, W. (1983). *Handlungsspielräume und Informationsverarbeitung.* Bern: Huber.

Reh, H. (1983). Denkstrukturen und deren Analyse. In: D.Dörner, H.W.Kreuzig, F.Reither, & T.Stäudel (Eds.), Lohhausen: *Vom Umgang mit Unbestimmtheit und Komplexität.* Bern: Huber.

Reither, F. (1979) *Über die Selbstreflexion beim Problemlösen.* Dissertation (unpublished). University of Gießen, West Germany.

Reither, F. (1981). Thinking and acting in complex situations. *Simulation & Games, 12,* 125–140.

Selz, O. (1913). *Über die Gesetze des geordneten Denkverlaufs.* Stuttgart: Spemann.

Sergejew, J. (1970). Psychologische Hintergründe großer Entdeckungen. *Bild der Wissenschaft, 7,* 546–553.

Schank, R., & Abelson, R. (1977). *Scripts, Plans, Goals, and Understanding.* Hillsdale NJ: Erlbaum.

Scheele, B. (1981). *Selbstkontrolle als kognitive Interventionsstrategie.* Weinheim, FRG: Beltz.

Scheier, M., & Carver, C.S. (1982). Cognition, affect and self-regulation. In H.S.Clark, & S.T.Fiske, (Eds.), *Affect and cognition.* Hillsdale, NJ: Erlbaum.

Stäudel, T. (1983). Die Denkprozesse. In: D.Dörner, H.W.Kreuzig, F.Reither, & T.Stäudel (Eds.), Lohhausen: *Vom Umgang mit Unbestimmtheit und Komplexität.* Bern: Huber.

Wiegand, J. (1981). *Besser Planen.* Teufen, Switzerland: Arthur Niggli.

Chapter 11

A Control-Systems Approach to the Self-Regulation of Action

Charles S. Carver and Michael F. Scheier

> "Beliefs are really rules for action, and the whole function of thinking is but one step in the production of . . . action."
>
> William James, The Pragmatic Method

Over the last several years we have been working on the development of a theoretical account of the self-regulation of behavior. Our approach derives from many sources, including Duval and Wicklund's (1972) self-awareness theory and the broader set of ideas known as control theory or cybernetics (e. g., MacKay, 1963, 1966; Powers, 1973a, 1973b; Wiener, 1948). Ours is a theory of the control of behavior, but not a theory of motor control per se. It is a theory of intentions and actions, but not a theory of cognition or comprehension. We believe, however, that the ideas that we have been using are eminently compatible with currently popular theories concerning motor control (see, e. g., Adams, 1976; Kelso, Holt, Rubin, & Kugler, 1981; Schmidt, 1976) and theories concerning cognition and comprehension (see, e. g., Anderson, 1980; Schank & Abelson, 1977). We also believe that the point of view we have adopted allows us to usefully address certain issues that traditionally have been approached from rather discrete and restricted theoretical perspectives. Thus, we suggest that the theory serves to pull together divergent ideas and research literatures in a way that is internally consistent, providing an integration that we view as highly desirable.

We begin this chapter by describing the assumptions of our model of self-regulation, and some of the research evidence that supports the model. We then discuss how the model can be applied to the analysis of a particular category of research and theoretical problems in personality-social psychology – responses to failure. In this context we compare our ideas with those of several other theorists who have approached the same phenomena from different directions. Finally, we discuss some of the implications of this approach for understanding ineffective self-management and the process of behavior change.

A Control-Systems Model of Self-Regulation

In brief, we believe that the self-regulation of behavior is meaningfully construed in terms of the principles of cybernetic control. (For more general statements on the utility of this perspective see, e.g., MacKay, 1963, 1966; Miller, Galanter, & Pribram, 1960; Powers, 1973 a, 1973 b.) More concretely, we assume that the self-regulatory efforts of the human being reflect an ongoing comparison of present behavior against salient behavioral standards, and the attempt to bring the one into correspondence with the other. We see these activities as nicely illustrating the functions of a negative, or discrepancy-reducing feedback loop (see Figure 11.1), the basic unit of cybernetic control. That is, there is an input function (awareness of one's present behavior or state), a comparison between this input and some reference value (salient behavioral standard), and an output function (adjustments in behavior, when necessary) that works to minimize discrepancies between the two.

We also suggest (after Powers, 1973 a), that the various behavioral qualities inherent in overt action are usefully conceptualized as forming a hierarchy, in terms of their complexity or abstractness. Though we will be focusing in what follows almost exclusively upon behavioral qualities that are relatively abstract (and thus high in the postulated hierarchy, see Fig. 11.1) our discussions assume the existence of an elaborate network of control systems underlying the system under discussion (see Carver & Scheier, 1981 a, or Powers, 1973 a, for greater detail).

Schematic Organization

Having made this summary statement, let us now add some conceptual and empirical depth to the picture. Exactly where to begin in discussing this perspective is somewhat arbitrary, in that all of the components of the feedback system con-

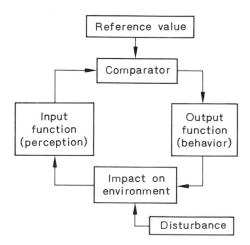

Fig. 11.1. Schematic of a negative feedback loop, the basic unit of cybernetic control

tribute to its successful functioning. One useful starting point, however, is to ask how to conceptualize behavioral standards and their utilization. How, for example, does a behavioral reference value become salient in any given situation? In order to discuss this, we must back up one more step, and talk a bit (in very general terms) about cognitive functioning.

We begin with the assumption that people impose order on their experience, based upon regularities that are encountered over time and accumulated events. This "order" takes the form of schematic organization. This basic assumption is commonly made by contemporary cognitive psychologists, who have developed a variety of theoretical approaches to conceptualizing the nature of such organizations of knowledge (e. g., Anderson, 1980; Anderson & Bower, 1973; Posner & Keele, 1968, 1970; Reitman & Bower, 1973; Rosch & Mervis, 1975). The notion that organization is imposed in memory is easily illustrated. For example, identifying a stimulus (e. g., an apple) on the basis of a few attributes (smooth red skin, rounded shape) commonly leads people to infer other attributes that have not been presented (e. g., seeds). Apparently the first few attributes are connected in memory to a broader constellation of information, which is activated by the initial identification.

Empirical demonstrations of the existence of such knowledge structures are not limited to studies conducted by cognitive psychologists. There is also evidence that people develop schematic representations of interpersonal stereotypes (e. g., Brewer, Dull, & Lui, 1981; Cantor & Mischel, 1977; Hamilton, 1979), environments (e. g., Brewer & Treyens, 1981), and even the self (e. g., Markus, 1977; Rogers, Rogers & Kuiper, 1979; Rogers, 1981). We assume that as people move through constantly varying social and nonsocial environments, they use the mental organization that has been developed over time to perceive and construe that environment. Similarly, as people focus on aspects of them*selves,* they use stored records of past experience of the self to recognize and implicitly categorize the nature of their present experience.

An important assumption built upon this foundation is that the knowledge structures under discussion can also incorporate information about *behavior.* Once again, this notion is consistent with assumptions made by cognitive psychologists, though workers in that area have rarely made a point of emphasizing it. We assume that many schemas are associated with, or incorporate, behavior-specifying information. Furthermore, if such a knowledge structure is activated in memory (via processing of information about one's present situation), we assume that the behavioral information becomes more accessible as well (cf. Collins & Loftus, 1975). This information may thereby become available for use as a reference value for a discrepancy-reducing feedback loop.

There is, in fact, evidence from recent research which suggests that activating interpretive structures in memory through relatively passive "priming" can activate behavioral information as well. In one research program bearing on this issue, Wilson and Capitman (1982) found that having male subjects read a boy-meets-girl story under a guise caused the behavioral qualities associated with that story to be displayed to a greater degree a few minutes later when an attractive girl entered the room. Similarly, Carver, Ganellen, Froming, and Chambers

(1982) found that activating the perceptual category of aggressiveness resulted in an increase in the intensity of shocks that subjects delivered to an ostensible co-subject in an unrelated context. In both of these cases, the behavioral quality emerged via the simple process of activating a mental structure that was related to the behavioral quality.

We should note that the attributes of categories – whether perceptual, conceptual, or behavioral – can vary in their abstractness. They can be concrete and restricted (e.g., "redness" in vision; a change in biceps constriction in behavior). Or they can be much more complex and abstract (e.g., the semantic information conveyed by a photograph of some event in vision; the abstract quality of "altruism" in behavior). There are at least two points to be made about this variation among attributes. First, information presumably is being processed by the nervous system simultaneously at many levels of the hierarchy of abstractness (cf. Norman, 1981). Thus, for example, monitoring of muscle tensions occurs more-or-less continuously while a person goes about behaving altruistically, despite the fact that monitoring of muscle tensions is normally outside conscious awareness during the course of an altruistic act.[1]

The second point to be made here amplifies upon a statement we made earlier: that our own interest will be primarily in what goes on at relatively high levels of abstraction. Three levels will be particularly important to us. The first is what Powers (1973a) termed "program" control. A program is roughly equivalent to what Schank and Abelson (1977) termed a "script". It represents a particular kind of event sequence in which multiple possibilities for action can be expected to occur, requiring decisions to be made. A script often used by Schank and Abelson for illustrative purposes is "dining out". Such an event clearly entails many behavioral choices within the framework of a partially predefined set of subevents. It should be quite obvious from this example that much of the flow of human action takes the form of events that are program- or script-like in their character.

The intentions underlying these activities are quite abstract. But even more abstract levels of control can be implicated in human action. Powers (1973a) postulated what he called "principle" control as a level that was directly superordinate to program control. This concept seems very similar to what Schank and Abelson (1977) termed "meta-scripts". Principles or meta-scripts are guidelines for action that are more general than programs. Said differently, a principle defines a quality that might be realized in a great many programs of behavior. Using a principle helps a person to decide what programs of action to undertake, and what decisions to make while engaged in a program. For example, the principle of frugality might cause one to pass up the purchase of a new car this year. Alter-

---

[1] Many people have argued that closed-loop control of action occurs primarily in the early stages of learning, and that once behavior has become automatic, monitoring diminishes or eventually ceases altogether (see, e.g., Greenwald, 1970; Norman, 1981; Stelmach, 1976). Though it does seem likely that monitoring is reduced as actions become familiar, there is also a basis for the position that it never ceases entirely (see Norman, 1981). Further discussion of this issue is beyond the scope of this chapter, inasmuch as we will be addressing almost exclusively behaviors that are not highly automated.

natively, given that the person truly needs to purchase a new car, the same principle might be observed by avoiding the more extravagant options and choosing
from among relatively inexpensive models.

The final level of abstraction that we will adress here is what Powers (1973 a)
termed "system concepts". The sense of continuity, personal wholeness, and ac-
tualization that comprise a person's ideal self is an illustration of what we have
viewed as system concepts. One might self-regulate with regard to other system
concepts – for example, a similar mental construction representing an idealized
image of a group which one holds to be particularly important. But the sense of
ideal self seems intuitively meaningful as a system concept, and it is one to which
we will return later in the chapter.

Self-regulation with regard to the system-concept self means attempting to be
who you think you should be. This is done by trying to live up to principles that
are specified by your image of who you should be. The principles, in turn, are realized in programs of action. In each case, a superordinate level of control acts
(output function, from Fig. 11.1) by specifying reference values to be used at the
next lower level (see Fig. 11.2). Progress toward goal attainment is monitored at
each level. The result is a hierarchy of action qualities, the end result of which is
overt behavior. (For a more complete description, with behavioral examples, see
our discussions elsewhere, e. g., Carver & Scheier, 1981 a, 1982 a, 1983).

We should make two final points concerning this hierarchical organization.
First, implicit in this approach to conceptualizing action is the notion that there

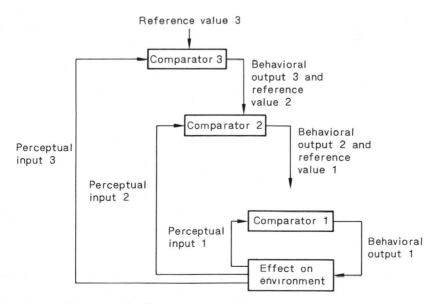

Fig. 11.2. A 3-level hierarchy of feedback systems, in which the behavioral output of a superordinate system constitutes the resetting of the reference value of the next lower system

are multiple paths to the attainment of any given goal value. This is particularly obvious with regard to relatively abstract goals. A specific principle, for example, can be realized in a great many ways in a great many behavioral domains. Similarly, progress toward being the sort of person one wants to be is monitored not just in one domain, but in many. This renders the model nicely compatible with analyses in which diverse actions serve a common function (see Wicklund & Gollwitzer, Chapter 4, this volume, for a theory that focuses on precisely such a notion).

Finally, we should point out that people often self-regulate with regard to more than one reference value at a time. In some cases the values are entirely compatible with each other (e.g., being frugal while also being austere). But in other cases the reference values that are salient are less easily satisfied simultaneously (e.g., being frugal while being a patron of the arts). The conflict embodied in this sort of circumstance can create problems in behavior.

Self-Directed Attention

Our discussion thus far suggests a basis for action. But there are several additional assumptions that need to be mentioned. First, we have argued that self-regulation of behavior depends in part upon the person's focus of attention. More specifically, we have suggested that a precondition for the operation of the comparator at the superordinate level of control is the focusing of attention inward to the self. When a behavioral standard is salient, self-focused attention is assumed to lead to an increased tendency to compare one's present behavior or state with that standard. The result of this comparison, in our view, is the engagement of the discrepancy-reducing feedback loop. (The functions just outlined are precisely those that were postulated by Duval & Wicklund, 1972, though we differ from those authors regarding how best to construe the functions.)

That self-focused attention increases the tendency to compare one's behavior with situationally salient standards has recently been demonstrated, though the evidence is somewhat indirect. In a series of four studies (Scheier & Carver, 1983a) subjects were given an opportunity to examine a source of concrete information that would be useful in determining the degree to which the subjects had been, would be, or were currently matching their behavior to standards at a higher level ("doing well" at an experimental task). In each case, higher levels of self-focus were associated with a stronger tendency to compare oneself with the standard.

Self-focus was operationalized in this research (as in most research on the effects of self-focus) in one of two ways: by exposing some subjects to manipulations that would remind them of themselves (e.g., a wall mirror), or by assessing individual differences in the disposition to be self-attentive, via the Self-Consciousness Scale (Fenigstein, Scheier, & Buss, 1975). This scale was developed as an explicit attempt to tap the same psychological states as were created by the various experimental manipulations. A large number of successful conceptual replications indicates that the attempt was highly successful (see Carver & Scheier, 1981 a, for a full review). This convergence between manipulation and disposition is only one of many sources of evidence of the validity of the self-attention construct (once again we refer readers to Carver & Scheier, 1981 a, for a discussion of this evidence).

As we noted earlier, when a behavioral standard is salient, directing attention to the self is presumed to engage a discrepancy-reduction process. The result of this process is that the self-focused person conforms more closely in his or her behavior to the salient standard. Evidence that this does occur was acquired in some of the earliest research on the effects of self-awareness (Wicklund & Duval, 1971). Since that time, the demonstrations have been many and varied. The behaviors in question have ranged from aggression (Scheier, Fenigstein, & Buss, 1974; Carver, 1974, 1975), to use of the equity norm (Greenberg, 1980), to the taking of Halloween candy by children (Beaman, Klentz, Diener, & Svanum, 1979). Indeed, this standard-matching or discrepancy-reduction consequence of self-focus is among the more completely documented phenomena in the recent literature of social psychology.

Aspects of the Self

We have been discussing the focusing of attention on the self. But the self quite obviously is a complex and multifaceted entity. When attention is self-directed, the person may be focusing on transient feeling states, on memories reflecting typical behavioral characteristics, on the actions presently being engaged in, or on other self-relevant information. How can these various aspects of self be sorted out in an orderly way? A simple answer is that the aspect of self taken as the object of scrutiny when attention is directed inward is whatever self-aspect is situationally salient (cf. Duval & Wicklund, 1972). Sometimes the context renders internal sensations salient as an aspect of self (e.g., Gibbons, Carver, Scheier, & Hormuth, 1979; Scheier, 1976; Scheier & Carver, 1977; Scheier, Carver, & Gibbons, 1979), sometimes the context points to stored knowledge about one's behavioral tendencies (e.g., Scheier, Buss, & Buss, 1978; Turner, 1978), and sometimes it points to one's present behavior and goals (e.g., Carver, 1974, 1975; Scheier et al., 1974; Wicklund & Duval, 1971).

Another way to address this question has proven to be quite useful, however. This approach proceeds from the assumption that some aspects of the self are covert, hidden, or "private", and that other aspects of the self are overtly social, or "public" (Fenigstein et al., 1975; see also Buss, 1980; Scheier & Carver, 1981, 1983 b). Sometimes attention is directed selectively to the private self, and sometimes it is directed selectively to the public self. This private-public distinction was first highlighted in the development of the Self-Consciousness Scale, which measures private self-consciousness and public self-consciousness with separate subscales. But the distinction has proven useful with regard to manipulations as well.

This difference in focus has profound behavioral implications, under certain circumstances. Specifically, if standards or reference values stemming from private considerations would induce behavior of one sort, but standards deriving from social or public considerations would induce behavior of an opposite sort, focusing on these two sides of the self should create opposite patterns of action.[2] Indeed, a number of studies have confirmed this reasoning. Froming and Carver (1981) found that attention to the private self was associated with a decrease in conformity to a unanimously incorrect group in an Asch-type setting; attention to the public self had a precisely opposite effect – an increase in conformity (see also Santee & Maslach, 1982). Carver and Scheier (1981b) found that when reactance was induced in a social-influence setting, focus on the private self was associated with a strong reactance effect – increased resistance – and focus on the public self was associated with a decrease in the reactance effect. In yet another context, Froming, Walker, and Lopyan (1982) found that focusing on the private self caused increased conformity to personal attitudes in administering punishment, whereas focusing on the public self caused increased conformity to what subjects thought "most people" believed was the appropriate level of punishment.

In some of this research, focus on public versus private self-aspects was varied by selecting subjects who differed on the relevant dimensions of the Self-Consciousness Scale. In other cases, experimental manipulations were used to vary the focus of attention. Based in part upon these and related data, there is a substantial basis for believing that the presence of a small mirror focuses attention on private self-aspects, and that manipulations such as the presence of TV cameras, audiences, or one's image on a TV monitor focus attention on public self-aspects. (See Scheier & Carver, 1981, 1983b, for more exhaustive surveys of research on the public-private distinction.)

Interruption and Disengagement

The processes outlined thus far account for behavioral conformity to salient goal values, goals which can vary from private, personal concerns to more social or self-presentational concerns. But there is one very important aspect of action control that we have not yet broached. Specifically, people sometimes encounter obstacles in their attempts to attain goals. Some of these obstacles can be dealt with, but others constitute sufficient impediments that they render further attempts useless. How are we to account for the ways in which people confront and respond to such obstacles?

In response to this very basic issue, we have assumed the existence of an expectancy assessment process, which occurs separately from the more basic dis-

---

[2] There obviously is some degree of conceptual overlap between the distinction under discussion here and the dimension of self-monitoring. That is, chronic focus on the private self seems a good deal like low self-monitoring; focus on the public self seems a good deal like high self-monitoring. Upon closer examination, however, it becomes apparent that the fit between constructs is less than perfect. Though discussion of the differences is beyond the scope of this chapter, interested readers will find the issue addressed elsewhere (Carver & Scheier, 1981a, pp. 322–325).

crepancy-reduction process. This assessment is induced when behavior is interrupted by such events as frustrated efforts on a task, or rising anxiety (cf. Simon,
1967). We construe this assessment as leading to a sort of psychological "watershed." That is, we see the assessment as prompting one of two subsequent responses (see Figure 11.3). Either the person returns to the attempt to match behavior to the salient reference value, or the person experiences an impetus to
disengage from that attempt. Virtually all responses would seem to fall into one
or the other of these categories. And despite the fact that there probably is a
range of effort intensities as one examines the entire spectrum of expectancies

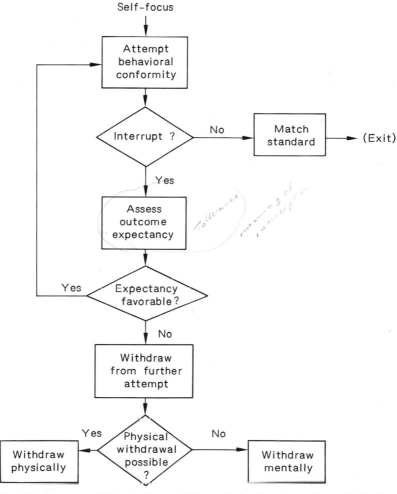

Fig. 11.3. Postulated sequence following from self-directed attention when a behavioral reference value has been evoked, including the possibilities of interruption and disengagement
(adapted from Carver, 1979)

we see some utility in emphasizing the watershed dividing the two response classes from each other. Just as self-focus leads to greater initial attempts to conform to salient standards, self-focus when an interruption has led to favorable expectancies leads to a more pronounced renewal of effort. If expectancies are unfavorable, self-focus leads to a more complete disengagement.

There are three points about this aspect of the model that require additional attention. Two of them stem from the fact that people cannot always disengage overtly from the behaviors which they have begun. When situational constraints prevent overt disengagement, we assume a mental or psychological disengagement. This may be reflected in task-irrelevant rumination, daydreaming, or the like. Such a condition is also likely to be reflected in behavior as a deterioration in performance.

The second point to be made here is that there are many reference values from which complete and permanent disengagement is simply not feasible – for example, doing well at one's chosen profession, or getting along well with the people with whom one repeatedly interacts. There may be a disengagement from effort when a person perceives that matching such a value is unlikely. But that disengagement sooner or later will be followed by a renewed confrontation with the reference value. The result in such a case is a repeated cycle of sporadic effort, doubt, and disengagement.

Finally, we have argued that this assessment process is associated with, or even gives rise to, the emotions elation and depression (Carver, 1979; Scheier & Carver, 1982a; see also Stotland, 1969). The implications of this notion for the cycle just described seem obvious. The confronting and reconfronting of one's inability to attain important goals in life is likely to be accompanied by substantial negative affect, a point to which we return in the context of clinical issues.

The watershed aspect of the model has received considerable support from research conducted over the last several years. In one study, subjects were selected as having moderate levels of fear toward nonpoisonous snakes (Carver, Blaney, & Scheier, 1979a). Subjects varied, however, in their self-reported expectancies of being able to pick up and hold a snake if asked to do so. Self-directed attention during an actual approach task led to enhanced perceptions of anxiety among all subjects, and presumably to interruption and outcome assessment at an earlier stage of the task (see also Scheier, Carver, & Gibbons, 1981). But the effect of self-focus on subsequent action depended upon subjects' preassessed levels of confidence. Doubtful subjects stopped the approach task sooner in the sequence when self-focus was high than when it was lower; but confident subjects tended to persist *longer* when self-focus was high than when it was lower.

Another project conducted shortly after this one studied the model's applicability to people's responses to failure. Subjects in this research (Carver, Blaney, & Scheier, 1979b) were told that they were working on items from an intelligence test. All subjects were given a pretreatment failure experience, followed by a manipulation of expectancy concerning a second task. The second task actually consisted of a single insoluble puzzle item. Subjects led to believe that they had a good chance to redress their failure on the second task persisted longer on that item when self-focus was high than when it was lower. And subjects led to hold unfavorable expectancies for redressing their failure were less persistent when self-focus was high than when it was lower.

These projects illustrate the kind of support obtained for the behavioral aspects of this portion of the model. Conceptual replications have also been reported in several other related studies (e.g., Brockner, 1979; Carver & Scheier, 1982b; Scheier & Carver, 1982c). Finally, evidence for our predictions concerning affect has also been gathered in one of these studies (Carver & Scheier, 1982b), evidence that dovetails nicely with data collected recently by Weiner and his colleagues (Weiner, 1982; Weiner, Russell, & Lerman, 1978, 1979).

Reassertion and Giving Up: Helplessness and Alternative Interpretations

The ideas outlined in the preceding section were originally intended to describe the general structure of behavioral responses to impediments to action. As a general statement, the reasoning appears to be applicable to fear-based self-regulation, to the effects of environmental obstacles, and to the effects of personal inadequacies. These ideas, however, also prove to have a good deal in common with recent theoretical accounts of the performance impairments that are often termed "helplessness." It would perhaps be instructive to examine more closely the relationships among current accounts of helplessness phenomena, alternative conceptualizations of those phenomena, and our own approach.

Helplessness

The helplessness concept had its empirical origin in the finding that exposing a dog to unavoidable aversive events made it more difficult for the animal to learn an avoidance or escape response later on (Overmier & Seligman, 1967; Seligman & Maier, 1967). A good deal of research subsequently was done to determine whether conceptually similar effects would occur among human subjects (for reviews see Abramson, Seligman, & Teasdale, 1978; Miller & Norman, 1979; Roth, 1980). The prototypical helplessness effect among humans is the finding that after exposure to a series of outcomes that are not contingent upon behavior, the subject displays impaired performance on a subsequent task (e.g., Hiroto, 1974; Hiroto & Seligman, 1975). This sort of finding is common in the helplessness literature, generalizing across types of tasks employed, and the way in which noncontingency is operationalized. But this finding does not inevitably occur. Indeed, sometimes exposure to noncontingency leads to performance enhancement instead of impairment (e.g., Hanusa & Schulz, 1977; Roth & Bootzin, 1974; Roth & Kubal, 1975).

This apparent paradox – that noncontingent pretreatment sometimes leads to worse performance and sometimes to better performance – stimulated additional theory construction. In the main, these theories have been much more cognitive than the initial account. Wortman and Brehm (1975), for example, accounted for the paradox by arguing that the effect of the pretreatment depends upon the expectancy that it leaves in its wake. If the person has an expectancy of no control, subsequent performance will suffer. If the person still has an expectancy of being able to exert control, on the other hand, a reassertion will take place, leading to performance improvement. Other theorists, even while emphasizing attributional parameters as determinants of helplessness (Abramson et al., 1978; Miller & Norman, 1979), concur with Wortman and Brehm in this regard. That is, the final pathway to helplessness in these models is an expectancy of no control, or an expectancy of noncontingency.

These expectancy constructs are quite similar to the outcome expectancy construct that we use in our own work. Thus, these accounts of learned helplessness appear to be assimilable to a more general statement on the self-regulation of action. In addition, the model of self-regulation outlined above appears to add important elements to these accounts of reassertion and giving up. Most obviously, we would suggest that no such theory is complete without taking into account the role of self-focus. Self-focus exaggerates both the performance decrements that follow from unfavorable expectancies and the renewed efforts that follow from favorable expectancies.

Secondly, we have suggested that an impulse to disengage behaviorally from the task attempt lies at the heart of the impaired performances displayed by "helpless" persons. This suggestion seems quite compatible with data from several sources in the helplessness literature (Diener & Dweck, 1978; Brunson & Matthews, 1981; see also Halisch & Heckhausen, 1977). Perhaps what has often been characterized by helplessness theorists as a learning deficit is really attributable to a mental disengagement from the task at hand, inspired by a desire to leave the testing context, a desire whose expression is prevented by social constraints.

Egotism

There has not been complete unanimity concerning the meaning of helplessness effects and the role of expectancies in their occurrence, however. An alternative argument has been advanced by proponents of a set of ideas concerning the protection and maintenance of self-esteem. Termed "egotism" (Snyder, Stephan, & Rosenfield, 1976, 1978), this perspective holds that people choose their actions so as to minimize threats to self-esteem. This line of reasoning is often used to account for the tendency to attribute successes internally and failures more externally (cf. Heider, 1958). It has been used to explain instances in which people appear to go out of their way to make success difficult (Berglas & Jones, 1978). And it is also applicable to so-called helplessness effects.

Frankel and Snyder (1978) argued that people in an ordinary helplessness paradigm are not learning that they have no control over outcomes. Rather, the pretreatment serves as a threat to their self-esteem. The second task constitutes an additional threat. In order to minimize that threat as much as possible, subjects create a self-esteem-protective attribution for potential failure. Specifically, they withdraw their effort from the second task. If they do poorly, then, it is because they did not try hard enough, not because they lack the necessary ability. Presumably the lack-of-effort attribution is less threatening to one's self-image than is the lack-of-ability attribution.

Frankel and Snyder reasoned further that if subjects were provided with a different self-protective attribution, there would be no need to protect self-esteem by withdrawing effort. The result should be an elimination of the alleged helplessness effect. In a test of this reasoning, Frankel and Snyder told some subjects that the second task was very difficult, thereby creating a face-saving attribution for a potential failure. As predicted, this manipulation eliminated the performance deficit. The finding has since been conceptually replicated (Snyder, Smoller, Strenta, & Frankel, 1981).

Do these findings refute the expectancy-based analyses of human responses to failure? Perhaps not. It should be noted that a manipulation of perceived difficulty does more than change the attributional contingencies surrounding one's performance. It also has an impact on the incentive value of a good performance. That is, doing well on a difficult task would seem to go farther toward compensating for an initial failure than would doing well at a task that is only moderately difficult.

And why is this important? We have not mentioned incentive until now. But it seems likely that the degree of incentive (or importance) inherent in a task will have an impact on how favorable one's expectancy must be in order for one's efforts to continue. Presumably people will continue the attempt to attain a very valuable goal even when its attainment appears relatively unlikely. Less-valued goals are likely to be abandoned even when subjective probability of their attainment is higher. In effect, changing the importance of a goal shifts the psychological watershed point.

This reasoning suggests that subjects in the Frankel-Snyder study who received the task-difficulty manipulation may have begun the second task more motivated, and thus trying harder, than the other subjects. The importance of this possibility lies in the fact that all subjects in the Frankel and Snyder study actually received as their second task a set of anagrams that were only moderately difficult. To the degree that high levels of incentive, and the effort to which they led, paid off in successful anagram solutions, subjects' expectancies of making up for their pretreatment failure should have become more favorable, resulting in even greater efforts.

But what if these subjects had actually been given items that were high in difficulty? What if there had been no early payoff for trying hard? If the effect of a task-difficulty attribution is simply to provide protection for self-esteem, there is no particular reason to assume that this effect diminishes over items. We, on the other hand, would suggest that even with high incentives there comes a point where efforts give way to disengagement. We therefore would predict that subjects in this circumstance would display a performance decrement.

In a study designed to test this reasoning (Scheier & Carver, 1982b), we replicated the procedures of Frankel and Snyder (1978), but divided the task-difficulty group into two groups. One of these groups received a set of anagrams that began with moderately difficult items; the other received anagrams that began with highly difficult items. The dependent measure was performance on the final 10 items, which were only moderately difficult for all subjects. As predicted, subjects who had the opportunity to do well initially displayed a reassertion effect, and those who had been given difficult items displayed the usual helplessness effect. Further, and also in line with our theoretical approach, both of these effects were most pronounced among persons high in dispositional self-consciousness. Finally, there was evidence that these variations in behavior across experimental groups was a product of variations in expectancy across trial blocks.

Thus expectancies do appear to matter. Indeed, there seems to be no other way for the egotism point of view to account for the fact that people sometimes respond to failure pretreatments with renewed efforts, even in the absence of a self-esteem-protective attribution (e. g., Hanusa & Schulz, 1977; Roth & Kubal, 1975). In a different context, Snyder et al. (1978) have acknowledged that the occurrence of egotistical behavior in general will depend in part upon the perceived probability that such behavior will successfully protect or enhance self-esteem (see also Bradley, 1978; Frey, 1978). As such, however, egotism theory then becomes very similar to our own account of self-regulation in at least one important way. Specifically, both approaches appear to provide for a psychological watershed among responses, based upon expectancies of being able to make up a discrepancy between one's present behavior and a standard of comparison.

There are two other points of comparison between egotism theory and our own approach that deserve some note. The first concerns the self-protective function of behavioral disengagement. We have discussed disengagement as a function that removes the person from impossible circumstances, that allows one, in effect, to back out of corners. Our emphasis on this aspect of the process tends to obscure the fact that it clearly is self-protective to disengage in cases where further attempts would be fruitless (see Janoff-Bulman & Brickman, 1982). Disengagement and withdrawal prevent one from confronting the discrepancy between where one is and where one wants to be, and thereby short-circuits the negative affect that accompanies expectations of little progress. Thus, a refusal to try is in fact quite self-serving.

The second point that we should make here stems from the notion of hierarchical organization, which we have adopted from Powers (1973 a). We suggested earlier in the chapter that a good example of what Powers termed a "system concept" is the abstract sense of the ideal self – who one thinks one ought to be. Attempting to *be* that person would appear to constitute an attempt to enhance or protect one's self-esteem. Thus, we would think of self-esteem-protective behavior as reflecting self-regulation in which the system-concept level of control is functionally superordinate.

When many people talk about helplessness, however, they implicitly are discussing behavior at the level of problem solving or task performance. We would tend to think of problem-solving behavior – if there are no higher order implications – as reflecting superordinate control at what Powers termed the program level. In a sense, then, the difference between contexts in which the egotism explanation is the more accurate and those in which the helplessness explanation is the more accurate may be this: Egotism occurs when people have been failing to be who they want to be, by virtue of failing at problems. Helplessness occurs when people have simply been failing at problems, problems that may have no relevance for self-esteem. Both of these models of the effects of failure would appear to be accurate under certain circumstances. But what determines which of these levels of control is functionally superordinate at a given time is a matter for resolution by future research.

Action and State Orientations

Kuhl (1981, 1982, Chapter 6, this volume) has proposed yet a different account of helplessness phenomena. His theory is based in an expectancy-value framework, but Kuhl argues for the importance of a third factor not incorporated into other such theories. This factor is termed the person's degree of "action orientation". A high level of action orientation is characterized by "cognitive activities focusing on action alternatives and plans that serve to overcome a discrepancy between a present state and an intended future one;" the opposite pole of this dimension, called "state orientation," is characterized by "cognitive activities that focus on the present, past, or future state of the organism" (Kuhl, 1981, p. 159). This theory holds that behavioral intentions are formulated on the basis of expectancy and value considerations. But an action orientation is indispensable for successful activity, because this orientation is required for the execution of an intention.[3]

This theory suggests the basis for two distinct types of performance decrements. The first derives from the condition postulated in other helplessness theorists – that is, a motivational or intentional deficit caused by low expectancies of control or success. Without the intention to seriously attempt a task, there is nothing for an action orientation to carry out. But the theory also suggests that the lack of an action orientation can by itself impede behavior, resulting in performance decrements (termed "functional helplessness"). That is, being motivated to perform is not sufficient; this motivation must have overt expression. Kuhl thus points out that performance deficits have many possible antecedents, rather than just one, and that lumping all instances of poor performances together may be quite misleading.

The distinction between state and action orientations emphasized by Kuhl appears to be one that deserves additional scrutiny independent of helplessness phenomena. Indeed, bringing this distinction to bear on our own work reveals points at which issues remain to be resolved. Throughout this chapter we have emphasized the notion that self-focus promotes a comparison of one's present behavior against salient reference values, followed by an active attempt to bring

---

[3] As an aside, we note that the action-state distinction might be a useful way of construing the difference between the "Type A, coronary-prone behavior pattern" and the opposite pattern, termed Type B (e.g., Glass, 1977; Matthews, 1982). That is, the Type A is usually characterized as achievement oriented, time urgent, and aggressive, with the Type B being less so. This sounds not unlike an action orientation among A's, and a state orientation among B's.

the one into line with the other. This appears to imply that self-attention induces an action orientation. Indeed, Kuhl's description of what occurs during action orientation is very similar in many respects to our own descriptions of the discrepancy-reduction process induced by self-focus.

But there are also circumstances in which self-focus leads simply to enhanced awareness of emotions and other personal states (e.g., Gibbons et al., 1979; Scheier & Carver, 1977; Scheier et al., 1979), and to a more accurate recollection of what one is like (Scheier et al., 1978; Turner, 1978). These latter cases would seem to reflect a state orientation. The mental disengagement that we have assumed follows when physical withdrawal is stifled might also be interpreted as a state orientation. We would argue, however, that such a "motivated" disengagement is different in important ways from a simple preoccupation with action-irrelevant information, although the two conditions may appear quite similar superficially. Thus the state-action distinction appears to be one that cuts across the attentional-focus distinction that we have emphasized, rather than paralleling it.

As a last comment in this regard, we would like to note that an analogy drawn from the realm of computers may be useful in thinking about the difference between state and action orientations. Many computer systems have two distinct modes of activity: an "edit" mode in which existing data and instructions for operation can be examined passively or examined and altered, and an "executive" or "control" mode in which the instructions specifying operations are carried out. A state orientation is somewhat like the edit mode – examination, preoccupation, but no overt behavior. An action orientation is somewhat like the executive mode – carrying out the processes specified by the program, or by the behavioral intention or standard. It is obvious how different these modes are in the computer. And the difference would seem equally important to human behavior. Just precisely what prompts one orientation or the other, and how best to conceptualize the meaning of the distinction for human behavior, are matters for further consideration.

Applications: Ineffective Self-Management, and Behavior Change

The preceding section outlined several theoretical orientations to the performance deficits associated with the term helplessness. Our primary reason for examining those theories was to show how they can be assimilated to the general approach to self-regulation outlined earlier. However, the preceding section also serves as a bridge to an additional area of theory and research. In particular, although our ideas were intended to represent an analysis of normal self-regulation, they also prove to be applicable to the understanding of problems in self-management. To illustrate this applicability, we begin by briefly outlining two representative areas of research in which data suggest the model's usefulness in examining particular problems in behavior. We then turn to more general questions pertaining to the domains of ineffective self-management and behavior change.

Test Anxiety

Test anxiety is a problem in self-management for many students. People who are test anxious experience unpleasant affect during evaluative testing situations, and often suffer impairments in test performance. Early theories of test anxiety were based on the assumption that the test anxious experienced overly high levels of arousal in test situations (e. g., Spence & Spence, 1966). More recent approaches, however, have placed greater emphasis on the cognitive processes that accompany the arousal (e. g., Sarason, 1975, 1978; Wine, 1971, 1982). This shift in orientation epitomizes the "cognitive revolution" that has taken place in understanding behavior problems. And the shift seems particularly appropriate in the case of test anxiety, for the following reason. At least two recent research projects have found that persons high in test anxiety do not experience more intense arousal during exams than do persons lower in test anxiety (Hollandsworth, Glazeski, Kirkland, Jones, & Van Norman, 1979; Holroyd, Westbrook, Wolf, & Badhorn, 1978). What appears to distinguish the groups from each other is not their levels of arousal, but rather how they *respond* to the arousal and to the test situation more generally.

It is often reported that persons high in test anxiety engage in self-deprecatory rumination when in highly evaluative circumstances (e. g., Meichenbaum, 1972; Mandler & Watson, 1966; Deffenbacher, 1978), and that they tend to neglect or misinterpret available task-related cues. Wine (1971, 1982) has attempted to integrate findings such as these by arguing that the test anxious become highly self-focused, when placed in evaluative settings. In Wine's view, self-focus impairs task performance because it diminishes the attention available for task focus.

We have argued elsewhere (Carver, Peterson, Follansbee, & Scheier, 1983; Carver & Scheier, 1983) that this view is too simple. For one thing, it fails to take into account the fact that self-focus sometimes facilitates performance, rather than impairing it (e. g., Carver & Scheier, 1981c; Wicklund & Duval, 1971). There appears to be no plausible way of accounting for these facilitation effects without postulating a more elaborate model. It is our position that what Wine and others (e. g., Sarason, 1975, 1978) have seen as representing self-focus is what we have discussed in the context of outcome assessment and psychological disengagement from task attempts (Figure 3, earlier in the chapter). That is, we see the test anxious as anticipating unpleasant affect and poor performances. As they assess their outcome expectancies, they experience an impetus to withdraw. Because the social context is one that does not sanction overt withdrawal, the result is cognitive disengagement. The test situation does not permit this disengagement to last long, however. Thus the person experiences a repeated cycle of concern about the possibility of doing poorly, negative self-centered ideation, off-task thinking, and sporadic efforts.

This analysis fits well with aspects of the literature of test anxiety. Most obviously, the phenomenology of the experience is likely to lead to data such as Wine has summarized. As additional examples, consider the following. Nottelman and Hill (1977) found that test-anxious children glanced away more frequently from a task on which they were working than did less test anxious children. This is easily interpreted as a kind of physical disengagement from the task attempt. More recently, Galassi, Frierson, and Sharer (1981) reported that the most frequent thought category experienced by highly test-anxious college students during an actual test concerned escaping from the test situation. These findings seem highly consistent with our reasoning.

More direct support for aspects of our reasoning has been obtained in recent research in which subjects' attentional focus was experimentally manipulated. Our analysis suggests that high levels of self-focus in an evaluative setting should be performance impairing among persons high in test anxiety. But it also suggests that the same initial state should be performance facilitating among persons who are not test anxious. Such an interactive effect would parallel the many interactions obtained in other circumstances between self-focus and outcome expectancies (e.g., Brockner, 1979; Carver et al., 1979a, 1979b; Carver & Scheier, 1982b; Scheier & Carver, 1982c). In a test of this prediction (Carver et al., 1983, Experiment 1), subjects high and low in test anxiety completed a set of moderate and difficult anagrams under stress-inducing conditions, while seated before a mirror or with no mirror present. As expected (see Fig. 11.4), mirror presence interacted with level of test anxiety, impairing those high in anxiety and facilitating those lower in anxiety.

Our theoretical analysis also makes another prediction that is not easily derived from alternative conceptualizations. Specifically, we have assumed that the performance impairments displayed among persons high in test anxiety have

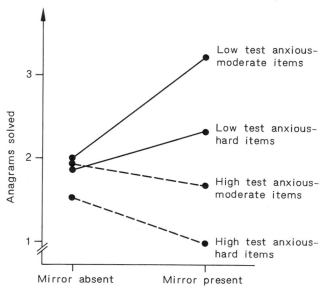

Fig. 11.4. Performance on moderately difficult and very difficult anagrams, as a function of level of test anxiety and mirror presence or absence (from Carver et al., 1983)

their root in an impulse to disengage physically from the task attempt. Presumably this impulse is stifled because the social context does not sanction physical withdrawal. If, however, a situation could be arranged in which physical removal of oneself from the behavioral attempt were permissable, we would expect persons high in test anxiety to display withdrawal under conditions of self-focus. In contrast to this reduced persistence, persons low in test anxiety should display increased persistence under conditions of enhanced self-focus.

This reasoning was tested (Carver et al., 1983, Experiment 2) by giving subjects a set of anagrams on separate cards, with the following instructions. Subjects were to take as much or as little time as they wished in attempting any given anagram, and the items could be attempted as often as subjects wished. But the anagrams had to be attempted in a specific order. In the event the subject wished to (temporarily) give up the attempt to solve a particular item, that item was to be placed at the bottom of the stack of cards. This instruction allowed the experimenter to determine when the subject gave up the attempt to solve the first item, which was insoluble. When the subject gave up on that item, the experimenter recorded the elapsed time, and the session was ended. Analysis of the data yielded an interaction between mirror presence and test anxiety that was very similar in form to the interaction obtained in the experiment just discussed (see Fig. 11.5). Self-focus led to greater persistence among persons low in test anxiety, and to reduced persistence among persons high in lost anxiety. Dispositional public self-consciousness exerted a similar and independent effect on persistence. Thus our behavioral predictions were supported by the data.

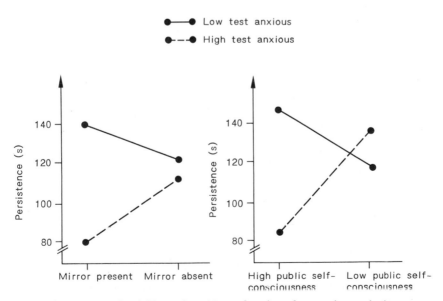

Fig. 11.5. Persistence on an insoluble test item (a) as a function of test anxiety and mirror presence or absence, and (b) as a function of test anxiety and level of public self-consciousness (from Carver et al., 1983)

In brief, we would suggest that the phenomenon of test anxiety is easily viewed as a special case of the more general processes discussed earlier, in which the anticipation of an undesired outcome leads to an impulse to disengage. This impulse may be executed behaviorally – perhaps as a continuous search for the elusive item that is easily answered. Or it may be reflected only in mental disengagement, to the degree that the setting permits it. In either case, the processes would seem to represent extreme cases of what were postulated earlier as normal functions of self-regulation.

Alcohol Use and Abuse

A second area in which data appear to be supportive of our general line of reasoning concerns the use of alcohol, and the functions served by alcohol consumption. This research was conducted by Hull and his colleagues, who have taken a perspective on the meaning of self-directed attention that differs somewhat from ours (see Hull, 1981). Despite this difference, however, the findings themselves seem easily assimilable to our approach to the control of behavior.

After reviewing a substantial portion of the literature discussing the effects of alcohol on various aspects of human functioning, Hull concluded that a major effect of alcohol ingestion was diminished self-awareness. Consistent with this point of view were, for example, findings that alcohol use was associated with less well-reasoned self-regulation, more impulsive behavior, and an inattention to salient standards (reviewed in Hull, 1981). Interestingly enough, these effects are very similar to those displayed under conditions of deindividuation. And there is independent evidence that the deindividuation experience is associated with reduced self-focus (Diener, 1979; Prentice-Dunn & Rogers, 1982). Hull and his colleagues subsequently gathered additional data to validate the self-awareness-reduction hypothesis. They found, for example, that alcohol intake decreased the frequency of self-reference during an impromptu speech (Hull & Young, 1983), an effect that is often regarded as reflecting diminished self-focus (cf. Carver & Scheier, 1978). There is also evidence that alcohol consumption interferes with access to self-relevant memory structures (Hull & Young, 1983).

The notion that alcohol reduces self-focus has far-reaching implications, particularly when taken in combination with an assumption advanced earlier in the chapter. The assumption in question is that the experience of self-focus is aversive when the person has unfavorable expectancies about being able to reduce salient discrepancies between present states and important reference values. If the person cannot disengage behaviorally from that reference value, cognitive withdrawal may ensue. Hull's line of reasoning suggests an additional possibility. If alcohol reduces self-focus, it will also thereby reduce the aversiveness of focusing on the absence of discrepancy reduction. In effect, drinking provides a very potent way of creating and maintaining mental disengagement. This theoretical statement, ironically enough, sounds very much like the common stereotype of the serious drinker as a person trying to escape from his problems.

Once again, this account is consistent with preexisting literature. Perhaps even more important, however, Hull and Young (1983) have reported two studies which provide additional evidence on the question. One of the studies was a laboratory experiment in which subjects first received a success or a failure on an ostensible intelligence test. Afterward they participated in a wine tasting session, which provided a measure of alcohol consumption. The combination of failure and high levels of dispositional self-consciousness was associated with the greatest consumption of alcohol. The data appear to suggest an active attempt on the part of these people to reduce their levels of self-focus, following the creation of a discrepancy that they had no possibility of reducing.

The final study we would like to mention in this domain stems from the argument that these laboratory findings have important ramifications for long-term patterns of alcohol consumption. More specifically, it would appear to be the case that high levels of self-consciousness, taken together with continued personal failure in other domains, would constitute a risk factor for alcohol abuse, or for relapse following detoxification. In a test of this reasoning (Hull & Young, 1983), subjects completed the Self-Consciousness Scale and a questionnaire assessing the quality of their life events, just as they were being released from a detoxification treatment program. They were recontacted three and six months later to determine current levels of alcohol consumption. As Hull and Young had predicted, subjects high in private self-consciousness who had had predominantly negative self-relevant events were more likely than any other group to resume their earlier levels of drinking within the first three months.

The apparent use of alcohol as a reducer of self-awareness is particularly intriguing in light of Kirschenbaum and Tomarken's (1982) recent suggestion that the only way a person can maintain the gains of a successful therapy experience is to develop an obsessive-compulsive pattern of carefully monitoring behavior in the domain in which the therapy was focused. Such monitoring requires self-directed attention. If alcohol reduces self-awareness, it thereby impedes the ability to self-regulate, including the capability of actively refraining from further alcohol consumption. The structure of this situation is such that there is very little middle ground for the person who confronts it. Though there doubtlessly are other contributing factors, this analysis suggests an important – and rather nonintuitive – basis for the common observation that one drink often leads to many, on the part of the person attempting to control a drinking problem.

Issues in Construing Problems in Self-Management

There seem to us to be important conceptual similarities between the two areas of research just discussed, despite their obvious differences. Indeed, these two are not the only areas of maladaptive behavior in which the conceptual similarities exist, and to which the elements of our theoretical model seem applicable.[4]

---

[4] This discussion of general issues in construing problems in self-management and the nature of the behavior-change process is a relatively brief overview of our thoughts on the matter. Interested readers will find a more extensive discussion elsewhere (Carver & Scheier, 1983).

Brockner (1979), for example, has made related arguments concerning the performance impairments associated with low self-esteem. Preliminary data also implicate the same structural elements in social anxiety (Burgio, Merluzzi, & Pryor, 1982). And there are data to suggest that the model is applicable to the understanding of depression, as well (Blaney, Strack, Ganellen, & Coyne, 1982).

In all of these cases there appear to be expectancies that things will go poorly. There is a phenomenology of negative self-relevant ideation. Efforts at attaining goals are halting and sporadic. Such sporadic, ineffectual efforts often result in confirmation of the expected failure. Indeed, based upon the more general finding that people seek out and attend primarily to hypothesis-confirming (as opposed to -disconfirming) information (e.g., Einhorn & Hogarth, 1978; Jenkins & Ward, 1965), there is every reason to believe that people experiencing significant problems in living will weight the failures they experience more heavily than their successes. Thus the expectancy of bad outcomes solidifies, and becomes the frame of reference for subsequent attempts. These structural elements (which are entirely consistent with our more general analysis of self-regulation) seem present in all of the cases cited above. What distinguishes those cases from each other would seem to be such parameters as the dimensions of the failure, the behavioral domains in which the failures occur, and perhaps the certainty with which failure is anticipated.[5]

And what of behavior change? How does one break out of this spiral of doubt, disengagement, and renewed doubt? Recent attempts to understand the behavior change process, and to develop more effective ways of changing maladaptive self-regulation have had a pronounced cognitive flavor. The cognitive-behavioral approach to behavior change emphasizes the self-imposition of behavioral standards, the monitoring or observation of one's actions, and the importance of expectancies concerning being able to change one's behavior (e.g., Bandura, 1977; Kanfer, 1977). These constructs are quite similar to those that we have been discussing throughout the chapter, though our discussion has focused on moment-to-moment self-regulation rather than on the self-regulatory processes involved in altering chronic behavioral tendencies. Indeed, Kanfer and Busemeyer (1982) have discussed the behavior-change process as a series of stages in which the client monitors progress toward continually evolving goals, guided by outcome feedback – a model that clearly embodies the characteristics of a dynamic control system (see also Bandura, 1978).

In more pragmatic terms, it is our feeling that the goals underlying most cognitive therapy techniques are easily discussed within the framework of our approach to self-regulation. To illustrate this, let us give our rather superficial im-

---

[5] Though a full consideration of this point is beyond the scope of this chapter, we note once again in passing that the assumption of a hierarchical organization among goals also has interesting implications for conceptualizing problems in behavior. For example, a given program of action (e.g. working long hours) may reduce a discrepancy regarding one higher-order value (attaining career success) while simultaneously enlarging a discrepancy regarding another higher-order value (maintaining a rich and fulfilling family life). Such conflicts can easily occur despite the fact that both higher-order reference values are viewed positively and contribute to an overall sense of self-worth.

pressions of three such techniques, phrased in terms of the processes we have postulated. Beck (1972, 1976) approaches therapy for depression (in part, at least) by attempting to alter clients' expectancies about the consequences of their actions. His emphasis on reality testing would seem to be a way of showing depressed clients that bad outcomes are not as likely as they might seem. This, in turn, should diminish negative affect, and increase the probability of subsequent engagement of effort.

Meichenbaum (1977), on the other hand, appears to emphasize the utility of short-circuiting the expectancy-assessment process altogether. Clients are taught to supply themselves with a steady stream of task-focused self-statements (cf. Brockner & Hulton, 1978), thereby inducing in themselves an orientation to action, rather than worrying about the long-term consequences of the action (cf. Kuhl, 1981, 1982, Chapter 6, this volume). Again, the result should be a return to a pattern of active striving. Provided the needed skills actually are present, active striving would seem to be the most important ingredient.

Finally, an important component of some cognitive therapies is the use of imagery, in particular the imaginal rehearsal of behaviors that the person finds problematic (Singer, 1974; Lang, 1977; Mahoney, 1979). Imaginal rehearsal would seem to represent a way of encoding behavioral reference values with a high degree of redundancy for later use. In addition, if rehearsal is employed shortly before the attempt to do the target behavior, it would seem to be a way of rendering the values more accessible for use, by virtue of their recent activation.

We recognize that this brief sketch does not begin to do justice to the complexity of the therapeutic techniques being addressed. But we believe that it does suggest a commonality of process underlying those techniques and the approach to understanding the control of action that we have advocated.

Conclusion

In this chapter we have attempted to accomplish four things. First, we presented a brief overview of our version of a control-systems approach to the conceptualization of behavior. Second, we outlined some of the research findings that are supportive of our assumptions. We should acknowledge that many of these findings are consistent with theoretical frameworks other than ours. Other findings, however, are less easily assimilated to those alternative frameworks (Carver & Scheier, 1981 c). From our point of view, then, when the data are considered as a whole, the approach outlined here provides the most internally consistent interpretation now available.

The other two goals behind the chapter represent attempts to apply the basic model to literatures other than the one from which it grew. The first was the literature of human helplessness. We briefly reviewed several perspectives on helplessness, and suggested that the phenomena documented by researchers in that area are easily understood as specific instances of the basic processes already discussed. The second application was to the broader domain of ineffective self-management. We discussed research on test anxiety and alcohol use as illustrations of the utility of the ideas presented earlier in the chapter. And we suggested that a range of more general issues in conceptualizing maladaptive behavior and its therapeutic change can be meaningfully addressed in similar terms.

We find this breadth of applicability of the concepts with which we have been working to be both intriguing and gratifying. Certainly we have obtained far less than complete closure on the problem of human behavior. And the effort to test the generality of these ideas in applied settings has really just begun. But we have observed an apparent similarity of processes across such diverse domains as motor control and the therapy process. This suggests to us that the ideas have a great deal of utility as a viewpoint on the self-regulation of action.

References

Abramson, L. Y., Seligman, M. E. P., & Teasdale, J. D. (1978). Learned helplessness in humans: Critique and reformulation. *Journal of Abnormal Psychology, 87,* 49–74.

Adams, J. A. (1976). Issues for a closed-loop theory of motor learning. In G. E. Stelmach (Ed.), *Motor control: Issues and trends.* New York: Academic Press.

Anderson, J. R. (1980). *Cognitive psychology and its implications.* San Francisco: Freeman.

Anderson, J. R., & Bower, G. H. (1973). *Human associative memory.* Washington, D.C.: V. H. Winston.

Bandura, A. (1977). Self-efficacy: Toward a unifying theory of behavioral change. *Psychological Review, 84,* 191–215.

Bandura, A. (1978). The self system in reciprocal determinism. *American Psychologist, 33,* 344–358.

Beaman, A. L., Klentz, B., Diener, E., & Svanum, S. (1979). Self-awareness and transgression in children: Two field studies. *Journal of Personality and Social Psychology, 37,* 1835–1846.

Beck, A. T. (1972). *Depression: Causes and treatment.* Philadelphia: University of Pennsylvania Press.

Beck, A. T. (1976). *Cognitive therapy and the emotional disorders.* New York: International Universities Press.

Berglas, S., & Jones, E. E. (1978). Drug choice as an internalization strategy in response to noncontingent success. *Journal of Personality and Social Psychology, 36,* 405–417.

Blaney, P. H., Strack, S., Ganellen, R. J., & Coyne, J. (1982). Attentional style and response deficits. Paper presented at the meeting of the American Psychological Association, Washington.

Bradley, G. W. (1978). Self-serving biases in the attribution process: A reexamination of the fact or fiction question. *Journal of Personality and Social Psychology, 36,* 56–71.

Brewer, M. B., Dull, V., & Lui, L. (1981). Perceptions of the elderly: Stereotypes as prototypes. *Journal of Personality and Social Psychology, 41,* 656–670.

Brewer, W. F., & Treyens, J. C. (1981). Role of schemata in memory for places. *Cognitive Psychology, 13,* 207–230.

Brockner, J. (1979). The effects of self-esteem, success-failure, and self-consciousness on task performance. *Journal of Personality and Social Psychology, 37,* 1732–1741.

Brockner, J., & Hulton, A. J. B. (1978). How to reverse the vicious cycle of low self-esteem: The importance of attentional focus. *Journal of Experimental Social Psychology, 14,* 564–578.

Brunson, B. I., & Matthews, K. A. (1981). The Type A coronary-prone behavior pattern and reactions to uncontrollable stress: An analysis of performance strategies, affect, and attributions during failure. *Journal of Personality and Social Psychology, 40,* 906–918.

Burgio, K. L., Merluzzi, T. V., & Pryor, J. B. (1982). The effects of self-focused attention and performance expectancies on social interaction. Paper presented at the meeting of the Eastern Psychological Association, Baltimore.

Buss, A. H. (1980). *Self-consciousness and social anxiety.* San Francisco: Freeman.

Cantor, N., & Mischel, W. (1977). Traits as prototypes: Effects on recognition memory. *Journal of Personality and Social Psychology, 35,* 38–48.

Carver, C. S. (1974). Facilitation of physical aggression through objective self-awareness. *Journal of Experimental Social Psychology, 10,* 365–370.

Carver, C. S. (1975). Physical aggression as a function of objective self-awareness and attitudes toward punishment. *Journal of Experimental Social Psychology, 11,* 510–519.

Carver, C. S. (1979). A cybernetic model of self-attention processes. *Journal of Personality and Social Psychology, 37,* 1251–1281.

Carver, C. S., Blaney, P. H., & Scheier, M. F. (1979a). Focus of attention, chronic expectancy, and responses to a feared stimulus. *Journal of Personality and Social Psychology, 37,* 1186–1195.

Carver, C. S., Blaney, P. H., & Scheier, M. F. (1979b). Reassertion and giving up: The interactive role of self-directed attention and outcome expectancy. *Journal of Personality and Social Psychology, 37,* 1859–1870.

Carver, C. S., Ganellen, R. J., Froming, W. J., & Chambers, W. (1983). Modeling: An analysis in terms of category accessibility. *Journal of Experimental Social Psychology, 19,* 403–421.

Carver, C. S., Peterson, L. M., Follansbee, D. J., & Scheier, M. F. (1983). Effects of self-directed attention on performance and persistence among persons high and low in test anxiety. *Cognitive Therapy & Research, 7,* 333–353.

Carver, C. S., & Scheier, M. F. (1978). Self-focusing effects of dispositional self-consciousness, mirror presence, and audience presence. *Journal of Personality and Social Psychology, 36,* 324–332.

Carver, C. S., & Scheier, M. F. (1981a). *Attention and self-regulation: A control-theory approach to human behavior.* New York: Springer-Verlag.

Carver, C. S., & Scheier, M. F. (1981b). Self-consciousness and reactance. *Journal of Research in Personality, 15,* 16–29.

Carver, C. S., & Scheier, M. F. (1981c). The self-attention-induced feedback loop and social facilitation. *Journal of Experimental Social Psychology, 17,* 545–568.

Carver, C. S., & Scheier, M. F. (1982a). Control theory: A useful conceptual framework for personality-social, clinical, and health psychology. *Psychological Bulletin, 92,* 111–135.

Carver, C. S., & Scheier, M. F. (1982b). Outcome expectancy, locus of attributions for expectancy, and self-directed attention as determinants of evaluations and performance. *Journal of Experimental Social Psychology, 18,* 184–200.

Carver, C. S., & Scheier, M. F. (1983). A control-theory model of normal behavior, and implications for problems in self-management. In P. C. Kendall (Ed.), *Advances in cognitive-behavioral research and therapy.* (Vol. 2) (pp. 127–194). New York: Academic Press.

Collins, A. M., & Loftus, E. F. (1975). A spreading-activation theory of semantic processing. *Psychological Review, 82,* 407–428.

Deffenbacher, J. L. (1978). Worry, emotionality, and task-generated interference in test anxiety: An empirical test of attentional theory. *Journal of Educational Psychology, 70,* 248–254.

Diener, C. I., & Dweck, C. S. (1978). An analysis of learned helplessness: Continuous changes in performance, strategy, and achievement cognitions following failure. *Journal of Personality and Social Psychology, 36,* 451–462.

Diener, E. (1979). Deindividuation, self-awareness, and disinhibition. *Journal of Personality and Social Psychology, 37,* 1160–1171.

Duval, S., & Wicklund, R. A. (1972). *A theory of objective self-awareness.* New York: Academic Press.

Einhorn, H. J., & Hogarth, R. M. (1978). Confidence in judgement: Persistence of the illusion of validity. *Psychological Review, 85,* 395–416.

Fenigstein, A., Scheier, M. F., & Buss, A. H. (1975). Public and private self-consciousness: Assessment and theory. *Journal of Consulting and Clinical Psychology, 43,* 522–527.

Frankel, A., & Snyder, M. L. (1978). Poor performance following unsolvable problems: Learned helplessness or egotism? *Journal of Personality and Social Psychology, 36,* 1415–1423.

Frey, D. (1978). Reactions to success and failure in public and private conditions. *Journal of Experimental Social Psychology, 14,* 172–179.

Froming, W. J., & Carver, C. S. (1981). Divergent influences of private and public self-consciousness in a compliance paradigm. *Journal of Research in Personality, 15,* 159–171.

Froming, W. J., Walker, G. R., & Lopyan, K. J. (1982). Public and private self-awareness: When personal attitudes conflict with societal expectations. *Journal of Experimental Social Psychology, 18,* 476–487.

Galassi, J. P., Frierson, H. T., Jr., & Sharer, R. (1981). Behavior of high, moderate, and low test anxious students during an actual test situation. *Journal of Consulting and Clinical Psychology, 49,* 51–62.

Gibbons, F. X., Carver, C. S., Scheier, M. F., & Hormuth, S. E. (1979). Self-focused attention and the placebo effect: Fooling some of the people some of the time. *Journal of Experimental Social Psychology, 15,* 263–274.

Glass, D. C. (1977). *Behavior patterns, stress, and coronary disease.* Hillsdale, N. J.: Erlbaum.

Greenberg, J. (1980). Attentional focus and locus of performance causality as determinants of equity behavior. *Journal of Personality and Social Psychology, 38,* 579–585.

Greenwald, A. G. (1970). Sensory feedback mechanisms in performance control: With special reference to the ideo-motor mechanism. *Psychological Review, 77,* 73–99.

Halisch, F., & Heckhausen, H. (1977). Search for feedback information and effort regulation during task performance. *Journal of Personality and Social Psychology, 35,* 724–733.

Hamilton, D. L. (1979). A cognitive-attributional analysis of stereotyping. In L. Berkowitz (Ed.), *Advances in experimental social psychology* (Vol. 12). New York: Academic Press.

Hanusa, B. H., & Schulz, R. (1977). Attributional mediators of learned helplessness. *Journal of Personality and Social Psychology, 35,* 602–611.

Heider, F. (1958). *The psychology of interpersonal relations.* New York: Wiley.

Hiroto, D. S. (1974). Locus of control and learned helplessness. *Journal of Experimental Psychology, 102,* 187–193.

Hiroto, D. S., & Seligman, M. E. P. (1975). Generality of learned helplessness in man. *Journal of Personality and Social Psychology, 31,* 311–327.

Hollandsworth, J. G., Jr., Glazeski, R. C., Kirkland, K., Jones, G. E., & Van Norman, L. R. (1979). An analysis of the nature and effects of test anxiety: Cognitive, behavioral, and physiological components. *Cognitive Therapy and Research, 3,* 165–180.

Holroyd, K. A., Westbrook, T., Wolf, M., & Badhorn, E. (1978). Performance, cognition, and physiological responding in test anxiety. *Journal of Abnormal Psychology, 87,* 442–451.

Hull, J. G. (1981). A self-awareness model of the causes and effects of alcohol consumption. *Journal of Abnormal Psychology, 90,* 586–600.

Hull, J. G., & Young, R. D. (1983). The self-awareness-reducing effects of alcohol consumption: Evidence and implications. In J. Suls & A. G. Greenwald (Eds.), *Psychological perspectives on the self* (Vol. 2). (pp. 159–190). Hillsdale, N. J.: Erlbaum.

Janoff-Bulman, R., & Brickman, P. (1982). Expectations and what people learn from failure. In N. T. Feather (Ed.), *Expectations and actions: Expectancy-value models in psychology.* (pp. 207–237). Hillsdale, N. J.: Erlbaum.

Jenkins, H. M., & Ward, W. C. (1965). Judgment of contingency between responses and outcomes. *Psychological Monographs: General and Applied, 79,* (1, Whole No. 594).

Kanfer, F. H. (1977). The many faces of self-control, or behavior modification changes its focus. In R. B. Stuart (Ed.), *Behavioral self-management: Strategies, techniques, and outcomes.* New York: Brunner/Mazel.

Kanfer, F. H., & Busemeyer, J. R. (1982). The use of problem solving and decision making in behavior therapy. *Clinical Psychology Review, 2,* 239–266.

Kelso, J. A. S., Holt, K. G., Rubin, P., & Kugler, P. N. (1981). Patterns of human interlimb coordination emerge from the properties of non-linear, limit cycle oscillatory processes: Theory and data. *Journal of Motor Behavior, 13,* 226–261.

Kirschenbaum, D. S., & Tomarken, A. J. (1982). On facing the generalization problem: The study of self-regulatory failure. In P. C. Kendall (Ed.), *Advances in cognitive-behavioral research and therapy* (Vol. 1). N. J.: Academic Press.

Kirschenbaum, D. S., Tomarken, A. J., & Ordman, A. M. (1982). Specificity of planning and choice applied to adult self-control. *Journal of Personality and Social Psychology, 42,* 576–585.

Kuhl, J. (1981). Motivational and functional helplessness: The moderating effect of state versus action orientation. *Journal of Personality and Social Psychology, 40,* 155–170.

Kuhl, J. (1982). The expectancy-value approach within the theory of social motivation: Elaborations, extensions, critique. In N.T. Feather (Ed.), *Expectations and actions: Expectancy-value models in psychology.* Hillsdale, N.J.: Erlbaum.

Lang, P. (1977). Imagery in therapy: An information processing analysis of fear. *Behavior Therapy, 8,* 862–886.

MacKay, D.M. (1963). Mindlike behavior in artefacts. In K.M. Sayre & F.J. Crosson (Eds.), *The modeling of mind: Computers and intelligence.* Notre Dame, Ind.: University of Notre Dame Press.

MacKay, D.M. (1966). Cerebral organization and the conscious control of action. In J.C. Eccles (Ed.), *Brain and conscious experience.* Berlin: Springer-Verlag.

Mahoney, M.J. (1979). Cognitive skills and athletic performance. In P.C. Kendall & S.D. Hollon (Eds.), *Cognitive-behavioral interventions: Theory, research and procedures.* New York: Academic Press.

Mandler, G., & Watson, D.L. (1966). Anxiety and the interruption of behavior. In C.D. Spielberger (Ed.), *Anxiety and Behavior.* New York: Academic Press.

Markus, H. (1977). Self-schemata and processing information about the self. *Journal of Personality and Social Psychology, 35,* 63–78.

Matthews, K.A. (1982). Psychological perspectives on the Type A behavior pattern. *Psychological Bulletin, 91,* 293–323.

Meichenbaum, D. (1972). Cognitive modification of test anxious college students. *Journal of Consulting and Clinical Psychology, 39,* 370–379.

Meichenbaum, D. (1977). *Cognitive behavior modification: An integrative approach.* New York: Plenum.

Miller, I.W., & Norman, W.H. (1979). Learned helplessness in humans: A review and attribution-theory model. *Psychological Bulletin, 86,* 93–119.

Miller, G.A., Galanter, E., & Pribram, K.H. (1960). *Plans and the structure of behavior.* New York: Holt, Rinehart & Winston.

Norman, D.A. (1981). Categorization of action slips. *Psychological Review, 88,* 1–15.

Nottelman, E.D., & Hill, K.T. (1977). Test anxiety and off-task behavior in evaluative situations. *Child Development, 48,* 225–231.

Overmier, J.B., & Seligman, M.E.P. (1967). Effects of inescapable shock upon subsequent escape and avoidance learning. *Journal of Comparative and Physiological Psychology, 63,* 28–33.

Posner, M.I., & Keele, S.W. (1968). On the genesis of abstract ideas. *Journal of Experimental Psychology, 77,* 353–363.

Posner, M.I., & Keele, S.W. (1970). Retention of abstract ideas. *Journal of Experimental Psychology, 83,* 304–308.

Powers, W.T. (1973a). *Behavior: The control of perception.* Chicago: Aldine.

Powers, W.T. (1973b). Feedback: Beyond behaviorism. *Science, 179,* 351–356.

Prentice-Dunn, S., & Rogers, R.W. (1982). Effects of public and private self-awareness on deindividuation and aggression. *Journal of Personality and Social Psychology, 43,* 503–513.

Reitman, J.S., & Bower, G.H. (1973). Storage and later recognition of exemplars of concepts. *Cognitive Psychology, 4,* 194–206.

Rogers, T.B. (1981). A model of the self as an aspect of the human information processing system. In N. Cantor & J.F. Kihlstrom (Eds.), *Personality, cognition and social interaction.* Hillsdale, N.J.: Erlbaum.

Rogers, T.B., Rogers, P.J., & Kuiper, N.A. (1979). Evidence for the self as a cognitive prototype: The "false alarms effect." *Personality and Social Psychology Bulletin, 5,* 53–56.

Rosch, E., & Mervis, C. (1975). Family resemblances: Studies in the internal structure of categories. *Cognitive Psychology, 7,* 573–605.

Roth, S. (1980). A revised model of learned helplessness in humans. *Journal of Personality, 48,* 103–133.

Roth, S., & Bootzin, R.R. (1974). Effects of experimentally induced expectancies of external control: An investigation of learned helplessness. *Journal of Personality and Social Psychology, 29,* 253–264.

Roth, S., & Kubal, L. (1975). The effects of noncontingent reinforcement on tasks of differing importance: Facilitation and learned helplessness effects. *Journal of Personality and Social Psychology, 32,* 680–691.

Santee, R. T., & Maslach, C. (1982). To agree or not to agree: Personal dissent amid social pressure to conform. *Journal of Personality and Social Psychology, 42,* 690–700.

Sarason, I. G. (1975). Anxiety and self-preoccupation. In I. G. Sarason & C. D. Spielberger (Eds.), *Stress and anxiety* (Vol. 2). New York: Wiley.

Sarason, I. G. (1978). The test anxiety scale: Concept and research. In C. D. Spielberger & I. G. Sarason (Eds.), *Stress and anxiety* (Vol. 5). New York: Halsted-Wiley.

Schank, R. C., & Abelson, R. P. (1977). *Scripts, plans, goals, and understanding.* Hillsdale, N. J.: Erlbaum.

Scheier, M. F. (1976). Self-awareness, self-consciousness, and angry aggression. *Journal of Personality, 44,* 627–644.

Scheier, M. F., Buss, A. H., & Buss, D. M. (1978). Self-consciousness, self-report of aggressiveness, and aggression. *Journal of Research in Personality, 12,* 133–140.

Scheier, M. F., & Carver, C. S. (1977). Self-focused attention and the experience of emotion: Attraction, repulsion, elation, and depression. *Journal of Personality and Social Psychology, 35,* 625–636.

Scheier, M. F., & Carver, C. S. (1981). Private and public aspects of the self. In L. Wheeler (Ed.), *Review of personality and social psychology* (Vol. 2). Beverly Hills Calif.: Sage.

Scheier, M. F., & Carver, C. S. (1982a). Cognition, affect, and self-regulation. In M. S. Clark & S. T. Fiske (Eds.), *Affect and cognition: The 17th annual Carnegie symposium on cognition.* Hillsdale, N. J.: Erlbaum.

Scheier, M. F., & Carver, C. S. (1982b). Learned helplessness or egotism: Do expectancies matter? Manuscript submitted for publication.

Scheier, M. F., & Carver, C. S. (1982c). Self-consciousness, outcome expectancy, and persistence. *Journal of Research in Personality, 16,* 409–418.

Scheier, M. F., & Carver, C. S. (1983a). Self-directed attention and the comparison of self with standards. *Journal of Experimental Social Psychology, 19,* 205–222.

Scheier, M. F., & Carver, C. S. (1983b). Two sides of the self: One for you and one for me. In J. Suls & A. G. Greenwald (Eds.), *Psychological perspectives on the self* (Vol. 2). (pp. 123–157). Hillsdale, N. J.: Erlbaum.

Scheier, M. F., Carver, C. S., & Gibbons, F. X. (1979) Self-directed attention, awareness of bodily states, and suggestibility. *Journal of Personality and Social Psychology, 37,* 1576–1588.

Scheier, M. F., Carver, C. S., & Gibbons, F. X. (1981). Self-focused attention and reactions to fear. *Journal of Research in Personality, 15,* 1–15.

Scheier, M. F., Fenigstein, A., & Buss, A. H. (1974). Self-awareness and physical aggression. *Journal of Experimental Social Psychology, 10,* 264–273.

Schmidt, R. A. (1976). The schema as a solution to some persistent problems in motor learning theory. In G. E. Stelmach (Ed.), *Motor control: Issues and trends.* New York: Academic Press.

Seligman, M. E. P., & Maier, S. F. (1967). Failure to escape traumatic shock. *Journal of Experimental Psychology, 74,* 1–9.

Simon, H. A. (1967). Motivational and emotional controls of cognition. *Psychological Review, 74,* 29–39.

Singer, J. L. (1974). *Imagery and daydream methods in psychotherapy and behavior modification.* New York: Academic Press.

Snyder, M. L., Smoller, B., Strenta, A., & Frankel, A. (1981). A comparison of egotism, negativity, and learned helplessness as explanations for poor performance after unsolvable problems. *Journal of Personality and Social Psychology, 40,* 24–30.

Snyder, M. L., Stephan, W. G., & Rosenfield, D. (1976). Egotism and attribution. *Journal of Personality and Social Psychology, 33,* 435–441.

Snyder, M. L., Stephan, W. G., & Rosenfield, D. (1978). Attributional egotism. In J. H. Harvey, W. Ickes, & R. F. Kidd (Eds.), *New directions in attribution research* (Vol. 2). Hillsdale, N. J.: Erlbaum.

Spence, J. T., & Spence, K. W. (1966). The motivational components of manifest anxiety: Drive and drive stimuli. In C. D. Spielberger (Ed.), *Anxiety and behavior.* New York: Academic Press.

Stelmach, G. E. (Ed.) (1976). *Motor control: Issues and trends.* New York: Academic Press.

Stotland, E. (1969). *The psychology of hope.* San Francisco: Jossey-Bass.

Turner, R. G. (1978). Consistency, self-consciousness, and the predictive validity of typical and maximal personality measures. *Journal of Research in Personality, 12,* 117–132.

Weiner, B. (1982). The emotional consequences of causal ascriptions. In M. S. Clark & S. T. Fiske (Eds.), *Affect and cognition: The 17th annual Carnegie symposium on cognition.* Hillsdale, N.J.: Erlbaum.

Weiner, B., Russell, D., & Lerman, D. (1978). Affective consequences of causal ascriptions. In J. H. Harvey, W. Ickes, & R. F. Kidd (Eds.), *New directions in attribution research* (Vol. 2). Hillsdale, N.J.: Erlbaum.

Weiner, B., Russell, D., & Lerman, D. (1979). The cognition-emotion process in achievement-related contexts. *Journal of Personality and Social Psychology, 37,* 1211–1220.

Wicklund, R. A., & Duval, S. (1971). Opinion change and performance facilitation as a result of objective self-awareness. *Journal of Experimental Social Psychology, 7,* 319–342.

Wiener, N. (1948). *Cybernetics: Control and communication in the animal and the machine.* Cambridge, Ma.: M. I. T. Press.

Wilson, T. D., & Capitman, J. A. (1982). The effects of script availability on social behavior. *Personality and Social Psychology Bulletin, 8,* 11–19.

Wine, J. D. (1971). Test anxiety and direction of attention. *Psychological Bulletin, 76,* 92–104.

Wine, J. D. (1982). Evaluation anxiety: A cognitive-attentional construct. In H. W. Krohne & L. C. Laux (Eds.), *Achievement, stress, and anxiety.* Washington, D. C.: Hemisphere.

Wortman, C. B., & Brehm, J. W. (1975). Responses to uncontrollable outcomes: An integration of reactance theory and the learned helplessness model. In L. Berkowitz (Ed.), *Advances in experimental social psychology* (Vol. 8). New York: Academic Press.

Chapter 12

From Cognition to Behavior: Perspectives for Future Research on Action Control

Julius Kuhl

The preceding chapters of this volume have presented theory and evidence regarding various processes intervening between cognition and action. We have seen how a traditional Expectancy × Value model of *reasoned action* can be elaborated to incorporate variables that moderate cognition-behavior correspondence (Ajzen, Chapter 2), and how an analysis of the processes underlying cognitive predictors of behavior can contribute to a fuller understanding of the cognition-behavior link (Kruglanski & Klar, Chapter 3). Moreover, whether or not a cognitive state results in action depends upon the type of goals activated at the time (Gollwitzer & Wicklund, Chapter 4), upon the efficiency of self-regulatory processes controlling the maintenance and protection of action-related cognitions (Chapters 5–8), and upon the processes underlying problem-solving and performance control (Chapters 9–11). Rather than summarizing the details of these chapters again (see Chapter 1 for a summary), we would like to reflect upon the more general lessons that can be learned from the analyses of various action-control processes.

What are the shortcomings of earlier attempts to explain behavior on the basis of what we know about cognition? What changes in the metatheoretical underpinnings of current research into cognition-behavior relationships might help stimulate future research into action control processes? In this, the concluding chapter, we will discuss five aspects of current research that one might consider changing to facilitate process-oriented research into action-control mechanisms. These aspects are (1) the current emphasis on *prediction* as opposed to explanation; (2) the one-sided focus on a *molar* as opposed to a *molecular level of analysis;* (3) the *neglect of ecological and psychological representativeness* of the behavior under study; (4) the implicit or explicit acceptance of *associationistic accounts* of human behavior; and (5) the theoretically naive use of the *concept of consistency.*

From Predictive to Explanatory Models

During the past three decades, many models of human action have been formulated in a predictive format. In these models, some observable aspect of behavior is described in terms of an algebraic function combining several cognitive predictor variables (Ajzen & Fishbein, 1973; Atkinson, 1957; Heckhausen, 1977; Vroom, 1964). Ajzen (Chapter 2) has shown how a predictive model of reasoned action can be applied and tested by regression analysis techniques. One measures the assumed cognitive variables, uses the model to predict some behavior, and tests the goodness of fit between predicted and observed behavior. Other authors have used similar models very fruitfully to deduce predictions that were then tested in laboratory experiments (e.g., Atkinson, 1974).

However, the prevalence of predictive models in the past may have contributed to the neglect of processes intervening between cognition and behavior. The illusion of a short-cut from cognition to behavior might have been avoided if past research had taken more into account the processes mediating behavior, rather than focusing on the measurement of cognitive predictors. Predictive models do not provide much explanatory depth. The common research strategy that consists of predicting behavior on the basis of subjects' rating of their expectations and values does not provide much information about the mental processes mediating that behavior.

Consider the study cited by Ajzen (Chapter 2, Table 1), in which the behavior predicted was voting choice in the 1976 presidential election (Ajzen & Fishbein, 1980). The multiple correlation between various beliefs and values regarding one candidate and the intention to vote for him was 0.83, and the intention-behavior correlation was 0.80. What inferences can be drawn from these results about the psychological processes mediating voting behavior? At first glance, one might conclude that the beliefs and values that voters have before election day determine their voting behavior in the way specified by the regression model. However, many voters in Ajzen and Fishbein's study might have formed their intention to vote for a particular candidate prior to the point in time when the investigators asked them to express their beliefs and values regarding the candidates. In those cases, the subjects' intentions may have determined their belief and value ratings rather than vice versa.

There is a second, more serious shortcoming of the exclusive reliance on predictive models. Aside from the problem of drawing any conclusions regarding causal directionality from correlational data, the correlational approach associated with these models does not provide any information regarding the psychological mechanisms underlying behavior. Do the regression weights for the various expectancy and value variables presumably affecting voting behavior (see Chapter 2, Table 1) indicate the extent to which *each* voter has taken the variable in question into account? Does, for instance, the regression coefficient of 0.64 obtained for the behavioral attitude as a predictor of the voting intention reflect the degree to which each voter has been affected by his/her personal attitudes rather than by social norms? Probably not. As Ajzen (Chapter 2) points out, these models may be very predictive on an aggregate level but fail to predict one single case. Analyses of *individual* decision-making processes reveal great individual differences in the type of information affecting individual decisions (Fischhoff, Goitein, & Shapira, 1982; Kruglanski & Klar, Chapter 3; Kuhl, 1982, in press). Some voters may attend primarily to their personal attitudes, while others may be guided by social expectations regarding their voting behavior. The relative weight assigned to personal attitudes or social expectations may even change over time within individuals. The application of algebraic models of human behavior suggests aggregating data across individuals and neglecting the study of the specific psychological mechanisms underlying individual behavior.

A similar point can be made regarding action control processes mediating the transition from cognition to behavior. Adding one or several "control" variables to the prediction equation (see Chapter 2) does not provide much information regarding the psychological processes mediating action control (cf. Chapter 6).

Do we mean to suggest that predictive models be discarded from psychological theorizing? Certainly not. We believe that psychologists should continue to utilize predictive models – where they are appropriate. The use of these models should be confined to the purpose for which they are constructed, i. e., *predicting,* as opposed to explaining, behavior. There are many applications for which predictive models are very useful (Ajzen & Fishbein, 1980). However, the utility of a predictive model as a tool to improve our understanding of the psychological processes mediating behavior is rather limited. Although empirical results obtained on the basis of a predictive model can have *some* explanatory value (see Ajzen, Chapter 2), a detailed analysis of the underlying processes requires the application of process-oriented models and experimental techniques.

The reason for the limited explanatory value of predictive models lies in the fact that they typically adopt a cruder level of analysis than explanatory models. A process-oriented (explanatory) model specifies the assumed sequence of the mental processes mediating the behavior in question on a rather fine-grained level of analysis. For most predictive purposes, it is more useful to assess the outcomes of the broadly defined processes rather than tracing the determinants of these processes into finer and finer levels of analysis. However, adopting the rather crude level of analysis of a predictive model entails the risk of overlooking the complex processes intervening between cognition and action. In the following section, we will elaborate this point in some detail.

From Molar to Molecular Levels of Analysis

Most research into cognition-behavior consistency focuses on rather molar summary constructs rather than on the specific mental processes mediating the transition from cognition to behavior. Let us discuss one typical experiment to illustrate this point. Snyder & Swann (1976) had male students formulate judgments of liability in a simulated sex-discrimination legal case. There was no covariation between the verdicts and previously measured attitudes ($r=0.07$) unless attitudes toward affirmative action were made salient experimentally ($r=0.58$). Salience is a rather global concept, which obviously had considerable predictive value but has little explanatory value. What are the psychological mechanisms that determine which cognitions become salient in a given situation? What mechanisms intervene between a salient cognition and the performance of an appropriate action?

The various chapters of this volume have addressed these questions. The key to answering the first question about the determinants of salience lies in the study of the specific action-related knowledge structures that individuals develop (Kruglanski & Klar, Chapter 3). Rating measures of subjects' attitudes are not precise measures of the propositional memory structures encoding attitude-related knowledge (cf. Kuhl, Chapter 6, this volume; Kuhl, in press). Suppose we want to explain why a subject's attitude toward affirmative action is activated in a particular situation. We need to understand two things to find an explanation. First, we need to know the specific memory structures encoding that individual's attitude-related knowledge. Second, we need a model of the memory mechanisms controlling the retrieval of those memory structures. Although some steps addressing these two points have been taken in this volume (see Kruglanski & Klar, Chapter 3; Kuhl, Chapter 6; Kluwe & Friedrichsen, Chapter 9), a lot remains to be done. Several techniques that cognitive psychologists have developed to assess complex knowledge structures (Chi, in press) and to understand complex pattern-matching mechanisms controlling retrieval processes (Anderson, 1983) provide promising perspectives for future research.

Even if we had a clear understanding of the processes mediating the salience of action-related cognitions, the specific processes intervening between the activation of those cognitions and the performance of appropriate actions would remain to be investigated. The molecular approach suggests an even finer-grained level of analysis than the one taken in various chapters of this volume. Future research should focus on the many information-processing mechanisms underlying action control that are still unexplored. How do different modes of representing action-related knowledge (i.e., propositional, pictorial, sequential, etc.) affect action-control processes? How do various modes of processing (e.g., parallel-holistic vs. sequential-analytic) interact with action-control processes? What is the functional significance of conscious awareness? How do various emotional states affect action control?

Although investigating these questions on a molecular level of analysis may considerably improve our understanding of action-control processes, we do not mean to suggest giving up the molar level of analysis altogether. We have argued in the preceding section that a molar level of analysis is more useful than a molecular one for many predictive purposes. In addition, the molar level is indispensable for maintaining contact with ecologically representative behavioral phenomena (see Herrmann & Wortman, Chapter 8). A one-sided focus on the molecular level entails the risk of becoming preoccupied with rather artificial behavioral phenomena that have no bearing on people's actions in everyday life. We believe that lack of generalizability is not the only risk associated with the molecular level of analysis. In the next section, we will argue that special precautions have to be taken to prevent molecular research from undermining generalizability of research findings.

From Simple Cases to Psychologically Representative Behavior

Investigators who emphasize the need to study behavior in ecologically valid settings are usually concerned about the lack of external validity of laboratory research. Experimental findings obtained in the laboratory may not be very generalizable to real-life situations (see Herrmann & Wortman, Chapter 8). In a typical laboratory experiment, the degree of freedom subjects have in choosing between action alternatives is considerably more restricted than in most real-life settings. However, as we have seen (Chapters 6 and 7), one of the functions of psychological processes intervening between cognition and behavior is to protect an intention against competing action tendencies. If no competing tendencies are aroused, action-control processes are not activated and cannot be investigated. If subjects can only choose between engaging in some attitude-consistent behavior and failing to do so, there is no need for them to invoke action-control processes that operate in the more complex everyday-life settings.

Field experiments may increase the generalizability of research findings to real-life situations (see Chapter 8). However, observing people in natural settings does not guarantee that the psychological processes that require study will actually be invoked. *Ecological* validity does not guarantee *psychological* validity. Many field experiments focus on simple cases, that is, behaviors that do not invoke the complex self-regulatory mechanisms that operate in more complex real-life settings. Ajzen and Fishbein's (1980) study on voting behavior is a case in point. While investigating voting behavior may be interesting in its own right, that particular behavior may be less than optimal for studying the general mechanisms underlying human behavior.

Specifically, self-regulatory processes may not have been invoked in that study because actually voting for the candidate one intends to vote for is not very problematic. In many real-life situations, intentions are considerably more difficult to enact than in typical experimental situations, and such difficulty is a necessary condition for the activation of self-regulatory processes (Ach, 1935; Kuhl, 1983-a, and Chapter 6, this volume). If we want to develop a better understanding of the processes mediating between cognition and action in everyday behavior, we need to develop methods that invoke those processes in an experimental situation. Obviously, we cannot develop methods of invoking processes without having at least a preliminary conception of what those processes are. We hope that the foregoing chapters have provided some useful steps toward developing a process-oriented model of cognition-behavior transitions. The more refined our models of action control become, the easier it will be to create *psychologically representative* experimental situations, that is, situations that invoke the psychological processes that occur in typical real-life settings.

From Associationistic to Dynamic Models

We believe that one reason why current models of human behavior tend to neglect many processes intervening between cognition and action derives from an
implicit or explicit emphasis on associationistic principles. Even the most sophisticated cognitive models of human behavior (e.g., Anderson, 1983) still reflect the behavioristic tradition that explains human behavior in terms of mental
"connections" between stimuli and responses. What goals an actor strives for
and what behavior a person performs depends, according to those models, primarily on the type of situation encountered. Goals and behavioral procedures
are assumed to be encoded in memory structures that specify the situational conditions under which these structures are activated. This emphasis on cognitive
associations between situational information and behavioral information leads
to a neglect of the functional significance of various emotional and motivational
states of the organism and of dynamic changes in these states over time. Depending on the emotional and motivational state a subject is in, exposing him or her
to a particular situation can activate quite different goals and behaviors (Atkinson & Birch, 1978; Izard, 1977; Kuhl & Atkinson, in press-b; Plutchik, 1980). At-
tributing this fact to associative connections between emotional states and behavioral information (Bower, 1981) does not fully explain the constructivistic
and organizational functions of those states. The functional significance of emotional states cannot be reduced to their potential for activating those memory
structures "connected" to them. Emotional and motivational states can lead to a
totally novel organization of incoming information, *creating* new percepts outside the repertoire of perceptual responses previously associated with the current
perceptual input (Bastick, 1982; Carr & Bacharach, 1976; Kuhl, 1983b). Moreover, changes in emotional and motivational states over time cannot be fully accounted for by concomitant changes in the environment (Izard, 1977; Kuhl &
Atkinson, in press-a; Kuhl & Blankenship, 1979).

The state of the organism resulting from the dynamics of emotional and motivational processes is probably considerably more complex than associationistic
principles suggest. According to such principles, a persisting *conflict* between action tendencies should be the exception rather than the rule. Within the associationistic framework, it is rather unlikely that several tendencies of nearly equal
strength should be associated with one situation. If this statistically unlikely case
does occur, priority rules associated with the perception of a state of conflict are
activated and lead to a rapid resolution of conflict (Anderson, 1983). According
to the dynamic view, action tendencies continuously change in strength even in a
constant environment (Atkinson & Birch, 1970). As a result, an organism's attempt to enact a behavioral tendency activated by some stimulus is constantly
jeopardized by the waxing and waning of competing action tendencies. Since,
nevertheless, individuals are usually able to adhere to an intention until the goal
has been reached a model of man as being completely at the mercy of the waxing
and waning of action tendencies seems inadequate. Instead, the resulting model
of man is unlike that of an automaton, which flawlessly responds to any situation with the appropriate action program, provided only that it is in the automa-

ton's repertoire. Instead, the dynamic view suggests the model of a complex self-organizing system, which continuously monitors multiple behavioral tendencies and attempts to maintain the tendency currently selected for action despite the waxing and waning of alternative tendencies. A straightforward enactment of one action tendency aroused in a given situation is expected only in the case of habitual behavior in which competing tendencies have little or no effect.

The dynamic view suggests that, in many real-life situations, the processes intervening between cognition and action are much more complex than is traditionally assumed. Assessing a subject's beliefs, attitudes, and values is important but not sufficient for closing the cognition-behavior gap (see Chapter 2). In several chapters of this volume, associationistic principles have been supplemented by dynamic principles (see Chapters 4, 6, 7, and 10). Despite these efforts toward a broadening of the theoretical framework underlying current models of action, we may be still too closely attached to traditional associationistic accounts. A shift from this traditional perspective to one that integrates associationistic and dynamic principles may promote a better understanding of complex human behavior as it occurs outside the laboratory.

From "Cognition-Behavior Consistency" to "Motivational Stability"

Implicit in the dynamic perspective discussed in the preceding section is a reformulation of the basic question regarding the cognition-behavior link. If the motivational state of the organism is continuously changing, the degree of cognition-behavior consistency depends on an arbitrary decision on the points in time at which one samples the cognitive and behavioral measures for estimating consistency. To the extent that all behavior is mediated by some cognitive event, cognition-behavior consistency should always be perfect, provided the behavior in question is matched with the "right" cognition (cf. Ajzen, Chapter 2). By this account, a failure to establish perfect consistency would be due to the investigator's failure to match an element of the "stream of behavior" to the corresponding element from the "stream of cognitions", rather than to the subject's failure to behave consistently.

However, looked at a different way, traditional measures of consistency such as intention-behavior correlations also reflect something about the actor. They indicate the actor's ability (and willingness) to *maintain* an intentional state over time despite continuous changes in competing action tendencies. This faculty is tantamount to what Ach (1910) called the "efficiency of the will," the organism's capacity to stabilize ("freeze") its intentional state over long periods of time (see Chapter 5).

We suggest that in future research, consistency measures be interpreted as stability measures. Apart from the fact that the term *stability* is theoretically more appropriate, it also has the advantage of having a relatively neutral evaluative connotation. Whether or not a given amount of stability is considered useful and adaptive depends on the criterion of evaluation one adopts (see Herrmann & Wortman, Chapter 8). Rigidly adhering to an intention despite substantial environmental changes can have very adverse consequences for an actor. On the other hand, yielding to each situationally aroused temptation and shifting one's intentions accordingly may be maladaptive as well.

The task of assessing the overall adaptiveness of a given degree of motivational stability is very difficult. Resolving this question is important to clinical psychologists who must decide whether to increase their clients' motivational stability or their motivational flexibility. However, future research into the action-control mechanisms mediating motivational stability can be conducted independent of the adaptiveness issue.

Conclusion

In our preface, we quoted Aristotle's remark that "it is not thought as such that can move anything, but thought which is for the sake of something and is practical" (Aristotle, 1975, p. 102). We feel that current cognitivistic approaches to a theory of human action often lag behind this insight by assuming a direct connection between cognition and action. Cases in which there is a direct, unswerving route from cognition to behavior may be rather exceptional. Aristotle's remark did not, of course, provide much of a clue as to the nature of the processes that make thought "practical." It is the task of psychological researchers to find such answers. Research will fail to find answers, however, unless it is undertaken in response to a question. We are confident that future research will reveal many new insights regarding action-control processes – provided we keep asking Aristotle's question: what it is "that makes thought move anything"?

References

Ach, N. (1910). *Über den Willensakt und das Temperament*. Leipzig: Quelle & Meyer.

Ach, N. (1935). Analyse des Willens. In E. Abderhalden (Ed.), *Handbuch der biologischen Arbeitsmethoden. Bd. VI*. Berlin: Urban & Schwarzenberg.

Ajzen, I., & Fishbein, M. (1973). Attitudinal and normative variables as predictors of specific behaviors. *Journal of Personality and Social Psychology, 27*, 41–57.

Ajzen, I., & Fishbein, M. (1980). *Understanding attitudes and predicting social behavior*. Englewood Cliffs, NJ: Prentice-Hall.

Anderson, J. R. (1983). *The architecture of cognition*. Cambridge, MA: Harvard University Press.

Aristotle. (1975). *Nicomachean ethics* (H. G. Apostle, Trans.). Boston: Reidel.

Atkinson, J. W. (1957). Motivational determinants of risk-taking behavior. *Psychological Review, 64*, 359–372.

Atkinson, J. W. (1974). The mainsprings of achievement-oriented activity. In J. W. Atkinson & J. O. Raynor (Eds.), *Motivation and achievement*. (pp. 13–41). Washington, D. C.: Winston.

Atkinson, J. W., & Birch, D. (1970). *The dynamics of action*. New York: Wiley.

Atkinson, J. W., & Birch, D. (1978). *Introduction to motivation*. New York: Van Nostrand.

Bastick, T. (1982). *Intuition: How we think and act*. New York: Wiley.

Bower, G. H. (1981). Mood and memory. *American Psychologist, 36*, 129–148.

Carr, T. H., & Bacharach, V. R. (1976). Perceptual tuning and conscious attention: Systems of input regulation in visual information processing. *Cognition, 4*, 281–302.

Chi, M. T. H. (in press). Representing knowledge and metaknowledge: Implications for interpreting metamemory research. In F. E. Weinert & R. H. Kluwe (Eds.), *Metacognition, Motivation and Understanding*. Hillsdale, NJ: Erlbaum.

Fischhoff, B., Goitein, G., & Shapira, Z. (1982). The experienced utility of expected utility approaches. In N.T.Feather (Ed.), *Expectations and actions: Expectancy-value models in psychology.* (pp.315–340). Hillsdale, NJ: Erlbaum

Izard, C. (1977). *Human emotions.* New York: Plenum.

Heckhausen, H. (1977) Achievement motivation and its constructs: A cognitive model. *Motivation and Emotion, 1,* 283–329.

Kuhl, J. (1982). The expectancy-value approach in the theory of social motivation: Elaborations, extensions, critique. In N.T.Feather (Ed.), *Expectations and actions: Expectancy-value models in psychology.* Hillsdale, NJ: Erlbaum.

Kuhl, J. (1983 a). *Motivation, Konflikt und Handlungskontrolle.* Heidelberg, FRG: Springer.

Kuhl, J. (1983 b). Emotion, Kognition und Motivation: II. Die funktionale Bedeutung der Emotionen für das problemlösende Denken und für das konkrete Handeln. *Sprache und Kognition, 4,* 228–253.

Kuhl, J. (in press). Motivation and information processing: A new look at decision-making, dynamic conflict, and action control. In R.M.Sorrentino & E.T.Higgins (Eds.), *The handbook of motivation and cognition: Foundations of social behavior.* New York: Guilford Press.

Kuhl, J., & Atkinson, J.W. (in press-a). Perspectives in human motivational psychology: A new experimental paradigm. in V.Sarris & A.Parducci (Eds.), *Perspectives in psychological experimentation: Toward the year 2000.* Hillsdale, NJ: Erlbaum.

Kuhl, J., & Atkinson, J.W. (Eds.) (in press-b). *Motivation, thought, and action.* New York: Praeger.

Kuhl, J., & Blankenship, V. (1979). The dynamic theory of achievement motivation: From episodic to dynamic thinking. *Psychological Review, 86,* 141–151.

Plutchik, R. (1980). *Emotion: A psychoevolutionary synthesis.* New York: Harper & Row.

Snyder, M., & Swann, W.B. Jr. (1976). When actions reflect attitudes: The politics of impression management. *Journal of Personality and Social Psychology, 34,* 1034–1042.

Vroom, V.H. (1964). *Work and motivation.* New York: Wiley.

Author Index

Subject Index

SSSP

Springer
Series in
Social
Psychology

Springer-Verlag
Berlin
Heidelberg
New York
Tokyo

T.M.Amabile
The Social Psychology of Creativity
1983. ISBN 3-540-90830-7

Attitudinal Judgment
Editor: J.R.Eiser
1984. ISBN 3-540-90911-7

J.R.Averill
Anger and Aggression
An Essay on Emotion
1982. ISBN 3-540-90719-X

Basic Group Processes
Editor: P.B.Paulus
1983. ISBN 3-540-90862-5

C.S.Carver, M.F.Scheier
Attention and Self-Regulation:
A Control-Theory Approach to Human Behavior
1981. ISBN 3-540-90553-7

Compatible and Incompatible Relationships
Editor: W.Ickes
1985. ISBN 3-540-96024-4

Directions in Soviet Social Psychology
Editor: L.H.Strickland
Translated from the Russian by E.Lockwood,
N.Thurston, I.Gavlin
1984. ISBN 3-540-90959-1

J.Brockner, J.Z.Rubin
Entrapment in Escalating Conflicts
A Social Psychological Analysis
1985. ISBN 3-540-96089-9

The Ethics of Social Research
Fieldwork, Regulation, and Publication
Editor: J.E.Sieber
1982. ISBN 3-540-90691-6

The Ethics of Social Research
Surveys and Experiments
Editor: J.E.Sieber
1982. ISBN 3-540-90687-8

SSSP

Springer Series in Social Psychology

Springer-Verlag
Berlin
Heidelberg
New York
Tokyo

Facet Theory:
Approaches to Social Research
Editor: **D.Canter**
1985. ISBN 3-540-96016-3

Gender and Nonverbal Behavior
Editors: **C.Mayo, N.M.Henley**
1981. ISBN 3-540-90601-0

K.J.Gergen
Toward Transformation in
Social Knowledge
1982. ISBN 3-540-90673-8

Language and Social Situations
Editor: **J.P.Forgas**
1985. ISBN 3-540-96090-2

Social Psychology of Aggression
From Individual Behavior to Social Interaction
Editor: **A.Mummendey**
1984. ISBN 3-540-12443-8

M.L.Patterson
Nonverbal Behavior
A Functional Perspective
1983. ISBN 3-540-90846-3

Personality, Roles, and Social Behavior
Editors: **W.Ickes, E.S.Knowles**
1982. ISBN 3-540-90637-1

The Social Construction of the Person
Editors: **K.J.Gergen, K.E.Davis**
1985. ISBN 3-540-96091-0

Sociophysiology
Editor: **W.M.Waid**
1984. ISBN 3-540-90861-7

Sports Violence
Editor: **J.H.Goldstein**
1983. ISBN 3-540-90828-5